Organic Coatings: Science and Technology

SPE MONOGRAPHS

Organic Coatings: Science and Technology

Volume I: Film Formation, Components, and Appearance

ZENO W. WICKS, JR.
North Dakota State University

FRANK N. JONES
Eastern Michigan University

S. PETER PAPPAS
Polychrome Corporation

A WILEY-INTERSCIENCE PUBLICATION

JOHN WILEY & SONS, INC.

New York • Chichester • Brisbane • Toronto • Singapore

Copyright © 1992 by John Wiley & Sons, Inc.

Library of Congress Cataloging in Publication Data:

Wicks, Zeno W.
  Organic coatings: science and technology / Zeno W. Wicks, Jr., Frank N. Jones, S. Peter
  Pappas.

     p. cm.—(SPE monographs)
  "A Wiley-Interscience publication."
  Includes index.
  Contents: v. 1. Film formation, components, and appearance
  ISBN 0-471-61406-8 (acid-free paper)
  1. Plastic coatings. I. Jones, Frank N., 1936–   . II. Pappas, S. Peter (Socrates Peter),
     1936–   . III. Title. IV. Series.
     TP1175.S6W53   1992
     667′.9–dc20
                                                                                    92-214

Printed in the United States of America

10 9 8 7 6 5 4 3 2 1

To the thousands of students at North Dakota State University, at Eastern Michigan University, and at industrial laboratories around the world whose interest and searching questions have shaped this book.

Foreword

The Society of Plastics Engineers (SPE) is pleased to sponsor *Organic Coatings: Science and Technology*. This volume combines a concise review of theory with numerous practical applications to produce a basic understanding of coatings technology.

While the book can be utilized as a basic text, it provides in addition a ready industrial reference tool for the engineer and manager. The outstanding organization and presentation reflect upon the extensive background of the three authors, all of whom are recognized experts in the field.

SPE, through its Technical Volumes Committee, has long sponsored books on various aspects of plastics and polymers. Its involvement has ranged from identification of needed volumes to recruitment of authors. An ever-present ingredient, however, is review of the final manuscript to insure accuracy of the technical content.

This technical competence pervades all SPE activities, not only in publication of books but also in other activities such as technical conferences and educational programs. In addition, the Society publishes periodicals— *Plastics Engineering*, *Polymer Engineering and Science*, *Polymer Processing and Rheology*, *Journal of Vinyl Technology*, and *Polymer Composites*—as well as conference proceedings and other selected publications, all of which are subject to the same rigorous technical review procedure.

The resource of some 37,000 practicing plastics engineers has made SPE the largest organization of its type worldwide. Further information is available from the Society at 14 Fairfield Drive, Brookfield, Connecticut 06804.

ROBERT D. FORGER

Executive Director
Society of Plastics Engineers

Technical Volumes Committee
Raymond J. Ehrig, Chairperson
Aristech Chemical Corporation

Preface

In teaching university courses in Coatings Science and condensed courses for industrial personnel, we have found that there is no adequate textbook available. The purpose of this book is to provide a text that combines a presentation of current scientific understanding in the field of organic coatings with a summary of the applied technology of the field. An important element of the text is to present the jargon of the field, so that people entering the field can begin learning its specialized language.

This first of two volumes covers primarily raw materials for coatings: resins, solvents, and pigments. The second volume covers primarily characteristics, application, and formulation of coatings.

Coatings technology has evolved to a large extent as an art. Over the course of the last few decades, partial scientific understanding of many of the principles has been gained. However, the complexities of the field are such that the formulator's art is still essential in developing and using coatings. Our conviction is that an increasing understanding of the underlying science can increase the effectiveness with which the formulator can operate, and that the scientist starting to work in the area can gain effectiveness by understanding the formulator's language and appreciating the thousands of variables involved in developing a single commercially satisfactory coating.

Better effectiveness will be needed. Impending environmental regulations will demand that radically new formulations be developed, tested, and placed in service faster than ever before while at least maintaining, and preferably improving, coatings performance.

The literature in the coatings field is of spotty value. There are many excellent papers published in the field but, unfortunately, there are also many whose authors did not fully understand the complexity of the field. We have included an appendix of reference sources that we have found useful. We also tried to select specific references for the various topics covered that are reliable and have omitted those that are not technically sound. In some chapters we have included general references that cover the material in more depth than allowed by the space limitations of an introductory text.

The text is written for people who have had college level chemistry courses at least through organic chemistry. It does not assume, however, that the reader will have had any course work in polymers.

The text was initially targeted to students in university courses at an upper undergraduate level and first year graduate students. However, we also strived to make the text of value to both scientists and technicians working in industrial laboratories. We hope it will be useful to technical personnel of coatings manufacturers and of companies supplying raw materials to the coatings industry and to personnel involved in selecting and applying coatings in commerical applications.

While written specifically for the coatings field, many of the principles involved are applicable to the related fields of printing inks, adhesives, and parts of the plastics industry.

No book can cover every aspect of coatings in enough detail for the specialist, but we hope to put the subject in perspective and to stimulate readers to ask fruitful questions. One of our objectives is to help people who specialize in one subfield to relate their work to the rest of the field. We believe that a broad understanding of the subject can help workers be more productive and can greatly increase their enjoyment of the fascinating and challenging field of coatings science.

We have found that learning about coatings is an engaging intellectual pursuit, and we hope this book will help others share the pleasure of learning about coatings. Most of all, we hope the readers will build on our experience.

We want to acknowledge the many people who have helped in the development of the material for this text. Among the hundreds of people who have helped us, Loren Hill, formerly a Professor of Polymers and Coatings at North Dakota State University, is undoubtedly the one person whose contributions stand out. Others who cooperated by reviewing individual chapters are Ronald Bauer, Josef Jilek, K. F. Lin, Tosko Misev, Loren Odell, Richard Roesler, Clifford Schoff, Douglas Wicks, John Williams, and Larry Williams. We appreciate their suggestions and corrections.

ZENO W. WICKS, JR.
FRANK N. JONES
S. PETER PAPPAS

Las Cruces, New Mexico
Ypsilanti Michigan
Carlstadt, New Jersey
August 1992

Contents—Volume I

18. PIGMENTS

Contents—Volume II

Symbols

A	Arrhenius preexponential term
C	Concentration—weight per unit volume of solution
c	Concentration—moles per liter
CPVC	Critical pigment volume concentration (content)
CRH	Critical relative humidity
°C	Degrees Celsius
E	Modulus
E'	Storage modulus (elastic modulus)
E''	Loss modulus
E	Relative evaporation rate
E_a	Thermal coefficient of reaction rate (Arrhenius activation energy)
F	Functionality of a monomer
$\overline{F}$	Average functionality of a monomer mixture
f	Functionality of a polymer (resin)
f_n	Number average functionality of a polymer (resin)
G	Free energy
G	Small's molar association constant
g	gram
g	Gravitational constant
H	Enthalpy
i	Angle of incidence
K	Kelvin temperature
K	Absorption coefficient
K_E	Einstein constant
k	Rate constant for a specific reaction
kg	Kilogram
KU	Krebs unit
L	Liter
M	Molecular weight
M_c	Molecular weight between cross-links
$\overline{M}_n$	Number average molecular weight
$\overline{M}_w$	Weight average molecular weight
MFT	Minimum film formation temperature
mL	Milliliter
N	Newton
N	Number of moles
n	Refractive index
NVV	Nonvolatile volume (volume percent solids)
NVW	Nonvolatile weight (weight percent solids)

OA	Oil absorption
P	Vapor pressure
P	Degree of polymerization
$\overline{P}_n$	Number average degree of polymerization
$\overline{P}_w$	Weight average degree of polymerization
p	Extent of reaction
p_g	Extent of reaction at gelation onset
Pa	Pascal
Pa·s	Pascal second = 10 poise
mPa·s	Millipascal second = 1 centipoise
PDI	Polydispersity index = $\overline{M}_w/\overline{M}_n$
PVC	Pigment volume concentration (content)
R	Gas constant
RH	Relative humidity
r	Angle of reflection or angle of refraction
S	Entropy
S	Scattering coefficient
s	Second
T	Temperature (K if not otherwise specified)
T_b	Brittle—ductile transition temperature
T_g	Glass transition temperature
T_m	Melting point
t	Time
tan δ	tan delta, loss tangent, E''/E'
V	Molar volume
V_i	Volume fraction of internal phase
VOC	Volatile organic compound
w	Weight fraction
X	Film thickness
x	Mole fraction
x	Optical path length
XLD	Cross-link density
γ	Surface tension
$\dot{\gamma}$	Shear rate
δ	Solubility parameter
δ	Phase shift in viscoelastic deformation
ε	Molar absorbance
ε	Absolute shear viscosity
η_e	External phase viscosity
η_r	Relative viscosity = η/η_s
η_s	Viscosity of solvent
η^*	Extensional viscosity
$[\eta]$	Intrinsic viscosity
$[\eta]_w$	Weight intrinsic viscosity
$[\eta]_\theta$	Intrinsic viscosity under theta conditions
θ	Contact angle
λ	Wavelength
ν	Kinematic viscosity

ν_e	Mole of elastically effective network chains per cm^3
ρ	Density
τ	Shear stress
τ_o	Yield value
ϕ	Packing factor
χ	Activity coefficient

All equations using logarithms are given in terms of natural logarithms (ln). Equations from the literature using base 10 logarithms have been restated in terms of ln.

SI Units are used throughout with conversion factors given at the site of first usage.

Organic Coatings: Science and Technology

CHAPTER 1

What Are Coatings?

Look around you. Coatings are everywhere. If you are inside, there are coatings on the walls, refrigerator, cabinets, and furniture; less obviously, coatings are on the wires of electrical motors, printed circuits inside television sets, cassette tapes, video tapes, and compact disks. If you are outside, coatings are on your house and car, as well as inside your car on the vinyl seats, under the hood, and on components of the automotive stereo and computer systems. Whether you drink beer or soft drinks, there are coatings on the inside and outside of the cans. Clearly, the functional and decorative requirements of coatings span a very broad spectrum. Less obviously, there is a corresponding diverse science and technology that supports the development, production, and use of coatings. As consumers we have come to expect high performance in coatings applications such as house and automobile paints, audio and video tapes, for example, but are generally less aware of the science and technology that makes that performance possible.

More important than an awareness of our present state of coatings technology is the recognition that there are important limitations in our understanding of coatings and of the many ingredients in their compositions. In particular, it is important for students to realize that coatings science is a relatively youthful field, which offers exciting challenges and career opportunities. Furthermore, there is the opportunity to contribute to the major thrusts of coatings development: of reducing emissions that lead to ozone generation in the atmosphere, reducing energy requirements, and protecting against corrosion.

1.1. DEFINITIONS AND SCOPE

Coatings may be described by their appearance (e.g., clear, pigmented, metallic, or glossy) and by their function (e.g., corrosion protective, abrasion protective, skid resistant, decorative, or photosensitive). Coatings may be distinguished as organic or inorganic, although there is overlap. For example, many coatings consist of inorganic pigments dispersed in an organic matrix (the binder).

A somewhat confusing situation results from multiple meanings of the term "coating." It is used to describe the material (usually a liquid) that is applied to a

1

substrate, the resultant "dry" film, and the process of application. Usually, the intended meaning of the word coating can be inferred from the context.

We will limit our discussion to coatings with organic binders, generally organic polymers, which are applied purposefully to a substrate. Many kinds of coatings will not be included. Porcelain enamels on kitchen ranges are coatings, but they do not have organic chemical binders. Electroplated copper, nickel, or zinc coatings are excluded from the scope of this book for the same reason. The auto mechanic gets a coating of grease on his/her hands, but not purposefully. We will further restrict organic coatings to those materials that can be historically traced back to paints. What is the difference between a coating and a paint? Not much—the terms are often used interchangeably, confusing multiple meanings included. However, it has become common practice to use *coatings* as the broader term and to restrict *paints* to the familiar architectural and household coatings and sometimes to coatings for bridges, tanks, and the like. We will follow this practice. Many people prefer to call sophisticated materials that are used to coat automobiles and computer components coatings, perhaps sensing that paint sounds too lowbrow. Still another common term that is essentially a synonym for coating and paint is *finish*; it is afflicted with similar multiple meanings.

In limiting the scope of this book to organic coatings that can be related to historic paints we will exclude many materials that could be called coatings. Printing inks, polymers applied during production of paper and fabrics, coatings on photographic films, decals and other laminates, and cosmetics are but a few examples. However, many of the basic principles that will be covered in this text are applicable to such materials.

Restrictions of scope are necessary if the book is to be kept within two volumes, but our restrictions are not entirely arbitrary. The way we are defining coatings is based on common usage of the term in worldwide business. It is close to the U.S. Department of Commerce, Bureau of Census definition of organic coatings for their statistical analyses of industrial output [1].

The Census Bureau estimated that shipments by U.S. manufacturers of paints and coatings in 1990 were slightly more than 1 billion gallons (3.8×10^9 L) with a value of $11.5 billion [1]. The Census Bureau defines three broad categories: (1) architectural coatings, (2) product coatings sold to original equipment manufacturers (OEM coatings), and (3) special-purpose coatings. The shipments reported for 1990 were broken down as shown in Table 1.1. The worldwide coatings market in 1989 was estimated at $37 billion [2].

In 1990 the volume of shipments decreased almost 6% under 1989 but the value increased 3%. The increase in value as compared to the decrease in volume is a

Table 1.1. United States Coatings Shipments, 1990

Coatings	Liters $\times 10^{-9}$	Dollars $\times 10^{-9}$
Architectural	1.96	4.90
Product—OEM	1.22	4.08
Special purpose	0.61	2.48
	3.8	11.5

continuation of a long-term trend caused by several factors: regulatory pressure to reduce volatile organic compound (VOC) emissions tends to reduce volume but increase value per unit volume; requirements for higher product performance tend to increase value; inflation increases the value figures from year to year; importation of manufactured goods usually shifts coatings production to the country where the goods are manufactured; overall economic growth tends to increase both volume and value; competitive pressure on prices decreases value.

Architectural coatings include the familiar paints and varnishes (transparent paints) used to decorate and protect buildings, outside and inside. They also include other paints and varnishes sold for use in the home and by small businesses for application to such things as cabinets, household furniture (not those sold to furniture factories), and the like. They are often called "trade sales paints." They are sold directly to painting contractors and *do-it-yourself* users through paint stores and other retail outlets. In 1990 in the United States, a little over one-half of the total volume of coatings was architectural paints; however, the unit value of these coatings is lower than for the other categories, so that they comprised about 43% of the total value. This market is the least cyclical of the three categories. While new construction drops during recessions, the resulting decrease in paint requirements tends to be offset by increased repainting of older housing, furniture, and so forth, during at least mild recessions. Latex-based coatings comprise by far the largest fraction of architectural coatings.

Product coatings, also commonly called industrial coatings or industrial finishes, are applied in factories on products such as automobiles, appliances, magnet wire, furniture, metal cans, chewing gum wrappers . . . the list is almost endless. This market is often called the OEM market, that is, the original equipment manufacturer market. In 1990 in the United States, product coatings were about 32% of the volume and 35% of the value of all coatings. The volume of product coatings depends directly on the level of manufacturing activity. This category of the business is cyclical, varying with OEM cycles. In most cases, the product coatings are designed for a particular customer's manufacturing conditions and the performance requirements of its products. Thus the number of products in this category is much larger than in the others and the research and development requirements tend to be significantly higher.

Special purpose coatings is a catchall category for coatings that do not fit the other two. It includes coatings for cars and trucks that are applied outside the OEM factory (usually in body repair shops), coatings for ships and aircraft and the familiar stripes of coating on highways and parking lots. Coatings for steel bridges, storage tanks, chemical factories, and the like are classified as "maintenance coatings" and are usually included in this category. In 1990 in the United States, special purpose coatings comprised about 16% of the total volume and 22% of the total value of all coatings.

Coatings are applied for one or more of three reasons: (1) for decoration, (2) for protection, and/or (3) for some functional purpose. The low gloss paint on the ceiling of a room fills a decorative need, but it also has a function—it reflects and diffuses light to help provide even illumination. The coating on the outside of an automobile adds beauty to a car and also helps protect it from rusting. The coating on the inside of a beverage can has little or no decorative value, but it protects the beverage from the can. (Contact with metal affects flavor.) In some cases the interior coating protects the can from the beverage. (Some soft drinks are so acidic they can dissolve the metal.) Other coatings reduce the growth of algae and barnacles on ship bottoms, protect optical fibers for telecommunications against abrasion, serve as the recording medium in audio and video tapes, and on and on. While the public most commonly thinks of house paint when talking about coatings, coatings are important throughout the economy, and they make essential contributions to most hi-tech fields.

Traditionally, coatings have changed relatively slowly in an evolutionary response to new performance requirements, new raw materials, and competitive pressures. An important reason for the relatively slow rate of change is the difficulty of predicting product performance on the basis of laboratory tests. It is less risky to make relatively small changes in composition and check actual field performance before making further changes. Since about 1965, however, the pace of change has been forced to increase substantially. A major driving force for change has been the need to reduce VOC emissions because of their effect on air pollution. Coatings have been second to the gasoline–automobile complex as a source of VOC pollutants that are responsible for excess ozone in the air of many cities on many days of the year. This has resulted in increasingly stringent regulatory controls on such emissions. The drive to reduce VOC emissions has also been fueled by the rapidly rising cost of organic solvents. Other important factors have also accelerated the rate of change in coatings. Increasing concern about toxic hazards has led to the need to change many raw materials that were traditionally used in coatings. Furthermore, the increasing cost of energy has resulted in pressure to reduce the time and temperature of baking ovens. Product performance requirements have tended to increase; most notably increased effectiveness of corrosion protection by coatings has been needed.

1.2. COMPOSITION OF COATINGS

Organic coatings are complex mixtures of chemical substances that can be grouped into four broad categories: (1) binders, (2) volatile components, (3) pigments, and (4) additives.

Binders are the materials that form the continuous film that adheres to the "substrate" (the surface being coated) on one side, binds together the other substances in the coating to form a film, and presents an adequately hard surface to the outside. The binders of coatings within the scope of this book are organic polymers. In some cases, these polymers are prepared and incorporated into the coating before application; in other cases, final polymerization takes place only after the coating has been applied to the substrate. The binder governs, to a large extent, the properties of the coating film.

Volatile components are included in a large majority of all coatings. They play a major role in the process of applying coatings—they are liquids that make the coating fluid enough for application and they evaporate during and after application. Until about 1945 almost all of the volatile components were low molecular weight organic solvents that dissolved the binder components. For historic reasons, they are often called "solvents." However, this term is potentially misleading because since 1945 many coatings have been developed in which the binder components are not fully soluble in the volatile components. Because of the great need to reduce VOC emissions, a major continuing drive in the coatings field is to reduce use of solvents by making the coatings more highly concentrated (higher solids coatings) or by using water as a major part of the volatile components (water-borne coatings). "Vehicle" is a commonly encountered term. It usually means the combination of the binder and the volatile components of a coating. Today most coatings, including water-borne coatings, still contain at least some volatile organic solvents. Exceptions are powder coatings and many radiation curable coatings.

Pigments are finely divided insoluble solids that are dispersed in the vehicle and remain suspended in the binder after film formation. Generally, the primary purpose of pigments is to provide color and opacity to the coating film. However, they also can have important effects on application characteristics and on film properties. While a majority of coatings contain pigments, there are many important types of coatings that contain little or no pigment, commonly called "clear coatings" or just "clears." Transparent varnishes are an example.

Additives are materials that are included in small quantities to modify some property of the coating. Examples are catalysts for polymerization reactions, stabilizers, and flow modifiers.

Most coatings are extremely complex mixtures. Many contain several substances from each of the four categories and each substance is usually a chemical mixture. The number of possible combinations is limitless. The number of different applications is also limitless.

The person who selects the components from which to make a coating for a particular end use is called a "formulator," and the overall composition he/she designs is called a "formulation" or a formula. An example of a formulation is given in Table 1.2. The proportions of components are stated in both weight and volume units. The reason is that many of the components are most conveniently and accurately measured by weight, but the properties of coatings are governed mainly by volume relationships among the components, not by weight relationships. It is difficult to overemphasize the importance of volume relationships in formulating and studying coatings. In Table 1.2 the total weight and volume of each component are tabulated, and then broken down to show the weight and volume of each major category in each component.

Coatings are applied by brush, roller, spray, dip, and other means as discussed in Chapter 22. After application they form adherent films in a process called film formation. This process, which is critical to the subsequent performance of the coating, will be discussed in Chapter 3.

Formulation of paints started millennia ago as an empirical art or craft. Succeeding generations of formulators built on the experience of their predecessors and formulated paints with better and better performance characteristics. Starting perhaps 50 or 60 years ago, formulators have been trying to understand the underlying scientific principles that control the performance of paints. Most coatings systems are so complex that our understanding today is still limited. Real progress has been made, but the formulator's art is still a critical element in developing high-performance coatings. Demands on suppliers of coatings to develop new and better coatings are increasing at an accelerating pace so that time is now too limited

Table 1.2. Formulation of an Automotive Thermosetting Acrylic Enamel

Component	Total (kg)	Total (L)	Pigment (kg)	Pigment (L)	Binder (kg)	Binder (L)	Volatiles (kg)	Volatiles (L)
TiO$_2$ Dispersion (60 wt% pigment)	94.3	52.7	56.7	13.5	19	15.9	18.7	21.6
Color pigment dispersions	14	10.6	2.3	0.9	6	4.9	6	6.4
Acrylic polymer (55 wt% solids)	153	150.2			84	71.2	69	78.7
Melamine–formaldehyde resin (55 wt% solid)	102	100			56	50.7	46	49
n-Butyl alcohol	13.5	15.1					13.5	15.1
Catalyst (50 wt% solid)	1.2	1.4			0.6	0.8	0.6	0.7
Xylene	29.8	34					29.8	34
Propylene glycol methyl ether acetate	13.7	14.5					13.7	14.5
	421.9	378.5	59	14.4	165.6	142.5	197.3	220

Source: Adapted from Ref. [3], Table 2.

to permit trial and error formulation. Understanding the basic scientific principles can help a formulator design better coatings faster. In the chapters ahead, we will try to present, to as great an extent as present knowledge permits, the current understanding of the scientific principles involved in coatings science.

We will also cover areas in which our basic understanding remains inadequate and discuss approaches to more efficient and effective formulation despite inadequate understanding. In some cases, where no hypotheses have been published to explain certain phenomena, we will offer speculations. Such speculations will be based on our understanding of related phenomena and on our cumulative experience acquired over several decades in the field. We recognize the risk that speculation tends to increase in scientific stature with passing time and may even be cited as evidence or adopted as an experimentally supported hypothesis. It is our intent rather that such speculations will promote the advancement of coatings science and technology by stimulating discussion that leads to experimentation designed to disprove or support the speculative proposal. Obviously, we believe that the latter purpose outweighs the former risk; and we will endeavor to identify the speculative proposals as such.

Cost is an essential consideration in formulation. Beginning formulators are inclined to think that the *best* coating will be the one that will last the longest time without any change in properties, but such a coating may be very expensive, and for some applications not be as *good* as another less expensive coating whose performance is adequate for the particular application. Furthermore, it is seldom possible to maximize all of the performance characteristics of a coating in one formula. Some of the desirable properties are antagonistic with others; formulators must balance many performance variables while keeping costs as low as possible.

REFERENCES

1. U.S. Department of Commerce, Bureau of Census, *Current Industrial Reports—Paint, Varnish, and Lacquers*, Issued monthly.
2. M. S. Reisch, *Chem. Eng. News*, **68** (38), 39 (1990).
3. B. N. McBane, *Automotive Coatings*, Federation of Societies for Coatings Technology, Blue Bell, PA, 1987, p. 25.

CHAPTER II

Introduction to Essential Concepts

This chapter is designed to give an overview of some basic concepts of polymer chemistry, rheology, surface tension, and colloid chemistry to provide a foundation for discussion of coating binders. Further details are discussed in subsequent chapters.

2.1. POLYMERS

Polymer science has emerged as one of the main branches of chemistry, and a vast body of knowledge of polymer science is being developed. In this book, there is only space to summarize the polymer science concepts that are most important in the coatings field. This chapter covers only general principles; further detail can be found in Chapters 3 through 13, as well as many other chapters. Additional readings in the field of polymers as a whole are listed in the general references of this chapter and the list of sources given in the Appendix.

A polymer is often defined as a material of high molecular weight derived by covalently bonding together small units called "mers" into chains. There is disagreement about how high the molecular weight has to be in order for a material to qualify as a polymer. Some people refer to materials with molecular weights of 1000 as polymers; others insist that only materials with molecular weights over 10,000 (or even 50,000) qualify. Still others set the boundary at the lowest molecular weight at which the bulk polymer behaves as if the chains of individual molecules were entangled—a less than ideal criterion because the critical molecular weight for entanglement varies widely with the polymer structure and with the method of measurement.

The term "oligomer," meaning "few mers," is often used for low molecular weight polymers. This additional term does not help the definition problem much because there is no clear-cut boundary between oligomer and polymer, but the term can be useful because it provides a name with which most can agree for materials containing 2 to about 20 mers.

Polymers occur widely in nature; familiar examples of *biopolymers* are proteins, starch, cellulose, silk, and enzymes. Synthetic, man-made polymers are a relatively recent development—the field was pioneered in the 1920s and 1930s. In coatings, we are concerned mainly with synthetic polymers, although some chemically modified biopolymers are also used.

Synthetic polymers and oligomers are prepared by a chemical process, called *polymerization*, in which small molecules become joined by covalent bonds. The small molecules are called *monomers*, and each ends up as a single mer. A polymer made from a single monomer is called a *homopolymer*. If it is made from a combination of monomers, it is often (not always) called a *copolymer*. An example of a homopolymer is provided by the polymerization of vinyl chloride.

$$CH_2{=}CHCl \longrightarrow X{-}(CH_2{-}CHCl)_n{-}Y$$

Vinyl chloride monomer   Poly(vinyl chloride)

In this example the $-(CH_2{-}CHCl)-$ unit is the mer, and n represents the number of mers joined together in the molecule. Note that the electron pairs in the double bonds of the monomers are utilized to form the new single bonds in the polymer. The X and Y represent groups of some kind on each end of the chain of mers; discussion of the end groups will be postponed.

Polymers can be made in a variety of structures; three important classes are shown in Figure 2.1. When the mers are linked in chains, the polymers are called *linear polymers*, a term that is potentially misleading because the large molecules seldom form a straight line—they twist and coil unless they are crystalline. If there are forks in the chains (Fig. 2.1*b*) they are called *branched polymers*. A class of great importance in coatings results from the bonding of chains with each other at several sites to form *cross-linked* or *network* polymers (Fig. 2.1*c*). In effect, these are branched polymers where the branches covalently bond to other molecules so that eventually the mass of polymer consists mainly of a single, interconnected molecule. The reactions that join polymer or oligomer molecules are called cross-linking reactions. Polymers and oligomers, which can undergo such reactions, are frequently called *thermosetting polymers*. Some confusion can result because the term thermosetting is applied not only to polymers that cross-link when heated but

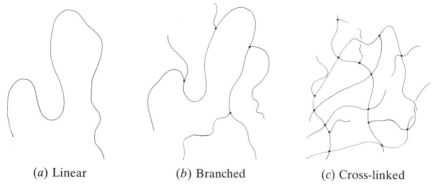

　　　(*a*) Linear　　　　　　　(*b*) Branched　　　　　　(*c*) Cross-linked

Figure 2.1. Schematic representations of three common classes of polymer structure.

also to those that can cross-link at ambient temperature. A polymer that does not undergo cross-linking reactions is called a *thermoplastic polymer*, because it usually becomes plastic (softens) when heated.

Copolymers of all three of the classes illustrated in Figure 2.1 are common. In linear copolymers, the different monomers may be distributed more or less at random throughout the chain (random copolymers), they may tend to alternate (alternating copolymers), or they may be grouped together (block copolymers). One can attach a polymer chain of one type of polymer to the side of a chain of a different polymer to form a type of branch copolymer called a *graft copolymer*. It is analogous to growing a graft of a limb of one kind of apple tree on the trunk of another kind of apple tree. Analogous complexities arise with cross-linked co-polymers.

Another term that is commonly but loosely used in the coatings field is "resin." This term overlaps to a major degree the meanings of polymer and oligomer. Originally, the term meant hard, brittle materials derived from tree exudates, such as, rosin, dammar, and elemi. A wide variety of these naturally occurring resins was used since prehistoric times to make coatings. Especially in the nineteenth and early twentieth centuries, these resins were cooked with drying oils to make *varnishes*.

The first synthetic polymers used in coatings were phenol–formaldehyde poly-mers, which could replace naturally occurring resins in many applications. Hence, it was natural to call them phenol–formaldehyde resins, or more briefly, phenolic resins. As further synthetic products were developed to replace resins, drying oils, and varnishes, it was natural for people to call them resins also. The term now is used in two senses—very broadly for polymeric materials that have not yet been fabricated into final form, as, for example, polyolefin resins to be made into milk bottles, and less broadly for polymers or oligomers used to make cross-linkable coatings, inks, and adhesives. The latter sense of the term seems most prevalent in the coatings field.

As in most situations where words do not have precise meanings, it is important to understand what a writer or a speaker means when he/she uses the words. It is all too common for people to assume, without much thought, that information that has been learned about high molecular weight polymers is also applicable to low molecular weight polymers or oligomers, because all are often called polymers. Many characteristics, however, depend on molecular weight. While much of the information that is available from studies of high molecular weight polymers can be useful in the coatings field, it must be used with caution, because the resins used in making coatings are commonly low molecular weight polymers or oligomers, even though they are usually called polymers.

2.1.1. Polymer Characterization

Characteristics of high molecular weight polymers differ in many ways from those of low molecular weight organic compounds. The characteristics of low molecular weight polymers and oligomers represent a transition stage between high polymers and small molecules. In the following subsections, we describe some of the key characteristics of synthetic polymers and oligomers.

2.1.2. Molecular Weight

For most pure organic compounds the concept of molecular weight is very straight-forward—each compound has a molecular weight. For synthetic polymers, how-ever, the situation is much more complex. All methods of synthesis lead to mixtures of molecules with different numbers of mers and, therefore, with different molec-ular weights. Even relatively simple thermoplastic homopolymers, such as poly-styrene or poly(vinyl chloride), contain molecules with thousands of different, although similar, structures. With copolymers the number of different molecules present can be many orders of magnitude higher. There will be some kind of a distribution of molecular weights within each synthetic polymer. In the simplest cases, the distribution of the number of molecules of each molecular weight re-sembles a skewed Gaussian distribution. In other cases the distribution may be quite complex. Thus the molecular weights can be defined only by some sort of statistical calculation. While many types of average molecular weight can be cal-culated, the two most widely used are "number" and "weight" average molecular weights.

Number average molecular weight $\overline{M}_n$ is the molecular weight average based on summing the products of the numbers of molecules and their molecular weights and dividing by the sum of the number of molecules in the sample. Mathematically it is expressed by the following equation, where M_1, M_2, and M_i are the molecular weights of the first, second, and ith species, and the N values are the numbers of molecules of each species present.

$$\overline{M}_n = \frac{\sum N_1 M_1 + N_2 M_2 + \cdots}{\sum N_1 + N_2 + \cdots} = \frac{\sum_i N_i M_i}{\sum_i N_i}$$

$$\overline{P}_n = \frac{\sum_i N_i P_i}{\sum_i N_i}$$

A similar equation is often used to represent the number average *degree of polymerization* $\overline{P}_n$, where P is the number of mers in a molecule. The term P_i is the number of mers in the ith polymer. For homopolymers $\overline{M}_n = \overline{P}_n$ times the molecular weight of each mer; for copolymers a weighted average molecular weight of the mers is used. The differing weights of end groups can be neglected in calculating $\overline{M}_n$ (or $\overline{M}_w$) of high polymers since the effect is negligible, but not for oligomers, where it can be appreciable.

Weight average molecular weight $\overline{M}_w$ is defined by the following equation in which w_1, w_2, and w_i are the weights of molecules of species 1, 2, and i. Since $w_1 = N_1 M_1$, $\overline{M}_w$ can also be calculated from the numbers of molecules of the different species, as shown.

$$\overline{M}_w = \frac{w_1 M_1 + w_2 M_2 + \cdots}{w_1 + w_2 + \cdots} = \frac{\sum_i w_i M_i}{\sum_i w_i} = \frac{\sum_i N_i M_i^2}{\sum_i N_i M_i}$$

It is evident that the weight average degree of polymerization $\overline{P}_w$ can be defined by analogous equations.

Figure 2.2 shows an idealized plot of the weight fraction of molecules of each molecular weight as a function of degree of polymerization for oligomers made from the same monomer by three different processes. In relatively simple distributions of molecular weights, such as shown in Figure 2.2, the value of $\overline{P}_n$ is at, or near, the peak of the weight fraction distribution curve. Because, by definition, $\overline{M}_w$ and $\overline{P}_w$ give extra weight to higher molecular weight molecules, they are always larger than $\overline{M}_n$ and $\overline{P}_n$.

The breadth of the molecular weight distribution can have an important effect on the properties of a polymer and is often critical in achieving satisfactory performance of a coating. The ratio $\overline{M}_w/\overline{M}_n$ is widely used as an index of the breadth of distribution. In the case of high molecular weight polymers, $\overline{M}_w/\overline{M}_n = \overline{P}_w/\overline{P}_n$, but in the case of oligomers, differences in end groups can be significant and affect the equality of the ratios. These ratios are called *polydispersity* (PD) or sometimes the *polydispersity index* (PDI). In this text, we only use the symbolisms $\overline{M}_w/\overline{M}_n$ and $\overline{P}_w/\overline{P}_n$. The ratios provide a convenient way to compare the molecular weight distributions of different polymers. However, one must be cautious in the use of a single value to describe a possibly complex distribution.

As shown in Figures 2.2 and 2.3, synthetic polymers commonly have broad distributions of molecular weights. Note that as the $\overline{M}_w/\overline{M}_n$ increases, the fractions of polymer at the extremes both above and below the number average molecular

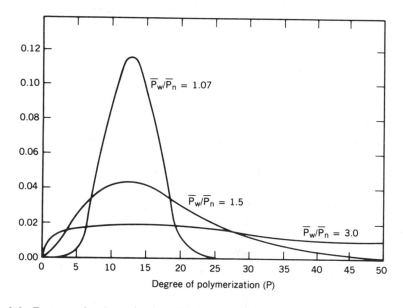

Figure 2.2. Degree of polymerization distribution plots calculated for three types of chain-growth polymers. $\overline{P}_w/\overline{P}_n = 1.07$ is for an ideal anionic polymerization, $\overline{P}_w/\overline{P}_n$ is 1.5 for an ideal free radical polymerization with termination by combination, and $\overline{P}_w/\overline{P}_n$ is 3.0 for a typical free radical polymerization. The $\overline{P}_n$ is 12 for all plots and $\overline{P}_w$ is 12.84, 18, and 36, respectively (from Ref. [1], with permission.)

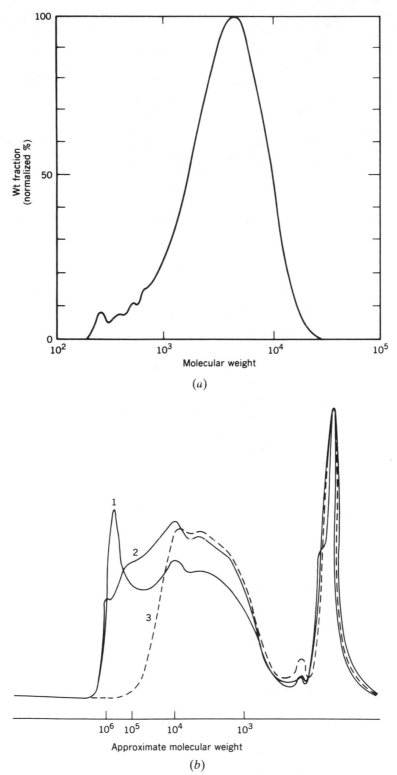

Figure 2.3. (*a*) Molecular weight distribution of a typical polyester resin (from Ref. [3] with permission). (*b*) Molecular weights of three alkyd resins as measured by gel permeation chromatography with a UV detector (from Ref. [4] with permission.)

weight increase. Even the oligomer with a number average of 12 mers and with $\overline{M}_w/\overline{M}_n = 1.07$ has substantial numbers of molecules containing 7–18 mers, and a more typical $\overline{M}_w/\overline{M}_n$ of 3 has molecules spanning several orders of magnitude of molecular weight.

The $\overline{M}_n$ is the molecular weight of most importance in relating stoichiometric ratios of reactants and in comparing certain physical properties. The $\overline{M}_w$ often proves more useful than $\overline{M}_n$ when considering the relationship between molecular weight and many physical properties of polymers including some of the properties that are crucial in coatings performance. Thus, at a minimum, both $\overline{M}_n$ and $\overline{M}_w$ must be considered in designing resins for coatings, and it is sometimes desirable to take other measures of molecular weight into account as well.

Measurement of molecular weight: Up until now we have described $\overline{M}_n$ and $\overline{M}_w$ as if they were mathematical abstractions. Molecular weight measurement has always been, and still is, one of the central problems of polymer science, and it is beyond the scope of this book to discuss it in detail (see Ref. [2] for a discussion of the various ways of determining $\overline{M}_w$ and $\overline{M}_n$). In practice, most coatings scientists use gel permeation chromatography (GPC), more properly called size exclusion chromatography (SEC), to measure molecular weights. In this method a dilute solution of oligomer or polymer is pumped at high pressure through a series of columns containing porous gels. The molecules are *sorted* by sizes, with the largest coming out first and the smaller ones, which are slowed by entering and leaving more of the pores, coming out later. The concentration of polymer in the solvent is analyzed as it leaves the column and is plotted as a function of time. Molecular weight is estimated by comparing retention times to those of standard polymers of known molecular weight. The $\overline{M}_n$, $\overline{M}_w$, and other types of average molecular weights can be calculated from the data. While GPC is convenient and is widely used, it is not dependably accurate; errors of $\pm 10\%$ can be expected, and much larger errors are possible. Errors can result because the molecular weight is not measured directly—the size of the polymer molecules in solution is measured—and from differences in detector response to different compositions. Despite its imprecision, GPC is extremely useful, especially for comparing polymers of similar structure.

The $\overline{M}_n$ of oligomers can be accurately measured by colligative methods such as freezing point depression and vapor pressure osmometry. However, the accuracy decreases as molecular weight increases and colligative methods are of little use above $\overline{M}_n = 50,000$.

Some polymers and oligomers have molecular weight distributions closely approaching the idealized kind of distributions shown in Figure 2.2, as illustrated by the GPC trace of a polyester oligomer in Figure 2.3a. However, many of the polymers used in coatings have complex distribution patterns as exemplified by the alkyds in Figure 2.3b. The $\overline{M}_w$ and $\overline{M}_n$ can be calculated for the entire trace or for portions of complex traces. But, polydispersity numbers obviously must be used with caution for complex traces. It is common that polymers showing *abnormal* distributions of molecular weight also show substantial differences in properties.

The molecular weight of resins is an important factor affecting the viscosity of coatings made with solutions of the resins. The higher the molecular weight, the greater the viscosity. Molecular weight is also an important factor controlling the strength of a film. In general terms, the higher the molecular weight, the higher the tensile strength of a film, at least up to a point. For example, in automotive acrylic lacquers, the acrylic polymer must have a $\overline{M}_w$ greater than about 75,000. The required molecular weight depends on polymer composition and the application. Film property considerations argue in favor of using high molecular weight polymers or resins in formulating coatings, but the viscosity considerations often favor low molecular weights. "High solids" formulations require the use of low molecular weight resins but this may limit film strength. As is so often the case in coatings, compromises are needed. It is often desirable to make resins with as narrow a range of molecular weight as possible since this will minimize the amount of very low and very high molecular weight resin. The very low molecular weight resin is generally undesirable from the standpoint of film properties; whereas the very high molecular weight resin increases resin solution viscosity disproportionately. However, there are exceptions—alkyd resins having very broad, complex molecular weight distributions may perform better than alkyds with similar compositions having narrow distributions [4].

2.1.3. Solubility and Miscibility

The concepts of solubility and miscibility are more complex in the case of polymers than with most small molecules. If we consider two small-molecule liquids, the two substances will be soluble in each other, that is, miscible, if the intermolecular interactions between the two different kinds of molecules are similar to the interactions with their own kind. In many cases there will be a fixed limit to the solubility of one substance in another, and in other cases the substances will be miscible in all proportions. The driving force for dissolution is most commonly an increase in entropy. However, in high polymers, because of the large size of the molecules and the relatively low number of molecules per unit volume, the entropy driving force is reduced. The polymer can dissolve only if the intermolecular interactions between polymer molecules and solvent are greater than the intermolecular interactions among the polymer molecules. This means that, in general, polymers with narrow molecular weight distributions and without significant differences in chemical composition from molecule to molecule will either be soluble or not soluble in some solvent rather than having some limited solubility. On the other hand, even though a polymer might not be soluble in some solvent, it is very possible, even probable, that the solvent will be soluble to a limited extent in the polymer. A further possibility is that the interaction between polymer and solvent might not be strong enough to make a true solution of individual molecules but could give a mixture in which clusters of small numbers of polymer molecules are distributed through the solvent matrix. Sometimes these systems will at least superficially resemble solutions.

The problem of solubility is further compounded because most polymers with which we deal have broad molecular weight distributions and frequently also have wide ranges of compositional variation. In such cases, it is common for a polymer to be soluble when a concentrated solution is made but to precipitate when the solution is diluted with a larger amount of the same solvent. It is a fairly safe assumption, implied in the preceding paragraph, that a polymer is more likely to be soluble in a given solvent as the molecular weight decreases. A concentrated solution of a mixed molecular weight polymer in some solvent can be thought of as a solution of the high molecular weight polymer fraction in a solution of its low molecular weight fraction in solvent. If more solvent is added, the low molecular weight material stays soluble but the more dilute solution of the low molecular weight polymer may no longer be a solvent for the high molecular weight polymer, which separates. In fact, one can fractionate some polymers into different molecular weight ranges by doing stepwise dilution, separating the insoluble fraction after each increment of solvent is added. The highest molecular weight materials are in the first precipitates and molecular weights of precipitates decrease as further dilution is carried out. Similarly, if a polymer consisting of molecules having differing degrees of polarity is dissolved in a relatively low polarity solvent, addition of further solvent can lead to separation of the more polar fractions.

Cross-linked polymers are not soluble in solvents, although any part of the material not chemically bound to the network may by soluble and can be extracted. On the other hand, solvent can still dissolve in the cross-linked polymer. In effect, solvent molecules can associate with the polymer network molecule but cannot separate the molecule so that it can dissolve. If the cross-links are far apart, that is, if the *cross-link density* (XLD) is low, a solvent will swell the polymer to a major degree. A familiar example is a rubber stopper swollen with solvent. If the cross-links are close together, less solvent can "fit in" and there will be less swelling. In fact, the extent of swelling can be used to determine the molecular weight between cross-links in a cross-linked polymer. The lower the swelling, the lower the molecular weight between cross-links and the higher the XLD.

A related matter, also of great importance, is the solubility of one polymer in another polymer. If two polymers are completely soluble in each other, they are said to be "compatible." This is not a common occurrence with high molecular weight polymers because the driving force of entropy in this case is small. It can happen when one polymer has chain segments that interact positively with chain segments of the other, for example, when one is acidic and the other is basic. Just as solubility of a polymer in solvent decreases as molecular weight increases, it is more difficult to find a polymer that is truly compatible with some polymer than it is to find a solvent for the polymer.

There have been many attempts to predict solubility from structures and properties, so far none has been completely successful. Solubility is discussed more fully in Chapter 14.

2.1.4. Morphology

Morphology is the study of the physical forms of materials. As with molecular weight and solubility, morphology is more complex with polymers than it is with small-molecule substances. Pure small-molecule compounds will generally freeze to crystals if the temperature is sufficiently low. In contrast, few synthetic polymers crystallize completely, and many do not crystallize at all. Noncrystalline materials that appear to be solids are called *amorphous solids*. There are at least two reasons why synthetic polymers are at least partly amorphous. In general, synthetic polymers are not pure compounds, so that it is more difficult to achieve the completely regular structure characteristic of a crystalline material. In addition, the molecules are so large that the probability of formation of perfect crystals is low and the process of forming a perfect crystal would be extremely slow. Part of a molecule can associate with part of a different molecule or with another part of the same molecule, greatly reducing the odds of perfect crystal formation. However, small crystalline domains are common in synthetic polymers; polymers with fairly regular structures, usually homopolymers, are most likely to crystallize. In these crystalline domains, fairly long segments of molecules associate with each other in a very regular way. The remaining parts of the same molecules have been unable to fit together so regularly and still remain amorphous. While polymers used in fibers and films (e.g., polyethylene and nylon) are often partly crystalline, the polymers used in coatings applications are, with few exceptions, completely amorphous.

Amorphous materials behave quite differently from crystalline materials. An important difference can be seen in Figure 2.4a and b, which schematically compares the changes in specific volume of crystalline and amorphous materials with temperature. In the case of a pure crystalline material, as temperature increases, initially, there is a slow increase in specific volume. The vibrations of the atoms and molecules become more intense as the temperature increases, they displace each other more and more and, therefore, the specific volume increases. Then, at a certain temperature, the substance melts. The melting point T_m is the lowest temperature at which the vibrational forces pushing molecules apart exceed the attractive forces holding them together in the crystals. With almost all substances (water is a notable exception), the molten compound occupies more volume at the same temperature than the crystals, because the molecules are freer to move and they "bounce" their neighbors out of the way leading to an abrupt increase in specific volume at T_m. Above T_m, the specific volume of the liquid slowly increases with a further increase in temperature. Physical chemists call T_m a first-order thermodynamic transition, that is, there is a sharp discontinuity in the change in volume as a function of temperature.

As seen in Figure 2.4b, an amorphous material behaves differently. Again, starting from a low temperature, there is a slow increase in specific volume as the

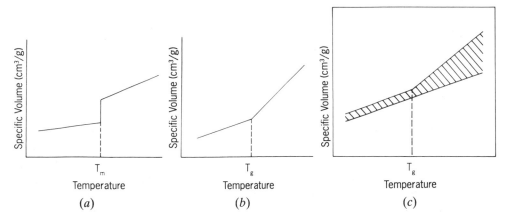

Figure 2.4. Specific volume as a function of temperature (a) for a crystalline material and (b) for an amorphous material. (c) Shows free volume within an amorphous material as a function of temperature. Units of specific volume are volume per mass (usually cm³/g). (Adapted from Ref. [1] with permission.)

temperature increases, but there is no temperature at which an abrupt change in volume occurs; in other words there is no melting point. Rather, there is a temperature at which there is a change in the rate of increase of specific volume with temperature. Above that temperature the thermal expansion coefficient is larger than it is below it. This change of slope is not a phase change; it is a second-order thermodynamic transition, that is, there is a discontinuity in a plot of the derivative of volume change as a function of temperature. The temperature at which it occurs is widely called the *glass transition temperature*, T_g. The T_g is properly defined as the temperature at which there is an increase in the thermal expansion coefficient. Unfortunately, T_g is often improperly defined as being the temperature below which a material is brittle and above which it is flexible. While this is true in many cases, we will see examples where this definition is misleading when we discuss mechanical properties of coatings in Chapter 24. Perhaps part of the reason for this misunderstanding is the connotation of the word glass, which we tend to associate with a brittle material. Like all amorphous materials, glasses undergo a second-order transition as shown. In fact, the phenomenon was first observed in the study of glasses, hence the name glass transition temperature. The idea seems to have gotten around that T_g is a phenomenon associated only with polymers. This is not true; many pure small-molecule compounds can be super-cooled without crystallization to form amorphous glasses and have T_gs. For example, the T_g of *m*-xylene is 125 K [5].

What is physically happening at T_g? As the amorphous material is heated, the atoms in the molecules vibrate with increasing energy, colliding with neighbors and shoving molecules apart for brief periods of time. At T_g, a few of the short-lived "holes" between the molecules become large enough that an adjacent molecule, or a segment of a polymer molecule, can fit between two molecules. Thus, T_g can be considered the lowest temperature at which segments of polymer molecules can move with some facility relative to neighboring segments. The increase in the coefficient of thermal expansion about T_g can be explained by the greater degree of freedom available to the molecule segments. The larger volume between molecules gives more degrees of freedom so that the same increase in temperature will give a greater increase in volume. As the temperature rises the specific volume increases but there is no more material, just the same material occupying more space. What is in this *extra* volume?—nothing. It is so-called "free volume," as shown schematically by the hatched area in Figure 2.4c. The molecular motions involved can be detected by spectroscopic techniques such as solid state NMR and can be seen to change as polymers are heated through T_g [6,7].

While it is difficult to overemphasize the importance of the concepts of T_g and free volume in coatings science, it must also be said that our understanding of these parameters and our ability to measure them is limited. The T_g values determined by different methods may not agree with each other; one must be careful in comparing T_gs to be sure that the various values are based on consistent test methods (see Ref. [8] for a discussion of various methods of measuring T_g).

The values for T_g determined by dilatometry (measurement of specific volume as a function of temperature) and most other methods depend on the heating rate. The faster the rate of heating during the determination, the higher the apparent T_g. When free volume is small, the rate of movement of molecules or segments must be slow. If the rate of heating is slow, there is more time for movement, and hence expansion, and the measured T_g will be lower.

The T_g can depend on the thermal history of the sample. If the sample was cooled very rapidly from a temperature above T_g to below T_g, free volume holes may be "frozen" into the polymer. The T_g of such a sample will be lower than it would have been if the sample had been cooled down to T_g slowly. On standing, even at temperatures somewhat below T_g, the molecules may move slowly resulting in "filling in" some of the free volume holes. This phenomenon is called "densification"; when it occurs, the T_g of the polymer increases. Since the physical properties change during densification, the process is sometimes called "physical aging." In view of these difficulties, some scientists feel that T_g is not a real thermodynamic point. They would say that if the determination of specific volume were done at a slow enough heating rate, no transition would be observed and that, rather than two straight lines as shown in Figure 2.4c, there would be a smooth curve.

Despite these difficulties, T_g is a very useful concept and is well understood in qualitative terms. Relationships between polymer structure and T_g are understood well enough that it is often possible to make reasonable predictions of T_g from knowledge of composition and $\overline{M}_n$. Important factors affecting T_g include the following:

Number Average Molecular Weight. The T_g increases with increasing $\overline{M}_n$, finally approaching a constant value at an $\overline{M}_n$ in the range of 25,000 to 75,000, depending on the polymer structure. It is logical that T_g would be related to $\overline{M}_n$ because $\overline{M}_n$ is inversely proportional to the fraction of chain ends to chain middles; since ends have much more freedom of movement than middles, they can be expected to have more effect on free volume.

Polymer Backbone Flexibility. The T_g is importantly affected by the ease of rotation about bonds in the polymer backbone. For example, the siloxane bond, Si—O—Si, rotates easily; the T_g of poly(dimethylsiloxane) is 146 K [9]. Aliphatic polyethers, such as poly(ethylene oxide) $+CH_2—CH_2—O\, \text{)}_n$, also have low T_g, generally in the range of 158 to 233 K, because there is considerable ease of rotation around the ether bond. The T_g of polyethylene is controversial. Most grades of polyethylene are partially crystalline; only the amorphous areas show a T_g. Also, although we commonly think of polyethylene as being just chains of methylene groups, the backbone is actually substituted to varying degrees with alkyl side chains like ethyl groups. However, all would agree that the T_g of a long linear aliphatic chain is low, perhaps less than 200 K. Substitution on the aliphatic chain by methyl groups, for example, polypropylene, increases the T_g since ease of rotation is decreased. Similarly, substitution of the chain with carboxymethyl groups, for example, poly(methyl acrylate) (PMA), gives a higher T_g, 283 K. And even higher T_gs result when two groups are substituted on alternate carbon atoms of the chain, for example, a methyl and a carboxymethyl group as in poly(methyl methacrylate) (PMMA), T_g 378 K. The presence of rigid aromatic rings in the polymer backbone substantially increases T_g.

Side Chains. Some polymers have side chains pendant from the backbone. If the side chains are several atoms long and are flexible, they will reduce T_g. An example is poly(n-butyl acrylate), which has $COOCH_2CH_2CH_2CH_3$ side chains; its T_g is 219 K. The T_g of poly(t-butyl acrylate) is much higher, 314 K, because its side chain is short and inflexible. The rigidity of the aromatic ring leads to a high T_g, 373 K, for poly(styrene).

Extensive lists of the T_gs of polymers are available [9]. Considerable care must be taken in comparing T_g values to be sure that the determinations have been done under comparable conditions and that the molecular weight is high enough to eliminate molecular weight effects.

Solutions of polymers in solvents and of solvents in polymers have T_gs intermediate between the T_g of the polymer and that of the solvent. The T_g of solutions increases with increasing polymer concentration. When small amounts of solvent are dissolved in the polymer, a simple mixing equation 2.1, where w_s is weight fraction of solvent and k is a constant, gives reasonable correlation between experimental and predicted results when w_s is less than about 0.2 [10].

$$T_{g(\text{solution})} = T_{g(\text{polymer})} - kw_s \qquad (2.1)$$

Over a wider range of concentrations, Eq. 2.2 gives poor correlations. In the case of solutions of oligomeric *n*-butyl methacrylate in *m*-xylene [5], Eq. 2.1 gave a good fit between observed and predicted data over the whole range from pure solvent to solvent-free oligomer. Here w_s and w_o are weight fractions and T_{g-s} and T_{g-o} are T_gs of solvent and oligomer, respectively. Equation 2.2 will have to be evaluated over a range of oligomer and polymer solutions to see if it is general.

$$\frac{1}{T_{g(\text{solution})}} = \frac{w_s}{T_{g-s}} + \frac{w_o}{T_{g-o}} + Kw_sw_o \qquad (2.2)$$

We will have many occasions throughout this text to discuss T_g. An excellent general review is available in Ref. [11]. We also refer frequently to the effects of availability of free volume on various aspects of coatings. Reference [12] provides a review of free volume considerations in coatings.

2.2. POLYMERIZATION

There are two major classes of polymerization reactions: *chain-growth* and *step-growth* polymerizations. The common denominator of chain-growth polymerizations is that the reactions are chain reactions. Frequently, this class of polymerization is called addition polymerization. The latter terminology is inadequate since, while all chain-growth polymerizations involve addition reactions, not all addition polymerizations involve chain-growth reactions—some are step-growth reactions. Terms such as *chain-growth addition* or *chain addition* polymerization seem needlessly cumbersome, so we have chosen to use the term chain-growth.

2.2.1. Chain-Growth Polymerization

Free radical initiated chain-growth polymerization, the most commonly used in the coatings field, is a complex process. Three types of chemical reactions, *initiation*, *propagation*, and *termination* are always involved, and a fourth type, *chain transfer*, often plays a significant role. In this chapter, we briefly describe the polymerization process; further detail will be found in Chapters 4, 5, and 7.

Initiation occurs when an *initiator* (I) reacts to form an initiating free radical

(I·) which, in turn, adds very rapidly to a monomer molecule to form a second free radical: Eqs. 2.3 and 2.4, respectively.

$$I \longrightarrow I\cdot \tag{2.3}$$

$$I\cdot \;+\; H_2C{=}\overset{\displaystyle H}{\underset{\displaystyle Y}{C}} \longrightarrow I{-}CH_2{-}\overset{\displaystyle H}{\underset{\displaystyle Y}{C}}\cdot \tag{2.4}$$

The polymer chain grows during the propagation reaction in which the monomer free radical adds to a second monomer molecule to extend the chain while forming a new free radical, Eq. 2.5.

$$I{-}CH_2{-}\overset{\displaystyle H}{\underset{\displaystyle Y}{C}}\cdot \;+\; H_2C{=}\overset{\displaystyle H}{\underset{\displaystyle Y}{C}} \longrightarrow I{-}CH_2{-}\overset{\displaystyle H}{\underset{\displaystyle Y}{C}}{-}CH_2{-}\overset{\displaystyle H}{\underset{\displaystyle Y}{C}}\cdot \tag{2.5}$$

Propagation reactions are extremely fast—so fast that a chain with hundreds of mers can grow in a small fraction of a second. At any moment, the concentrations of monomer and polymer greatly exceed the concentration of growing polymer molecules ($\sim 10^{-6}M$). With free radical initiated polymerizations, the rate of propagation, Eq. 2.5, is much faster than the rate of initiation, which is limited by the rate of Eq. 2.3.

The final stage is termination of the growing chain. Two common types of termination reactions are *combination*, Eq. 2.6, and *disproportionation*, Eq. 2.7. Theoretical calculations show that for high molecular weight polymers, under ideal circumstances, the lowest $\overline{M}_w/\overline{M}_n$ attainable in free radical polymerization with termination by combination is 1.5 and the lowest attainable with termination by disproportionation is 2.0.

$$\text{(P)}{\sim}CH_2{-}\overset{\displaystyle H}{\underset{\displaystyle Y}{C}}\cdot \;+\; \cdot\overset{\displaystyle H}{\underset{\displaystyle Y}{C}}{-}CH_2{\sim}\text{(P)} \longrightarrow$$

$$\text{(P)}{\sim}CH_2{-}\overset{\displaystyle H}{\underset{\displaystyle Y}{C}}{-}\overset{\displaystyle H}{\underset{\displaystyle Y}{C}}{-}CH_2{\sim}\text{(P)} \tag{2.6}$$

$$\text{(P)}{\sim}CH_2{-}\overset{\displaystyle H}{\underset{\displaystyle Y}{C}}\cdot \;+\; \cdot\overset{\displaystyle H}{\underset{\displaystyle Y}{C}}{-}CH_2{\sim}\text{(P)} \longrightarrow \text{(P)}{\sim}\overset{\displaystyle H\;H}{\underset{\displaystyle Y}{C}{=}C}$$

$$+\; H{-}\overset{\displaystyle H}{\underset{\displaystyle Y}{C}}{-}CH_2{\sim}\text{(P)} \tag{2.7}$$

As would be expected, when highly reactive species like free radicals are involved, side reactions also occur. Among the most important are chain-transfer reactions, in which the free radical on the end of the propagating polymer chain abstracts a hydrogen atom from some substance, X—H, that is present in the polymerization reaction mixture, as shown in Eq. 2.8.

$$X-H \;+\; \underset{Y}{\overset{H}{\underset{|}{\overset{|}{C}}}}-CH_2 \text{ℳ} \text{(P)} \longrightarrow X\cdot \;+\; H-\underset{Y}{\overset{H}{\underset{|}{\overset{|}{C}}}}-CH_2 \text{ℳ} \text{(P)} \qquad (2.8)$$

The net effect of chain transfer is to terminate the growing chain while generating a free radical that may start a second chain growing. X—H may be a molecule of the solvent, a molecule of polymer, or a *chain-transfer agent* deliberately added to modify the molecular weight. When it is a solvent or a chain-transfer agent, molecular weight is reduced.

Note that the structures of the propagating polymer chains show substitution on alternate carbon atoms. This results from the favored addition of the free radicals to the CH_2 end of the monomer molecules, so-called "head-to-tail addition." With most monomers, this is the predominant reaction but some small fraction of the additions take place to the other end of the monomer molecule, so-called head-to-head addition. This leads to polymer chain segments with substitution on adjacent carbon atoms. The effect of a small fraction of head-to-head structure is generally negligible, but it sometimes has significant consequences for weatherability and thermal stability.

Initiators, sometimes incorrectly called catalysts, are used in low concentration (usually in the range of 1 to 4 wt%). A variety of free radical sources has been used. Two classes of compounds are the most generally used: azo compounds such as azobisisobutyronitrile (AIBN) and peroxides such as benzoyl peroxide (BPO).

Azobisisobutyronitrile is fairly stable at 0°C, but when heated to 70–100°C it decomposes relatively rapidly to generate free radicals, as shown in Eq. 2.9. A substantial fraction of the resulting radicals initiates polymerization. The half-life of AIBN is about 5 h at 70°C and about 7 min at 100°C.

$$N\equiv C-\underset{CH_3}{\overset{CH_3}{\underset{|}{\overset{|}{C}}}}-N=N-\underset{CH_3}{\overset{CH_3}{\underset{|}{\overset{|}{C}}}}-C\equiv N \longrightarrow 2\; N\equiv C-\underset{CH_3}{\overset{CH_3}{\underset{|}{\overset{|}{C}}}}\cdot \;+\; N_2 \qquad (2.9)$$

AIBN

Benzoyl peroxide decomposes at similar temperatures—its half-life is about 20 min at 100°C. The benzoyl free radical generated initially may initiate polymerization; also it may further dissociate (rapidly at higher temperatures such as 130°C) to yield a highly reactive phenyl free radical and CO_2, as shown in Eq. 2.10.

$$\text{⬡}-\overset{O}{\overset{\|}{C}}-O-O-\overset{O}{\overset{\|}{C}}-\text{⬡} \longrightarrow 2\;\text{⬡}-\overset{O}{\overset{\|}{C}}-O\cdot$$

BPO

$$\longrightarrow 2\;\text{⬡}\cdot \;+\; 2\,CO_2 \qquad (2.10)$$

A wide range of monomers is capable of propagating a radical initiated chain reaction. Almost all are alkenes: methyl acrylate (MA) and methyl methacrylate (MMA) are examples.

$$H_2C=\overset{\overset{\displaystyle H}{|}}{\underset{\underset{\displaystyle COOCH_3}{|}}{C}} \qquad\qquad H_2C=\overset{\overset{\displaystyle CH_3}{|}}{\underset{\underset{\displaystyle COOCH_3}{|}}{C}}$$

$$\text{MA} \qquad\qquad\qquad \text{MMA}$$

The two processes for free radical chain-growth polymerizations of most interest to coatings scientists are *solution polymerization* and *emulsion polymerization*. Solution polymerization is discussed in Section 4.1.1 and Chapter 7; emulsion polymerization is discussed in Chapter 5. Another very complex example related to chain-growth polymerization of considerable importance in coatings is the *autoxidation polymerization* involved in the cross-linking of drying oils and drying oil derivatives, discussed in Chapter 9.

Chain-growth polymerization may also occur by *anionic polymerization*. In this case, polymerization is initiated by addition of an anion to a monomer molecule to form a new anion, which leads to polymer propagation. Anionic polymerization of methacrylates (but not acrylates) can be effected by initiating polymerization with strong bases such as alkali metal alkoxide in the presence of limited concentrations of alcohol and minimal concentration of water. Anionic polymerization has the potential advantage over its free radical initiated counterpart of permitting relatively narrow molecular weight distribution: very low $\overline{M}_w/\overline{M}_n$ is theoretically possible and $\overline{M}_w/\overline{M}_n$ values of 1.1 to 1.3 are attainable in practice. Theoretical $\overline{M}_w/\overline{M}_n$ values attainable by free radical and anionic polymerization are quite different because the processes follow different kinetics, as explained in polymer textbooks such as Odian's (see general reference section).

Group-transfer polymerization and other relatively new processes for chain-growth polymerization are briefly discussed in Section 7.2.

2.2.2. Step-Growth Polymerization

The second type of polymerization of central importance in coatings is *step-growth polymerization*. It results from the reaction of functional groups with each other and, as the name indicates, the polymer is built up a step at a time. The term "condensation polymerization" has also been widely used for this process because early examples involved *condensation reactions*—reactions in which a small molecule byproduct such as water is eliminated. While both terms are still used, *step-growth* seems the more appropriate because some of the reactions used today are not condensation reactions.

Step-growth polymerization reactions are used in two ways in coatings applications. One is for the preparation of resins for use as coatings vehicles, and the other is the use of such reactions for cross-linking after the coating has been applied to the substrate.

Esterification is one class of reaction that can be used in step-growth polymerization to prepare resins. Of the many reactions that form esters, three are commonly used in making step-growth polymers and oligomers for coatings: direct esterification of an acid with an alcohol, Eq. 2.11.

$$R-\overset{O}{\overset{\|}{C}}-OH \ + \ HO-R' \ \underset{\longleftarrow}{\overset{Cat.}{\longrightarrow}} \ R-\overset{O}{\overset{\|}{C}}-O-R' \ + \ H_2O \qquad (2.11)$$

transesterification of an ester with an alcohol, Eq. 2.12, and

$$R-\overset{O}{\overset{\|}{C}}-O-R'' \ + \ HO-R' \ \underset{\longleftarrow}{\overset{Cat.}{\longrightarrow}} \ R-\overset{O}{\overset{\|}{C}}-O-R' \ + \ HO-R'' \qquad (2.12)$$

reaction of an anhydride with an alcohol. Eq. 2.13.

$$R-\overset{O}{\overset{\|}{C}}-O-\overset{O}{\overset{\|}{C}}-R \ + \ HO-R' \ \longrightarrow \ R-\overset{O}{\overset{\|}{C}}-O-R' \ + \ HO-\overset{O}{\overset{\|}{C}}-R \qquad (2.13)$$

When one of the reactants undergoing esterification is monofunctional, polymer cannot form. If, however, all of the reactants have two or more functional groups, a polymer can be made. When all monomers are difunctional, linear polymers will form. Linear step-growth polymers are commonly used in fibers, films, and plastics, but there are few examples of their use in the coatings field. Most *polyester resins* used in coatings have relatively low molecular weights and are usually branched resins made using at least one monomer with three or more functional groups.

When a difunctional acid AA reacts with a difunctional alcohol BB in a direct esterification reaction, the molecular weight builds up gradually. Under ideal conditions polymer chains averaging hundreds of mers per molecule can be made, but this can occur only if (a) the reactants, AA and BB, contain no monofunctional impurities, (b) exactly equimolar amounts of AA and BB are used, (c) the reaction is driven virtually to completion, and (d) side reactions are negligible. If one of the reactants is present in excess, terminal groups of the excess monomer will predominate. Molecular weight of the completely reacted system will be progressively lower as the difference from equal equivalents is increased. For example, if 7 mol of dibasic acid are completely reacted with 8 mol of a dihydroxy compound (a diol), the average molecule will have terminal hydroxyl groups as shown in the following equation. (Here, for convenience, AA and BB represent both the reactants and the mers in the polymer.)

$$7AA \ + \ 8BB \ \longrightarrow \ BB(AA-BB)_6AA-BB \ + \ 14H_2O$$

The course of step-growth polymerization reactions is quite different from that of chain-growth polymerization. For example, if a chain-growth polymerization is stopped before it has gone to completion, the reaction mixture will consist of high molecular weight polymer and unreacted monomer. However, if a step-growth polymerization is stopped at a similar stage, the reaction mixture will contain mainly relatively low molecular weight reaction products and monomer but very little high molecular weight polymer. In the case of high molecular weight linear polymers, under ideal conditions, the $\overline{M}_w/\overline{M}_n$ obtained in step-growth polymerizations is 2.

Most step-growth coatings resins are branched. After application of the coating, the terminal groups on the branch ends are reacted with a cross-linker to form the final binder. Branching is built into step-growth polymers by using at least one monomer having more than two functional groups. The symbol F is used for the functionality of monomers. The term F is simply the number of reactive groups per molecule, for example:

$$\underline{F = 2} \qquad \underline{F = 3} \qquad \underline{F = 4}$$

$\begin{array}{c} CH_3 \\ | \\ HOCH_2-C-CH_2OH \\ | \\ CH_3 \end{array}$	$\begin{array}{c} H_2C-OH \\ | \\ HC-OH \\ | \\ H_2C-OH \end{array}$	$\begin{array}{c} CH_2OH \\ | \\ HOCH_2-C-CH_2OH \\ | \\ CH_2OH \end{array}$
Neopentyl Glycol	Glycerol	Pentaerythritol

Phthalic Anhydride   Trimellitic Anhydride

$$HO-\overset{\overset{O}{\|}}{C}\mathbf{+}CH_2\mathbf{)}_4\overset{\overset{O}{\|}}{C}-OH$$

Adipic Acid

Note that the anhydride groups in phthalic anhydride and trimellitic anhydride count as two functional groups because each anhydride group can form two ester groups during polymerization.

A second type of functionality is the average functionality of a mixture of monomers, $\overline{F}$. When the monomer mixture contains equal equivalents of hydroxyl and carboxyl groups, $\overline{F}$ is calculated from

$$\overline{F} = \frac{\text{total equivalents}}{\text{total moles}}$$

Most coating polyester resins are hydroxy-functional and are, therefore, made using monomer mixtures with excess hydroxyl groups. In this case, some of the hydroxyl groups have no carboxyl groups to react with so the equation must be modified to reflect only the total number of equivalents that can react. In a resin with excess hydroxyl groups, prepared from dicarboxylic acids ($F = 2$ only), the total equivalents that can react correspond to twice the number of equivalents of carboxylic acid groups.

$$\overline{F} = \frac{\text{total equivalents that can react}}{\text{total moles}} = \frac{2(\text{equiv of COOH})}{\text{total moles}}$$

As an example, consider a simple formulation for a polyester oligomer:

Adipic acid	0.9 mol	1.8 equiv
Phthalic anhydride	0.9 mol	1.8 equiv
Neopentyl glycol	1 mol	2 equiv
Glycerol	1 mol	3 equiv
Total	3.8 mol	8.6 equiv

$$\overline{F} \quad = \quad 2(3.6)/3.8 = 1.89$$

A further type of functionality of importance in designing resins to be used in cross-linking coatings, is the functionality of the resin. To distinguish this functionality from that of the monomers and the monomer mixture, the symbol f is used. Since almost all coatings resins are made using some triol or tetrol, one must usually use a number average functionality, $\bar{f}_n$.

$$\bar{f}_n = \frac{\text{number of functional groups in a sample}}{\text{number of molecules in the sample}}$$

The value of $\bar{f}_n$ can be calculated from $\overline{M}_n$ and the number of functional groups per sample weight obtained by analysis. A weight average functionality could, in principle, be calculated but would be difficult to measure.

Polyesters are just one class of step-growth polymers used in coatings. They and a wide range of others, are discussed at length in Chapters 6, 8, and 10 through 13.

2.3. RHEOLOGY

Rheology is the science of flow and deformation of materials. Rheological properties critically affect the application and performance of coatings. In the application of coatings, flow of liquids controls the formation and appearance of the film. Chapter 19 covers this topic in detail. A major factor affecting the performance of applied coatings films is their resistance to mechanical failure during deformation. Chapter 24 covers this important topic. In this section, we present an introduction, so as to have before the reader some of the rheological considerations that are important in the design of resins for coatings.

Liquids flow in various ways. The type that is most commonly measured in the coatings field is *shear flow*. Shear flow can be visualized by imagining a liquid between two solid plates; one fixed and one movable. When the movable plate is pulled or pushed, the liquid is subjected to *shear stress*, force per unit area, which is applied at some *shear rate*, velocity per unit thickness of the layer of liquid. *Viscosity* (dynamic shear viscosity in this case) is a measure of resistance to flow and is defined as the shear stress divided by the shear rate. The SI units of viscosity are pascal seconds (Pa·s); in older literature the CGS unit, poise (P), is often used (1 Pa·s = 10 P). When flow is driven by gravity, *kinematic viscosity* is measured. The SI units of kinematic viscosity are square meters per second ($m^2\ s^{-1}$), formerly stokes (St): $1\ m^2\ s^{-1} = 10^4$ St. Kinematic viscosity can be converted to dynamic viscosity by multiplying kinematic viscosity by density.

In the case of ideal liquids, the ratio of shear stress to shear rate, and hence the viscosity of the liquid, is constant for any shear stress. If the force per unit area is doubled, the rate of flow will double. Such ideal liquids are said to exhibit Newtonian flow. Almost all low molecular weight liquids, solutions of resins in good solvents, and dispersions of unassociated rigid particles in Newtonian fluids exhibit Newtonian flow. On the other hand, *solutions* of resins in poorer solvents in which there are clusters of resin molecules, or dispersions of liquids in liquids (emulsions), or dispersions in which there are particles that are attracted to each other usually exhibit non-Newtonian flow. In most of these cases that are important in coatings, viscosity at high shear rate is lower than it is at low shear rates. Such liquids are called *shear thinning fluids*. In some shear thinning fluids, the extent of shear thinning at any given shear rate is dependent on shear history and duration of application of shear stress—these systems are said to exhibit "thixotropy." Further details on these and other systems can be found in Chapter 19.

The viscosity of solutions of resins in good solvents increases with increasing concentration. In the range of 0.01 to 10 Pa·s, the range of greatest importance in coatings, the log of the viscosity increases approximately proportionally with the concentration. [Over a wider range of concentrations, the dependence is somewhat greater than proportional (see Section 19.4.2).] In good solvents, in this same range of viscosity, the log of the viscosity increases with the square root of the molecular weight of the resin. While applicable only over this relatively narrow range, Eq. 2.11 provides a useful approximation of the relationship of viscosity, η, with concentration C and molecular weight M where η_s is the viscosity of the solvent and K is a constant.

$$\ln \eta = \ln \eta_s + KCM^{1/2} \tag{2.14}$$

The effect of temperature on viscosity is important. Viscosity almost always decreases as temperature increases. In more general terms, viscosity of Newtonian liquids decreases as the free volume in the liquid increases. Since free volume is related (to a large degree) to the difference between the temperature and the T_g of the fluid, an important factor controlling viscosity is $(T - T_g)$. Higher T_g fluids will generally have higher viscosities at a given temperature. Therefore, it follows that the factors that govern T_g, discussed in Section 2.1.4, also affect viscosity. Structural factors such as chain rigidity and increasing molecular weight increase viscosity. An important difference, however, is that the influence of molecular weight on T_g levels off at high molecular weight, but the influence of molecular weight on viscosity does not level off.

The binders in the final coating films are almost always amorphous, as noted in Section 2.1.4. The viscosity of amorphous materials depends on the availability of free volume. Referring back to Figure 2.4b, we can see that free volume depends on $(T - T_g)$ and on the difference in the thermal coefficients of expansion above and below T_g. Mathematically, the dependence of viscosity on temperature of thermoplastic amorphous materials can be expressed by the empirical Williams, Landel, and Ferry (WLF) equation (2.15), where η_{T_g} is the viscosity at T_g, A is a constant related to the difference in thermal expansion coefficients, and B is a constant for a particular amorphous material corresponding to the number of degrees below T_g at which the viscosity would be infinite [13].

$$\ln \eta = \ln \eta_{T_g} - \frac{A(T - T_g)}{B + (T - T_g)} \tag{2.15}$$

The WLF equation is very useful and we will refer to it on many occasions, especially in Chapters 3 and 19. Its value results from using an extensive knowledge of the factors that affect the T_g of systems to estimate effects of changes in composition on viscosity.

As can be seen from Eq. 2.15, at temperatures well above T_g, amorphous thermoplastic materials are obviously liquid, but at temperatures nearer to or below T_g, they flow only slowly. At T_g, viscosity for most amorphous materials is about 10^{12} Pa·s. In the temperature range near T_g, amorphous materials show *viscoelastic* behavior, that is, they behave partly like an elastic material and partly like a viscous material. An elastic material can be deformed by applying a stress: however, when the stress is released, a perfectly elastic material will return immediately to its original shape. On the other hand, a nonelastic, viscous material deformed by stress keeps whatever shape it has at the moment the stress is released. When stress is applied and released in the case of a viscoelastic material, the material partially, but not completely, recovers its original shape.

Pure elastic deformation and recovery are virtually instantaneous, but viscous flow takes time. Therefore, the amount of deformation and the degree of recovery of a viscoelastic material will be dependent on the rate of application of stress and on the length of time that a stress is applied. At temperatures well below T_g, the response will be completely elastic. Of course, there will be some limit to the amount of deformation that such a material can undergo without breaking. If the amount of this deformation before breaking is small (as it generally is under such conditions), we call the material *brittle*. In the temperature range from somewhat below T_g to approximately 30°C above T_g, amorphous materials are viscoelastic. If the stress is applied rapidly, they will behave like elastic materials, that is, they will be brittle. If the stress is applied more slowly, there will be time for viscous flow and a better chance that the material will resist the deformation without breaking. Further information on viscoelasticity can be found in polymer texts such as Refs. [8] and [12] and in Chapter 24.

In the case of Newtonian systems with a dispersed phase, the volume fraction of the dispersed phase V_i affects the viscosity. In systems where the particles are rigid and where there is no particle–particle interaction, viscosity follows the Mooney equation 2.13, where K_E is a shape factor, 2.5 in the case of spheres, η_e is the viscosity of the *external* (continuous) phase, and ϕ is the packing factor [14].

$$\ln \eta = \ln \eta_e + \frac{K_E V_i}{1 - V_i/\phi} \qquad (2.16)$$

As can be seen in Eq. 2.16, when V_i is small, the effect of the dispersed phase on viscosity is small and the viscosity depends primarily on the viscosity of the external phase. At the other extreme when V_i equals ϕ, the particles are close-packed; no more particles will fit in, and the viscosity will be infinite. When V_i approaches ϕ, the viscosity increases very rapidly with small increases in V_i. The packing factor depends on the shape of the particles. In the case of monodisperse spheres that are randomly close-packed, the value of ϕ is 0.637. With monodisperse spheres, the packing factor is independent of particle size. However, in the case of spherical particles with mixed diameters, the packing factor depends strongly on particle size distribution. The broader the distribution, the larger the packing factor; at the same volume fraction, the viscosity of a dispersion of spheres of uniform size will be higher than when the spheres have a range of diameters. The effect of particle size distribution on viscosity becomes particularly important in latex systems (see Chapters 5 and 35) and in considering the effect of pigments on viscosity and film properties (see Chapters 20 and 21).

When the internal phase is not rigid, the dispersed droplets can be distorted when shear is exerted on them. The distortion results in a decrease of the shape constant and an increase in the packing factor; the viscosity decreases with shear. Such systems are shear thinning and often thixotropic. If there are clusters of particles with particle–particle interactions such that the attractive forces between the particles are weak enough to be overcome by low to moderate shear stress, the system is said to be "flocculated." Flocculated systems are also shear thinning and often thixotropic.

2.4. SURFACE TENSION

Surfaces of liquids behave differently from their bulk. Since coatings are applied as liquid films with a high ratio of surface/bulk, it is not surprising that the characteristics of the surfaces can become very critical in the application of coatings. As schematically shown in Figure 2.5, the molecules inside the bulk of the liquid can be thought of as being equally attracted in all directions by the surrounding molecules, but molecules at the surface are attracted more strongly by the nearby molecules in the liquid than they are by the molecules of the gas at the air interface. The result is a force that strives to minimize the surface area of the liquid. We call this force surface tension, the most widely used symbol is γ. Its dimensions are force exerted in the surface perpendicular to a line of length, newtons per meter in SI units, or more commonly, millinewtons per meter, (mN m^{-1}). [Older literature expresses surface tension in dynes per centimeter (1 mN m^{-1} = 1 dyn cm^{-1}).] Alternatively, one can consider surface tension as resulting from surface molecules having higher energy than bulk molecules, which experience greater stabilization by attractive interactions with surrounding molecules. Surface tension can also be expressed as energy per unit area, the units of which simplify to force per length.

In the case of solids, the same type of surface orientation effects are present but the material cannot flow to minimize surface area. Still the surface of a solid is under tension, and it has a surface tension.

The forces of nature work so as to decrease the surface free energy of all liquids and solids. The higher the surface tension of a liquid, the higher the surface free

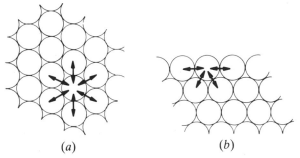

(a)   (b)

Figure 2.5. Schematic representation of forces acting on a molecule (a) in the bulk of a liquid and (b) at the surface of a liquid.

encrgy between the liquid and air. Therefore, the force of surface tension works to draw liquids into spheres since a sphere has the minimum ratio of surface area/volume. In a spaceship, liquid droplets assume an essentially perfect spherical shape; on earth, the force of gravity distorts the spheres.

If a layer of relatively high surface tension liquid is placed on the surface of a solid or immiscible liquid having lower surface tension, the liquid will draw up into droplets since that will minimize the overall surface free energy. The droplet seeks the shape that represents the lowest free energy balance between the forces exerted by surface tension and gravity. If, on the other hand, a liquid of relatively low surface tension is placed on a surface having higher surface tension, the liquid will spread spontaneously to cover the high surface tension surface so as to minimize surface free energy. If the surface of a layer of liquid on the solid is uneven, the surface tension of the liquid will work to make the liquid flow to a layer with a smooth surface, since the smooth surface will have less interfacial area with the air than the uneven surface.

One can also consider the surface effect in terms of the fact that the surface molecules are attracted back into the liquid by their neighboring molecules. In other words, it takes work to remove the surface layer of molecules and the higher this amount of work, the higher the surface tension. Liquids whose molecules interact strongly with each other have higher surface tensions than those where the interactions are weaker. The lowest surface tension solvents that we normally encounter are aliphatic hydrocarbons. Methyl groups give lower surface tensions than methylene groups, therefore, the surface tensions of straight-chain hydrocarbons increase as chain length increases. Oxygenated solvents generally have higher surface tensions than hydrocarbons, but the difference is less than might be expected because the molecules orient so that their least polar groups face the surface. Thus, surface tensions of alcohols are low and increase in the homologous series MeOH < EtOH < n-PrOH < n-BuOH, perhaps because the surface concentration of OH is low in each case and the methylene groups have higher surface tension than methyl groups. Surface tensions of solutions of resins in almost all cases increase with increasing concentration. Surface tensions also depend on temperature, decreasing as temperature increases.

Water, whose molecules interact very strongly, has the highest surface tension of any *solvent* used in coatings. The surface tension of liquids, especially water, can be lowered by adding *surfactants*. A surfactant is a substance with a long water immiscible aliphatic chain (hydrophobic) segment and a highly polar (hydrophilic) segment. The surface tension of water is reduced by addition of a surfactant since the surfactant orients at the surface with its hydrophobic segments towards the air and its hydrophilic segments inwards towards the water. There are three broad classes of surfactants: *anionic*, in which the hydrophilic segment is a negative ion such as a carboxylate or sulfate; *cationic*, in which the hydrophilic segment is a positive ion such as an amine salt; and *nonionic*, in which the hydrophilic segment is a chain of electrically neutral, water-miscible units such as poly(ethylene oxide).

Additives can reduce the surface tension of organic solvents and solutions as well. For example, addition of poly(dimethylsiloxane) to an organic solvent reduces surface tension below that of any straight-chain hydrocarbon. The chain of a siloxane molecule has great freedom of rotation and approaches linearity so that molecules of siloxane orient easily at the surface with all methyl groups facing the interface. Certain perfluorocarbons also substantially reduce the surface tension of organic solutions. It is important to remember that it is not possible to design an additive that will increase the surface tension of a solution.

Another surface tension flow effect that is of great importance in the coatings field is the *Marangoni effect*. If there is a differential in surface tension on the surface of a liquid, that part of the liquid with a lower surface tension will flow over the surface with a higher surface tension. This surface tension differential-driven flow can be important even when the differentials are small. Such differentials often result from uneven evaporation of solvent from different parts of a liquid coating after application. Some consequences of this phenomenon are discussed in Chapter 23.

Other effects of surface tension are addressed in succeeding chapters such as Chapter 20 on Pigment Dispersion, Chapter 23 on Film Defects, and Chapter 26 on Adhesion.

2.5. COLLOIDS

Colloids are dispersions of small particles in a continuous medium. Colloidal particles range in size from 1 nm to about 10 μm. There are many examples of colloidal dispersions involved in the coatings field; some examples are latexes and pigment dispersions. We can only summarize those aspects of colloids science of greatest importance in the coatings field.

Dispersed particles are continually subjected to the impact of molecules of the continuous phase and exhibit *Brownian movement*. These collisions help keep the dispersed particles suspended. However, most dispersions will separate over time. Almost all dispersions that are important in the coatings field are of materials the density of which is higher than that of the continuous phase in which they are dispersed. Eventually, the higher density particles settle. The rate of settling increases with increasing particle size and with increasing differential of density between the particles and the continuous phase. The rate of settling decreases with increasing viscosity of the continuous phase.

Colloidal dispersions are not thermodynamically stable and must be stabilized in some way so that the particles will not aggregate within the useful life of the dispersion. If the particles approach each other closely, van der Waals' forces will tend to hold them together; the particles "flocculate." As noted in Section 2.3, flocculation of a dispersed phase system leads to changes in flow properties; viscosity increases and flow becomes shear thinning. In the case of pigment dispersions, flocculation leads to reduction in color strength and hiding as is discussed in Chapter 20. Properties of films made from coatings with flocculated pigments and/or flocculated latex particles can be substantially different from those with the same composition but with stable dispersions. Also, gloss of films from coatings with flocculated pigments is generally lower.

2.5.1. Charge Repulsion

There are two general mechanisms by which stabilization of colloidal dispersions, important in coatings, is effected: charge repulsion and repulsion by adsorbed, solvated resin layers. In each case, the term repulsion refers to a force that causes the particles to repel each other when they approach each other, so that there is little observable deterioration of the dispersion over the lifetime.

In the case of charge repulsion, the surfaces of the particles are designed so that there will be an excess of one electrostatic charge. In dispersions used in coatings, this charge is most commonly negative. For example, a dispersion of a polymer in water—a latex—can be stabilized by having an anionic surfactant adsorb on the surfaces of the polymer particles. The surfactant molecules orient with the long hydrophobic, hydrocarbon tails in the polymer and the hydrophilic salt groups on the periphery associated with the water. As a result, the surface of the particles can be covered with anions; each anion will have an associated cation. Thus the surfaces of the particles will have a layer of negative charges with an associated layer of positive charges, called a "Stern layer." The Stern layer is quite rigid and behaves as if it were physically part of the particle. Its presence induces a second diffuse layer of anions to surround the particle giving a negatively charged so-called "double layer." When two particles approach each other, their diffuse, negatively charged layers electrostatically repel each other thus stabilizing the latex. Similar stabilizing layers can be applied to surfaces of pigments to stabilize dispersions of pigments in water. While somewhat less effective in solvent media, charge repulsion can also be utilized to stabilize dispersions of pigments in resin solutions.

The stability of charge repulsion stabilized dispersions in water can be affected by the addition of salts to the system. The stability is particularly sensitive to the addition of multivalent ions of the opposite charge of the stabilizing charge. Hence deionized water is generally used in aqueous dispersions.

2.5.2. Repulsion by Adsorbed, Solvated Resin Layers

The second mechanism for stabilization of dispersions is by repulsion resulting from adsorbed, solvated resin layers on the surface of the colloidal particles. Three terms are widely used in the literature to characterize this type of stabilization: *steric repulsion*, *entropic repulsion*, and *osmotic repulsion*. A layer of resin (or other material) swollen with solvent is adsorbed on the surface of the particles. If the layer is thick enough, the particles will not be able to approach each other sufficiently closely to flocculate. Rehacek [15] has done the seminal work on the stabilization of pigment dispersions by absorbed layers. He and several other workers have found that, if there is an adsorbed layer of resin swollen with solvent with an average thickness greater than 8–9 nm, the pigment dispersion will be stable. If the adsorbed layer is less than 8–9 nm in thickness, flocculation will occur. The work is discussed in Section 20.1.3.

Figure 2.6 shows an adsorbed layer schematically; the wavy lines represent partly adsorbed polymer molecules and S represents solvent molecules. Often polar groups such as OH and COOH of the polymer are adsorbed leaving less polar parts of the chain projecting out from the pigment surface in "loops and tails." The adsorbed layer is dynamic; the polar groups can adsorb and desorb, but since each polymer molecule has multiple polar groups, there is little probability of entire molecules desorbing. The polymer molecules can exist in a very large number of conformations; some average layer thickness will be present at equilibrium. In general, maximum film thickness is obtained with high molecular weight resins having a

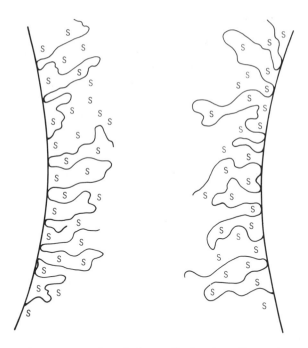

Figure 2.6. Schematic representation of a layer of solvent-swollen resin on a pigment surface.

limited number of adsorbing groups spaced along the backbone chain. Solvent molecules can move in and out of the adsorbed layer; there will be some average concentration present at equilibrium.

When two particles with adsorbed layers approach each other, the layers are compressed. If compression occurs, the number of conformations that the resin molecules can assume is reduced. As a result, the system would become less random, which requires a reduction in entropy. Resistance to the reduction in entropy leads to repulsion, hence the term "entropic repulsion." Some authors prefer to focus on the reduction in the amount of solvent in the adsorbed layer if compression occurs. There will be tendency for the solvent to return to the equilibrium concentration in the adsorbed layer; some consider this to be analogous to osmosis and call the result "osmotic repulsion." Since the amount of solvent in the layer was at equilibrium, expulsion of solvent would also require a decrease in entropy, resulting in a further element of entropic repulsion. Some authors prefer to avoid the controversy over terms by simply calling the repulsion, "steric repulsion." None of the terms are fully satisfactory. We will use the term entropic repulsion, but it should be understood that factors other than entropy may well be involved.

Entropic repulsion can also be applied to the stabilization of dispersions in water such as latexes, as discussed in Chapter 5. The stability of entropically stabilized dispersions against the effects of added ionic materials is greater than with charge repulsion stabilized dispersions. If the adsorbed layer is thick, there is sufficient increase in the volume of the internal phase, that the viscosity will be higher than that obtained with either thinner adsorbed layers or with charge repulsion stabilized dispersions. Since such thick adsorbed layers can be distorted, dispersions with thick adsorbed layers show shear thinning flow.

GENERAL REFERENCES

F. W. Billmeyer, Jr., *Textbook Of Polymer Science*, 3rd ed., Wiley-Interscience, New York, 1984.

G. Odian, *Principles of Polymerization*, 3rd ed., Wiley-Interscience, New York, 1991.

P. Munk, *Introduction to Macromolecular Science*, Wiley-Interscience, New York, 1989.

L. H. Sperling, *Introduction to Physical Polymer Science*, Wiley-Interscience, New York, 1986.

REFERENCES

1. L. W. Hill and Z. W. Wicks, Jr., *Prog. Org. Coat.*, **10**, 55 (1982).
2. H. G. Elias, "Structure and Properties," in *Macromolecules*, Plenum, New York, 1984, pp. 301–371.
3. C. J. Sullivan, D. C. Dehm, E. E. Reich, and M. E. Dillon, *J. Coat. Technol.*, **62** (791), 37 (1990).
4. J. Kumanotani, H. Hironori, and H. Masuda, *Org. Coatings Sci. Technol.*, **6**, 35 (1984).
5. Z. W. Wicks, Jr., G. F. Jacobs, I. C. Lin, E. H. Urruti, and L. G. Fitzgerald, *J. Coat. Technol.*, **57** (725) 51 (1986).
6. L. C. Dickinson, P. Morganelli, C. W. Chu, Z. Petrovic, W. J. Macknight, and J. C. W. Chien, *Macromolecules*, **21**, 338 (1988).
7. L. J. Mathias and R. F. Colletti, *Polym. Preprints, Am. Chem. Soc. Div. Polym. Chem.*, **30** (1), 304 (1989).
8. R. J. Roe, *Encyclopedia of Polymer Science and Engineering*, 2nd ed., Vol. 7, Wiley, New York, 1987, pp. 531–544.
9. W. A. Lee and R. A. Rutherford, "The Glass Transition Temperatures of Polymers," in J. Brandrup and E. H. Immergut, Eds., *Polymer Handbook*, 2nd ed., Wiley, New York, 1975, pp. III-139–III-192.
10. J. D. Ferry, *Viscoelastic Properties of Polymers*, 3rd ed., Wiley, New York, 1980, p. 487.
11. A. Eisenberg, "The Glassy State and the Glass Transition," in J. E. Mark, A. Eisenberg, W. W. Graessley, L. Mandelkern, and J. L. Koenig, Eds., *Physical Properties of Polymers*, American Chemical Society, Washington, 1984, pp. 55–95.
12. Z. W. Wicks, Jr., *J. Coat. Technol.*, **58** (743), 23 (1986).
13. M. L. Williams, R. F. Landel, and J. D. Ferry, *J. Am. Chem. Soc.*, **77**, 3701 (1955).
14. L. E. Nielsen, *Polymer Rheology*, Marcel Dekker, New York, 1977, p. 133.
15. K. Rehacek, *Ind. Eng. Chem., Prod. Res. Dev.*, **15**, 75 (1976).

CHAPTER **III**

Film Formation

Most coatings are liquids with a viscosity appropriate for the application method to be used, generally in the range of 0.05 to 1 Pa·s at high shear rates. After application, the liquid is converted into a *dry*, *solid* film. In powder coatings, the powder is liquefied after application and then converted to a solid film.

If the applied coating were crystalline, there would be no difficulty defining a solid film. The film would be considered solid if the temperature of the material was below its freezing point; however, binders of coating films are almost always amorphous, with no melting point and no sharp demarcation between a liquid and a solid. A useful definition of a solid film is that it does not flow significantly under the pressures to which it is subjected during testing or use. Thus, one can define whether a material is a solid under a set of conditions by stating the minimum viscosity required so that flow will not exceed a specified extent under specified pressure for a specified time. A film will be called dry or solid when it passes such a test; for example, it is reported that a film will be *dry to touch* if its viscosity is greater than about 10^3 Pa·s [1]. However, if the definition of dry is that the film will resist *blocking*, that is, sticking together, when two of the coated surfaces are put against each other for 2 s under a pressure of 1.4 kg cm^{-2} (20 psi), which is a more stringent definition than dry to touch, the viscosity will have to be greater than about 10^7 Pa·s.

For thermoplastic binders, we can use this information to predict polymer structures that could meet such tests. Using a modified form of the WLF equation mentioned in Section 2.3, we can make an estimate of the binder T_g that will be required so that the binder will have a viscosity such that it will not flow under some set of circumstances. Using so-called "universal constants" and assuming that viscosity at T_g is 10^{12} Pa·s, the WLF equation becomes

$$\ln \eta = 27.6 - \frac{40.2(T - T_g)}{51.6 + (T - T_g)}$$

Using this equation, we can estimate the approximate $(T - T_g)$ value required for a film to be dry to touch; that is, to have a viscosity of 10^3 Pa·s. The calculated $(T - T_g)$ value is 54°C, which corresponds to a T_g value of -29°C for a film to be dry to touch at $T = 25$°C. The T_g calculated to be necessary for block resistance (at 1.4 kg cm^{-2} for 2 s) at 25°C (i.e., for a viscosity of 10^7 Pa·s) is 4°C. Because the universal constants are only approximations, these T_g values are not exact but rather are estimates of the T_g required.

Since we have a reasonable idea of the relationships between structure and T_g (see Sections 2.1.1.3 and 4.1), we can approximate the structural requirements to make a binder that will approach the viscosity necessary to pass some test. If, for example, the coating has to pass a test at a higher temperature than 25°C, the T_g of the binder must be higher since the free volume dependence is on $(T - T_g)$. If the pressure to which the film is to be subjected is higher or the time under pressure is to be longer, the T_g must be higher. We can get a rough estimate of the required T_g using the WLF equation.

3.1. FILM FORMATION BY SOLVENT EVAPORATION FROM SOLUTIONS OF THERMOPLASTIC BINDERS

Films can be formed in a variety of ways. Perhaps the most obvious method is to make a coating by dissolving a polymer in solvent(s) at sufficiently low concentration for the viscosity to meet the application requirements, apply the coating, and then allow the solvent to evaporate. Let us use a solution of a vinyl chloride copolymer for illustration purposes; similar solutions were used to line beer and soft drink cans for many years until air pollution regulations required substantial reduction in solvent usage. A copolymer of vinyl chloride, vinyl acetate, and a hydroxy-functional vinyl monomer with an $\overline{M}_n$ of 23,000 is reported to give coatings with good mechanical properties without the need for cross-linking [2]. The T_g of the copolymer is reported to be 79°C.

While a can lining would be formulated with a stabilizer and a mixture of solvents, let us consider a simplified formulation with the polymer dissolved in only one solvent, methyl ethyl ketone (MEK). The viscosity of about 0.1 Pa·s for spray application would permit only about a 19 NVW (nonvolatile weight, i.e., wt % solids) solution of the resin in MEK and a NVV (i.e., vol % solids) about 12. MEK has a high vapor pressure at room temperature and would evaporate rapidly from a film. In fact, a sizable fraction of the MEK would evaporate from the atomized droplets of spray between the time when they left the spray gun and reached the walls of the inside of the can. As the solvent evaporates from the film, the viscosity increases, and the film would be dry to touch soon after application. Also in a short time, the coating would not block under the conditions mentioned earlier. Nevertheless, if the film were formed at 25°C, the dry film would contain several percent of retained solvent. Why?

In the first stage of solvent evaporation from the film, the rate of evaporation is essentially independent of the presence of polymer. Evaporation rate depends on the vapor pressure at the given temperature, the ratio of surface area to volume of the film, and the rate of air flow over the surface. However, as the solvent evaporates from the coating, the viscosity increases, T_g increases, free volume decreases, and the rate of loss of solvent from the film becomes dependent not on how fast the solvent will evaporate but, rather, on how rapidly the solvent molecules can reach the surface of the film so that they can evaporate. The rate of solvent loss becomes controlled by the rate of diffusion of the solvent through the film.

The solvent molecules must jump from free volume hole to free volume hole to reach the surface. As solvent loss continues, T_g increases further, $(T - T_g)$ decreases, free volume decreases, and the rate of solvent loss decreases further. When the T_g of the remaining polymer solution approaches the temperature T at which the film is being formed, the rate of solvent loss will slow even more. If the film is being formed at 25°C from a solution of a polymer which, when solvent free, has a T_g greater than 25°C (in this example, 79°C), the film will retain considerable solvent even though it is a hard, *dry* film. Solvent will slowly leave such a film but it has been shown experimentally that 2–3% of solvent remains in such films after several years at ambient temperature.

Beer drinkers would be up in arms if their beer was packaged in cans with MEK remaining in the interior coating linings. Although MEK alone evaporates very rapidly from a thin layer at 25°C, complete removal of all the MEK from a polymer film in a short time requires heating to a temperature significantly higher than the T_g of the polymer. Beer cans are baked at temperatures of the order of 200°C for up to 2 min. Solvent loss from films will be discussed in more detail in Section 15.1.4.

3.2. FILM FORMATION FROM SOLUTIONS OF THERMOSETTING RESINS

A major drawback of solution thermoplastic polymer based coatings is that the high molecular weights required for desirable film properties necessitate high solvent levels (on the order of 80 to 90 vol %) to achieve the viscosity for application. Substantially less solvent is needed if one makes a thermosetting, cross-linkable coating with low molecular weight resins, and then polymerizes the resins further after application of the coating to a substrate and evaporation of the solvent. Many cross-linking systems have been developed for use in thermosetting coatings and are discussed in Chapters 6 through 13. In this section, we consider the general principles involved.

3.2.1. Factors Affecting Mechanical Properties

A critical aspect of the design of any coating is the selection of components that will give the required mechanical properties. The mechanical properties of the cross-linked film depend strongly on many factors; two of the most important are the lengths of the segments between cross-links and the T_g of the cross-linked resin. Segment length depends on the average equivalent weight, the number average functionality $\bar{f}_n$ of the components, and the fraction of cross-linking sites actually reacted. The size of the segments between cross-links is often expressed as the average molecular weight between cross-links, $\overline{M}_c$. Except in ideal systems this is an oversimplification; $\overline{M}_c$ is more properly defined as the weight per mole of elastically effective network chains. An elastically effective chain is one that connects two segments in a network—not a cyclical chain or dangling end.

The effect of segment length differences can also be expressed in terms of *cross-link density* (XLD). The term cross-link density is sometimes used loosely indicating that when XLD is higher, the cross-links are closer together. Sometimes XLD is defined as the mole fraction of monomer units present as cross-links [3]. Cross-link density is best defined as the number of moles of elastically effective network chains per cubic centimeter of film, ν_e. In reading papers, one must be very careful to note how cross-link density is being defined.

For cases where stoichiometric amounts of reactants are used and the reaction is carried to completion, XLD can be calculated from the equivalent weight of reactants (corrected for the loss of any volatile condensation products), the stoichiometry of the reaction, and the film density. Everything else being equal, the higher the XLD, the higher the modulus; that is, the harder the film. Completeness of reaction can be an important variable. If the reaction is not carried to completion, the XLD will be lower than would be the case with a fully reacted coating. The situation is further complicated, especially in baking coatings, by reversibility of some cross-linking reactions. In such cases, an equilibrium condition may or may not be reached under the time and temperature conditions used. If, as can be true, side reactions occur, it is possible that extended heating may lead either to further increases in XLD or to reduction in XLD.

The T_g of cross-linked polymers is controlled by four factors and their interactions: T_g of the segments of polymer between the cross-links, the cross-link density, the presence of dangling ends, and the possible presence of cyclic segments [3]. While generalized equations showing the relationships of these factors with T_g have been developed, a full understanding of the complex interrelationships remains to be elucidated. The T_g of the polymer segments between cross-links is governed by the chemical structures of the resin and the cross-linking agent and by the ratio of these components. The factors discussed in Section 2.1.4 apply in their effects on the T_g of the segments of the cross-linked polymer chains. Since cross-links restrict segmental mobility, T_g increases as cross-link density increases. On the other hand, T_g decreases with increasing proportion of dangling ends; that is, chain segments with unreacted cross-link sites. The effect of cyclization on T_g has not been directly measured, but would be expected to restrict chain mobility and hence tend to increase T_g.

In the initial stages of cross-linking of low molecular weight resins, the molecular weight, and XLD increase while the fraction of dangling ends decreases resulting in an increase in T_g. As the cross-linking reaction continues, a *gel* will form; that is, the extent of cross-linking will have reached the point where the molecular weight of part of the polymer has become infinite. Reaction does not stop at the gelation stage but continues as long as there are functional groups to react, and there is sufficient mobility in the matrix to permit the reactive groups to move into positions for reaction. As the reaction continues, modulus (related to stiffness or resistance to deformation) above T_g increases. As cross-linking proceeds, the solubility of the film in solvent decreases finally giving films that are insoluble. Solvent can still dissolve in a cross-linked film leading to swelling. The extent to which a film will swell with solvent decreases as the distance between cross-links decreases; swelling is used as a method for experimental determination of XLD (see Chapter 24 on mechanical properties for further discussion).

As noted in Section 2.2.2, because of the difficulty of maintaining exact stoichiometric ratios in coatings, it is not appropriate to use only components with functionalities of 2. If a final film with low modulus above T_g is desired, XLD should be low; higher modulus films require higher XLD. The $\bar{f}_n$ required to make a final film of a desired XLD is strongly dependent on the molecular weight of the resins.

For high solids coatings, lower molecular weight resins must be used; to achieve the same XLD as with a similar conventional solids coating, the equivalent weight must also be lower in order to provide additional functionality, which is needed to build up the network. Although the equivalent weight is lower (i.e., less weight per functional group), the $\bar{f}_n$ is also lower as a consequence of the lower molecular weights. In most coatings, the binder consists of two or more resins; it is important to consider the $\bar{f}_n$ and $\bar{M}_n$ of each component not just the overall average functionality and molecular weight of the mixture.

In some coatings based on low molecular weight thermosetting acrylic resins (discussed in Section 7.2), the minimum $\bar{f}_n$ of the acrylic resin may well be affected by how high a molecular weight is required to assure that substantially all of the molecules have at least two reactive groups.

In other cases, such as oxidizing type alkyd resins (see Section 10.1), the rate of drying will depend on $\bar{f}_n$. If $\bar{f}_n$ is low, the concentration of functional groups is low and most of them may have to react to give a *dry* film. Since the rate decreases with the concentration and conversion of functional groups, the drying time will be slow. Conversely, if the average functionality is high, the concentration of functional groups will be high, and a dry cross-linked film will be formed in a shorter time. However, a higher proportion of unreacted functional groups may remain "frozen" in the cross-linked film in the latter case.

For polyisocyanates (Chapter 12) and polyamines for reaction with epoxy resins (Chapter 11), toxic hazards dictate the use of higher molecular weight cross-linkers with $\bar{f}_n$ greater than 2.

The possibility that steric hindrance will limit the number of groups that can undergo cross-linking reactions, a factor that has been seriously considered, has, in at least the case of melamine–formaldehyde (MF) cross-linked acrylic resins, been disproven (see Section 6.3.1.2) [4]. This is not to say that steric hindrance does not affect reaction rates—obviously it does—but the effect is built into the rate constants and it would now appear that the effect of steric hindrance for reaction of polymers is the same as for the reaction of similar groups on nonpolymeric molecules.

3.2.2. Package Stability versus Cure Rate

One problem with all thermosetting systems is the relationship between coating stability during storage and the time and temperature required to cure the film after application. Generally, it is desirable to be able to store a coating for many months or even years without a significant increase in viscosity that would result from reaction occurring during the storage period. On the other hand, one would like to have the cross-linking reaction proceed rapidly and at the lowest possible temperature after application.

More reactive combinations can be used in so-called "two-package coatings," where one package contains a resin with one of the reactive groups and the second package contains the component with the other reactive group. Alternatively, the second package could contain a catalyst for the reaction. In either case, the packages are mixed shortly before use. Two package coatings are used on a large scale commercially but are not generally desirable; they take extra time, material is usually wasted, they are generally more expensive, and there is chance of error in mixing. Even in two-package coatings, there is an analogous problem of "pot life"; that is, how long after the two packages have been mixed does the viscosity stay low enough for application? It has become increasingly common to refer to two-package coatings as "2K coatings" and single-package coatings as "1K coatings." The K stands for the German word for component, Komponent.

What controls the rate of a reaction? We can consider the question broadly using a general example of a cross-linking reaction between two functional groups represented by the symbols A and B. These two react with each other to form a covalent cross-link, A—B.

$$A \; + \; B \longrightarrow A{-}B$$

In the simplest cases, one can express the rate of the reaction (r) of group A with group B by Eq. 3.1, where k is the rate constant for the reaction between A and B at a specified temperature, and [A] and [B] represent the concentrations of functional groups in the coating in terms of equivalents per liter. The rate constant (or specific rate) is the reaction rate when [A] $\times$ [B] equals 1 eq^2 L^{-2}.

$$r = k[A][B] \tag{3.1}$$

[A] and [B] are reduced by dilution with solvent and increase as the solvent evaporates; therefore, other factors being equal, cross-linking will initially be faster in the applied film after solvent evaporation than during storage. As coating formulators shift to higher solids coatings to reduce VOC emissions, as well as to save cost by using less solvent, there will be higher concentrations of functional groups and greater difficulty in formulating storage-stable coatings. The problem results not only from the presence of less solvent but also from the lower molecular weight resins needed to make higher solids coatings that must have lower equivalent weights to achieve a cross-linked film of acceptable XLD. Both factors increase the concentration of functional groups in the stored coating. It also increases the concentration in the film after application and evaporation of solvent but, since the molecular weight is lower, more reactions must occur to achieve the desired cross-linked film properties.

In order to minimize the temperature required for curing while maintaining adequate storage stability, it is desirable to select cross-linking reactions for which the rate is strongly dependent on temperature. The dependence of the rate on temperature is reflected in the rate equation by the rate constant k. It is commonly taught, as a rule of thumb, in introductory organic chemistry classes that rate constants double with each 10°C rise in temperature. That generalization is true for only a limited number of reactions within a narrow temperature range near room temperature. A better estimate of the dependence of k on temperature is given by the empirical Arrhenius equation, Eq. 3.2, where A is called the preexponential term, E_a is the thermal coefficient of reactivity, commonly labeled *activation energy*, R is the gas constant, and T is temperature (in kelvin).

$$\ln k = \ln A - (E_a/RT) \tag{3.2}$$

Reaction rate data that follow the equation will give a straight line when $\ln k$ is plotted versus $1/T$ as illustrated in Figure 3.1. As seen in plot a of competing Reactions (1) and (2) where $A(1) = A(2)$ and $E_a(1) > E_a(2)$, the temperature dependence of rate constants increases with increasing activation energies. The larger the E_a value, the smaller the rate constant at any temperature. This trend can be counteracted by selecting a reaction with a higher A value, as shown in plot (b) where $A(3) > A(1)$ and E_a for the two competing reactions is equal. If A and E_a are both sufficiently greater for one reaction than for another, the rate constant at storage temperatures could be smaller while the rate constant at a higher tem-

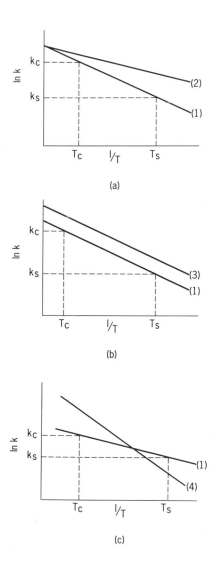

Figure 3.1. Arrhenius plots for competing reactions: (a) $A(1) = A(2)$, $E_a(1) > E_a(2)$; (b) $A(3) > A(1)$, $E_a(1) = E_a(3)$; (c) $A(4) > A(1)$, $E_a(4) > E_a(1)$. (from Ref. [7], with permission.)

perature could be larger than for the other reaction, as shown schematically in plot (c).

The preexponential term is controlled predominantly by entropic factors; more specifically by changes in randomness or order as the reaction proceeds to the activated complex in the transition state. Three important factors are (1) molecularity; unimolecular reactions tend to exhibit larger A values than do reactions of higher molecular order, (2) type of reaction; for example, ring-opening reactions tend to have high A values; whereas ring-closing reactions tend to have low A values, and (3) polarity changes; reactions in which reactants become less polar tend to exhibit larger A values. In addition, the kinetic parameters may be substantially affected by the solvent and solvent selection can have a significant effect on storage stability (see, Chapters 11 and 12 for examples).

Summarizing, the most desirable reactions for maximizing A values are unimolecular dissociations with ring opening and conversion of polar reactants to less polar activated complexes. However, cross-linking reactions are necessarily bimolecular.

$$A + B \longrightarrow A{-}B$$

One potential way around this dilemma is to use a "blocked" reactant (BX) that will release a reactant (B) thermally by a unimolecular reaction, most desirably with ring opening and decreasing polarity, followed by cross-linking between A and B.

$$BX \rightleftharpoons B + X$$

$$A + B \longrightarrow A{-}B$$

Another approach is to use a blocked catalyst (CX), where C catalyzes the cross-linking of A and B.

$$CX \rightleftharpoons C + X$$

$$A + B \xrightarrow{\ C\ } A{-}B$$

An important consideration is that the cross-linking reaction, which follows unblocking, should be faster than the reverse reaction, which regenerates the blocked reactant or catalyst. The kinetics of cross-linking reactions utilizing blocked reactants and catalysts have been analyzed—for further discussion see Ref. [5].

While one often encounters "threshold" or "unblocking" temperatures for reactions in the literature, such minimum reaction temperatures do not exist in the kinetics of reactions. Reactions will proceed at some rate at any temperature. Threshold or unblocking temperatures are actually the temperatures at which some observable extent of reaction has occurred within a specified time interval.

These concepts are helpful in understanding the large differences in storage stability we see, but another important reason for understanding these kinetic considerations is that they can be used to predict whether any chemical reaction will ever be found to meet some combination of stability and cure schedule that might be desired for a coating. Pappas and Hill [6] carried out a series of calculations to permit such predictions. They made a set of reasonable assumptions about reactive group concentrations during storage, and required extents of reaction during the curing step. Using these assumptions, they were able to calculate the ratio of the rate constants needed to permit any time of storage with any time for curing. In turn, this allows calculation of E_a and A values as a function of any combination of storage and curing temperatures. Kinetic parameters calculated for hypothetical unimolecular reactions, such as for blocked reactant systems, that proceed 5% over a 6-month period at 30°C and 90% in 10 min at various curing temperatures are listed in Table 3.1 [7].

Rate constants and kinetic parameters are independent of reactant concentrations for unimolecular reactions, but not for bimolecular reactions. Kinetic parameters calculated for bimolecular (i.e., second order) reactions (using concentrations corresponding to high solids coatings) were found to be of similar magnitude to those for unimolecular reactions. Although the values in Table 3.1 represent *order of magnitude calculations*, they provide useful guidelines for avoiding developmental projects aimed at impossible goals and also provide important insights for the design of thermosetting coatings.

Kinetic parameters are known for a large number of chemical reactions. As points of reference, a reasonable upper limit of A values for unimolecular reactions is 10^{16} s^{-1}, which corresponds to an upper value for the frequency of simple vibrations. For bimolecular reactions, A values tend to be less than 10^{11} L mol^{-1} s^{-1}, which represents an upper limit for the rate constant of diffusion, which must precede reaction. However, as shown in Table 3.1, a coating stable at 30°C would require an A value of 10^{24} s^{-1} if it were to cure within 10 min at 100°C. No such reaction is known or even conceivable. Users would like to have package-stable coatings that will cure in a short time at 80°C, a convenient temperature for low-pressure steam heat, but it is pointless to pursue this goal via kinetic control. This is not to say that it is impossible to make a storage stable coating that will cure in 10 min or less at 80°C; it merely says that the problem must be solved by some approach other than kinetic control.

Storage life could be lengthened by storing coatings under refrigeration. Users are seldom willing to incur this expense. As noted earlier, 2K systems may be used. The two packages are mixed just before application; the pot life of the mixed

Table 3.1. Kinetic Parameters as a Function of Cure Temperature

T (°C)	A (s^{-1})	E_a (kJ mol^{-1})[a]
175	10^{10}	109
150	10^{12}	121
125	10^{17}	146
100	10^{24}	188

[a]1 kJ = 0.239 kcal.

Source: See Ref. 7.

system must be long enough so that the viscosity of the mixture does not increase excessively before application. Pappas and Hill [6] made similar calculations of A and E_a values for the shorter storage times involved.

There are many possible approaches to increasing the package stability of coatings while permitting cure at ambient temperature or at moderately increased temperature. These approaches will be included in discussions of almost all the cross-linking systems covered in later chapters. The following list gives some approaches to solving this *impossible* problem:

1. Use of a radiation-activated cross-linking reaction instead of a thermally activated cross-linking reaction.
2. Use of a cross-linking reaction requiring an atmospheric component as a catalyst or reactant. Reactions involving oxygen or water vapor in the air are examples. The same principle is involved in passing a coated article through a chamber containing a volatile catalyst vapor.
3. Use of a volatile inhibitor so that the inhibitor works when the coating is stored in a closed container but volatilizes after application as a thin film, permitting the reaction to proceed. Examples include use of a volatile antioxidant in a system that dries by oxidation and use of oxygen as an inhibitor in a composition that cures anaerobically.
4. Use of a cross-linking reaction that is a reversible condensation reaction involving the loss of a volatile reaction product; some of the monofunctional volatile reactant can be used as a solvent in the coating. The equilibrium of the reaction will favor the noncross-linked side during storage but will shift to the cross-linked side after application when the solvent evaporates. A similar approach can be used in conjunction with blocked reactants and blocked catalysts in which the blocking group is volatile.
5. Use of an encapsulated reactant or catalyst with the capsules being ruptured during application. This approach has been quite effective in adhesives but has not yet been very useful in coatings because the residual capsule shells interfere with appearance and/or performance of the coating.
6. Use of systems in which a reactant undergoes a phase change. While no threshold temperatures for kinetic reactions exist, phase changes can occur over narrow temperature ranges. A crystalline blocked reactant or catalyst, insoluble in the vehicle, could give an indefinitely stable coating; heating over the melting point would melt the crystals permitting the unblocking reaction to occur releasing a soluble reactant or catalyst. Over a somewhat wider range of temperature, the same principle can be used if the blocked reactant is amorphous with a T_g of about 50°C above the storage temperature and perhaps 30°C below the cure temperature.

3.2.3. Mobility Effect on Reaction Rates

Another factor can be involved in the cross-linking of coatings, namely, the potential effect of availability of free volume on reaction rates and reaction completion. For reactions to occur, the reacting groups must be adjacent. If the reaction is occurring at a temperature well in excess of T_g, free volume will be large and the rate of the reaction will be controlled by concentrations and kinetic parameters. If, however, the temperature is well below T_g, free volume will be so limited that the polymer chain segment motions needed to bring unreacted functional groups close together will be at most exceedingly slow. At intermediate temperatures, the reaction may be able to proceed but the reaction rate will be controlled by the rate of movement through the matrix rather than by the kinetic parameters—the reaction rate becomes mobility controlled rather than kinetically controlled.

Since cross-linking often starts with comparatively low molecular weight components, the T_g will increase as the reaction proceeds. If the initial reaction temperature is well below the T_g of the solvent-free system, it is possible that little or no reaction will occur after solvent loss and that one will have a *dry* film merely due to the solvent evaporation without cross-linking. The usual result would be a weak, brittle film. If the reaction temperature is well above the T_g of the fully reacted system, then there will be no mobility effect on the reaction. However, if, as is often the case in ambient temperature cure thermoset coatings, the initial T_g is below ambient temperature and the ultimate T_g is above ambient temperature, the reaction rate will become mobility controlled as cross-linking proceeds. Furthermore, after additional reaction takes place cross-linking may well essentially stop before the reaction is complete.

This phenomenon has only been recognized recently and has not been widely studied, especially since the time frame for such studies has to be very long. At one time it was thought reactions ceased when T_g increased to the temperature at which the reaction was being carried out [8]. The term vitrification is commonly applied to this situation; it simply means glass formation. A more recent paper reports that reaction rate constants drop by about three orders of magnitude when T_g becomes equal to T, but that the reaction continues at a reduced rate [9]. This paper indicates that the reaction ceases when the T_g increased to $T + 50°C$. It is interesting to compare this value to the 51.6 value for the universal B constant in the WLF equation given above. The term $(T_g - B)$ is the temperature at which viscosity goes to infinity and free volume theoretically approaches zero.

Further study is required. However, pending additional data, it seems to be a reasonable speculation that reactions will slow down at temperatures perhaps 10°C or less above T_g and get progressively slower until T_g is about 50°C above the curing temperature, where the rate may be so slow that reaction will essentially cease. Caution is required since T_g values are very dependent on the method of determination and the rate of heating used. For this type of comparison, the most appropriate data to use are values obtained at very slow heating rates and at low rates of application of stress. Further discussion is available in Ref. [10].

One must be careful in defining what is meant by a dry film, especially when dealing with ambient temperature cure coatings. One consideration is whether the film is dry to handle. This stage could be reached with little or no cross-linking if the T_g of the solvent-free binder is high enough. Another consideration is whether some required degree of cross-linking has been achieved. This will probably have to be tested in some way other than hardness, perhaps most easily by determining resistance to dissolution or by the extent of solvent swelling.

3.3. FILM FORMATION BY COALESCENCE OF POLYMERIC PARTICLES

Thus far, we have been discussing film formation starting with vehicles made with soluble resins. Another important group of coatings utilizes vehicles in which the polymer, or resin, is not in solution but rather is present as a dispersion of insoluble polymer particles. In these cases, after application and loss of the volatile components, the particles must coalesce (fuse) to a continuous film.

3.3.1. Latexes

A latex is a dispersion of high molecular weight polymer particles in water. Although they are sometimes called emulsions, the terminology is poor and should be avoided. For many applications, latexes offer major advantages when compared with resin solutions such as those discussed in the preceding sections. Low viscosity, relatively high solids vehicles with high molecular weight polymers, commonly $\overline{M}_w$ $> 10^6$, are possible because the viscosity of the dispersion is independent of the molecular weight of the polymer and, as discussed in Section 2.3, depends primarily on the volume fraction of the internal phase and the packing factor of the particles. If such a high molecular weight polymer were in solution, the solids would have to be very low to attain application viscosity. The high molecular weight polymers provide sufficient strength in the coated films for many applications without the need for cross-linking; furthermore, the volatile material in a latex is primarily water, and the amount of VOC needed in a fully formulated paint is generally lower than in coatings made with solvent-soluble resins.

Latexes are stabilized as dispersions of fine particles in water by two mechanisms: charge repulsion and entropic repulsion [11], as discussed in Section 2.5. When a dry film is prepared from a latex, the stabilization of the dispersion must be overcome and the particles must coalesce into a continuous film. As water and water-soluble solvents evaporate, the latex particles come closer and closer together. It has been proposed that as the particles approach each other, the space between them acts as the equivalent of a capillary so that capillarity may provide the force required to overcome the stabilizing repulsions and permit coalescence [12]. It is estimated that such forces may generate as much as 3.5-MPa pressure where the latex particles touch. While capillarity may be a factor, Croll [13] points out that the time span during which the forces would be high is very short. His data on drying rate support a proposal of Kendall and Padget [14] that the major driving force for coalescence is surface energy reduction. The surface area of the coalesced film is only a small fraction of the surface area of all the latex particles, so that the driving force resulting from this reduction in surface area must be large. Eckersley and Rudin [15] showed that both forces are involved and, consistent with the differences in surface area, small particle size latexes form films at somewhat lower temperatures than do larger particle size ones of the same composition.

Neither capillarity nor surface energy forces can cause coalescence unless the polymer molecules in the latex particles are free to diffuse into neighboring particles so that the individual particles disappear in the coalesced film. The movement of the polymer molecules requires that there be a sufficient number and size of free volume holes in the latex particles to permit the movement of the polymer molecules. In other words, the T_g of the latex particles must be lower than the temperature at which film formation is being attempted.

The T_g of the latex particles can be affected by several variables. One important variable is the composition of the polymer. The T_g of poly(methyl methacrylate) (PMMA) is about 105°C and one cannot form a film from a PMMA latex at room temperature; instead, one gets a layer of material that powders readily. Many polymers and copolymers of other acrylate esters have T_gs well below room temperature and coalesce to a continuous film when applied at room temperature. The T_g of latex particles is also affected by the presence of other materials in the latex, such as stabilizing surfactants and water. One can reduce the T_g of a latex polymer by emulsifying a plasticizer into the latex. The plasticizer will dissolve in the latex particles, lowering their T_g and permitting film formation at a lower temperature.

It is common to refer to the minimum film formation temperature (MFT) of a latex. This is the lowest temperature at which a film will form with a latex under specified conditions. While films form rapidly from latexes when the temperature is above the T_g of the polymer, complete coalescence is a relatively slow process. The rate is strongly affected by $(T - T_g)$. It is generally desirable to have fairly rapid coalescence, which would require latexes with T_gs significantly below the temperature at which the film is to be formed. Usually, architectural paints are formulated so that film formation will occur fairly readily at temperatures as low as 5°C. To achieve film formation at 5°C, the T_g of the latex particles must be lower. However, as discussed in the introductory paragraphs of this chapter, $(T - T_g)$ will also affect whether the film obtained by coalescence will be a solid film. It was estimated that for a film to withstand the relatively mild blocking test described, $(T - T_g)$ would have to be of the order of 21°C. If the film is to be exposed to such a blocking test when its temperature is 50°C (not an unreasonable expectation during direct exposure to summer sun), the T_g should be about 29°C or higher. Thus, the paint formulator is faced with a difficult challenge to design a system that can form a film when applied at a temperature as low as 5°C, and yet resist blocking at a temperature of 50°C.

There have been two major approaches to this seemingly impossible challenge. First, one can add a "coalescing solvent" to the latex formula. Coalescing solvents must be soluble in the latex polymer and have low, but appreciable, evaporation rates. The coalescing solvent dissolves in the latex particles and acts as a plasticizer to lower the T_g, permitting film formation at a lower temperature. Then, after the film has formed, the coalescing solvent can slowly diffuse to the surface of the film and then evaporate. Since the free volume in the film is relatively small, the rate of loss of coalescing solvent will be slow. Furthermore, as the solvent evaporates, the T_g will increase so the rate of solvent loss slows further. Though such films feel dry, they will still block for several weeks after application.

A second approach is to design latex particles so there is a gradient of T_g from relatively high T_g in the center of the particles to relatively low T_g at the outer periphery of the particles [11]. The low T_g of the outer shell permits film formation at low temperature. Over time, the T_g of the coalesced film will approach the average T_g of the total polymer system. This higher T_g reduces the probability of blocking (see Section 5.2). Of course, one can also use coalescing solvents with such a latex.

Since pigments and other components also have significant effects on film formation, this introductory presentation is over-simplified; formulation of latex paints will be discussed at considerable length in Chapter 35.

3.3.2. Other Coalescing Systems

Several other types of coating systems are used in which film formation involves coalescence of particles, including nonaqueous dispersions, plastisols, organosols, and *water-reducible* resins. These coating systems will be discussed in future chapters.

Powder coatings form films by coalescence. They consist of finely divided powdered solid material. They are applied to metal objects, generally by an electrostatic process, and then coalesced by heat to form a continuous film. The T_g of the powder must be high enough ($\sim 55°C$) so that sintering (partial fusing) does not occur during storage and shipment of the powder. However, to minimize the oven temperature required for coalescence and leveling, the T_g should be as low as storage stability permits. In many cases, powder coatings are thermosetting. In these cases, it is important that coalescence and flow occur before a significant degree of cross-linking has taken place. Further discussion of powder coatings is presented in Chapter 31.

GENERAL REFERENCE

Z. W. Wicks, Jr., "Film Formation," Federation of Societies for Coatings Technology, Blue Bell, PA, 1986.

REFERENCES

1. H. Burrell, *Official Digest*, **34** (445), 131 (1962).
2. W. P. Mayer and L. G. Kaufman, *XVII FATIPEC Congress*, Book I, 110 (1984).
3. H. Stutz, K.-H. Illers, and J. Mertes, *J. Polym. Sci. B Polym. Phys.*, **28**, 1483 (1990).
4. L. W. Hill and K. Kozlowski, *J. Coat. Technol.*, **59** (751), 63 (1987).
5. W. J. Muizebelt, *J. Coat. Technol.*, **57**, (725), 43 (1985).
6. S. P. Pappas and L. W. Hill, *J. Coat. Technol.*, **53** (675), 43 (1981).
7. S. P. Pappas and H.-B. Feng, *Org. Coatings*, **8**, 139 (1986).
8. M. T. Aronhime and J. K. Gillham, *J. Coat. Technol.*, **56** (718), 35 (1984).
9. H. E. Bair, *Polym. Prepr.*, **26** (1), 10 (1985).
10. Z. W. Wicks, Jr., *J. Coat. Technol.*, **58** (743), 23 (1986).
11. K. L. Hoy, *J. Coat. Technol.*, **51** (651), 27 (1979).
12. D. P. Sheetz, *J. Appl. Polym. Sci.*, **9**, 3759 (1965).
13. S. G. Croll, *J. Coat. Technol.*, **58** (734), 41 (1987).
14. K. Kendall and J. C. Padget, *Int. J. Adhesion Adhesives*, **2** (3), 149 (1982).
15. S. T. Eckersley and A. Rudin, *J. Coat. Technol.*, **62** (780), 89 (1990).

CHAPTER IV

Solution and Dispersed Thermoplastic Polymers

Thermoplastic polymers, dissolved in organic solvents, were formerly widely used in coatings for automobiles, wood furniture, and corrosion protection coatings. Such coatings, which dry by solvent evaporation alone, are generally called lacquers. As discussed in Chapter 2, the application solids are low because of the relatively high molecular weight required to achieve good film properties. The molecular weight required varies with polymer structure; in the case of acrylics, the $\overline{M}_w$ must be above about 75,000, which leads to application solids of about 12% nonvolatile volume (NVV) depending on the color.

Use of lacquers has declined significantly because of VOC regulations and the increasing cost of solvents. Nevertheless, lacquers retain some importance; this chapter will cover the three major classes of thermoplastic polymers in commercial use in coatings: acrylics, halogenated polymers, and cellulose derivatives. Some other types are also used; phenoxy resins, relatively high molecular weight bisphenol A–epichlorohydrin copolymers, are sometimes used without cross-linking and are mentioned in Chapter 11; linear polyurethanes are briefly discussed in Chapter 12.

4.1. THERMOPLASTIC ACRYLIC POLYMERS

From the mid-1950s through 1971, acrylic lacquers were used on all General Motors cars produced in the United States. Due to the high levels of solvent required, the use of acrylic lacquers in original equipment manufacturing (OEM) of automobiles was phased out over the next 15 years. Acrylic lacquers are still important in refinishing automobiles, but as VOC regulations become more restrictive, they too will probably be phased out. The major reasons for use of acrylic lacquers on automobiles are their suitability for making brilliant *metallic colors* and their excellent exterior durability, especially gloss retention. Since they are thermoplastic coatings, the dry films retain their solubility in solvents. As long as one avoids spilling polar solvents like nail polish remover on the films, this is not a serious drawback. In fact, the continued solubility of the films is a distinct advantage in one way. Repair of damaged lacquer films is easy since the solvent in the repair lacquer dissolves in the old film providing for adhesion of the new film to the old film. Nearly invisible repairs can be made on small areas of the surface. Other important considerations are discussed in Chapters 33 and 36, which cover OEM and aftermarket automotive coatings.

In the case of OEM thermoplastic acrylic coatings, an intermediate step in reducing solvent usage was the use of nonaqueous dispersions (NADS) of polymers in so-called dispersion lacquers. Although NADs are no longer used for automotive coatings, some discussion is included since there may be future applications.

4.1.1. Solution Thermoplastic Acrylic Polymers

Solution thermoplastic acrylic polymers (TPAs) are prepared by chain-growth polymerization. A wide range of comonomers is used; azo initiators, such as AIBN (see Section 2.1.2), are almost always used for exterior coatings since the terminal groups introduced are less likely to have a negative effective on exterior durability. In the equations below, the initiating free radical from AIBN is represented by $I\cdot$. The principal reactions involved can be illustrated using methyl acrylate (MA) and methyl methacrylate (MMA).

Control of molecular weight in the synthesis of TPAs is critical. Film strength increases with molecular weight, although above $\overline{M}_w$ of about 90,000, the change in properties with increasing molecular weight is small. The upper end of the molecular weight range is limited because solutions of acrylic polymers with $\overline{M}_w$ greater than about 100,000 exhibit "cobwebbing" on spraying. Rather than atomizing to small droplets from the orifice of a spray gun, the lacquer comes out as threads. This can look very nice when spraying "angel's hair" decoration on Christmas trees but is unacceptable when spraying automotive top coats.

As molecular weight increases, viscosity of solutions also increases. Consequently, the solids at which the lacquers can be spray applied decreases. The viscosity is particularly affected by the amount of very high molecular weight polymer present. Therefore, it is critical to minimize the fraction of high molecular weight polymer by controlling the molecular weight distribution $\overline{M}_w/\overline{M}_n$) to as narrow a range as possible. Commercial TPAs have $\overline{M}_w$s of 80,000 to 90,000 and $\overline{M}_w/\overline{M}_n$ of 2.1 to 2.3. The solids at application viscosity are in the range of 11 to 13 NVV, depending on the level of pigmentation.

There are four major factors that control the molecular weight in solution free radical chain-growth polymerization with the same initiator, monomer, and solvent:

Initiator Concentration. The higher the initiator concentration, the lower the molecular weight. When initiator concentration is higher, more initiating free radicals are generated to react with the same total amount of monomer. More chains are initiated and terminated; therefore, the $\overline{M}_n$ and $\overline{M}_w$ of the polymer produced must be lower.

Temperature. At higher temperature more initiator is converted into initiating free radicals in a given time. As above, more chains are initiated. With more propagating chains present the probability of the free radicals on the ends of two chains approaching each other to undergo a termination reaction increases, leading to lower $\overline{M}_n$ and $\overline{M}_w$.

Monomer Concentration. The higher the monomer concentration, the higher the $\overline{M}_n$ and $\overline{M}_w$. The highest molecular weight would be obtained in a solvent-free reaction mixture. With the same concentration of growing free radical ends, higher monomer concentration increases the probability of chain growth relative to termination.

Solvent Concentration. The higher the solvent concentration, the lower the $\overline{M}_n$ and $\overline{M}_w$ for two reasons. Higher solvent concentration means lower monomer concentration, reducing molecular weight as described above. Furthermore, most solvents undergo, to varying extents, chain-transfer reactions, which is discussed later in this section. Chain-transfer to the solvent can significantly compete with propagation reactions, reducing the molecular weight.

When these factors change during a polymerization, $\overline{M}_n$ and $\overline{M}_w$ change throughout the course of the polymerization resulting in a broader molecular weight distribution. Close control of these variables is required to produce relatively narrow distribution polymers.

Furthermore, even with all four of these variables held constant, different monomers will give different molecular weight distributions. For example, consider the differences between our illustrative monomers, MA and MMA. Since the free radicals at the ends of growing chains of poly(methyl methacrylate) (PMMA) are sterically hindered, termination by combination is impeded and termination by disproportionation predominates. On the other hand, with MA, the major fraction of the termination reactions is by combination. Theoretical calculations show that under ideal circumstances the lowest $\overline{M}_w/\overline{M}_n$ attainable in free radical polymerization to high molecular weight with termination by combination is 1.5, and the lowest attainable with termination by disproportionation is 2.0. (Lower $\overline{M}_w/\overline{M}_n$s are attainable by other processes to be covered later.) If there were no other factors involved MA would give polymers of lower $\overline{M}_w/\overline{M}_n$ than MMA.

However, other factors are involved. A side reaction, chain transfer to polymer, must be considered. This reaction occurs to some degree in the polymerization of MMA, but is much more important in the polymerization of MA. The hydrogen on the carbon to which the carboxymethyl group is attached is more susceptible to abstraction by free radicals than any other type of hydrogen in PMA or PMMA, for example, by a growing chain R·.

$$R\cdot \;+\; -CH_2-\underset{\underset{COOCH_3}{|}}{\overset{\overset{H}{|}}{C}}-CH_2-\underset{\underset{COOCH_3}{|}}{\overset{\overset{H}{|}}{C}}- \;\longrightarrow\; RH \;+\; -CH_2-\underset{\underset{COOCH_3}{|}}{\overset{\overset{\cdot}{}}{C}}-CH_2-\underset{\underset{COOCH_3}{|}}{\overset{\overset{H}{|}}{C}}-$$

The new free radical on the PMA chain can now add to a monomer molecule initiating the growth of another chain as a branch on the original polymer molecule. The result is a polymer containing branched as well as linear molecules and having a larger $\overline{M}_w/\overline{M}_n$ than theoretically predicted for ideal linear polymerization. In extreme cases, chain transfer to polymer results in very broad distributions of molecular weight and ultimately to formation of gel particles through cross-linking.

Branching can also result from the abstraction of hydrogen atoms by initiating free radicals. The highly reactive phenyl free radical, formed by sequential decomposition of benzoyl peroxide (see Section 2.2.1), is a particularly effective hydrogen abstracter. The free radicals produced by decomposition of AIBN, while very reactive, are much less reactive than phenyl free radicals. Therefore, use of benzoyl peroxide as an initiator, especially at temperatures in excess of 130°C, leads to substantial degrees of branching. Indeed, homopolymerization of methyl acrylate, especially with benzoyl peroxide, can result in formation of a polymer containing many insoluble gel particles. Since branching causes broader molecular weight distribution it is generally undesirable in thermoplastic polymers for coatings; this is one reason that initiators such as AIBN are often preferred to initiators like BPO. Another is that polymer end groups derived from AIBN are thought to have less adverse effect on exterior durability.

The molecular weight and molecular weight distribution also depend on solvent structure. For example, substituting xylene for toluene and keeping other variables constant will lead to a decrease in average molecular weight. Since each xylene molecule has six abstractable hydrogen atoms relative to the three on toluene, the probability of chain transfer is increased; average molecular weight is reduced. If one wants to prepare a low molecular weight resin, one can add a compound that will undergo very facile hydrogen abstraction as a chain-transfer agent. If the hydrogen atoms are sufficiently easily abstracted, addition of even relatively low concentrations can lead to substantial reduction in molecular weight. Mercaptans (RSH) are widely used as chain-transfer agents.

$$\text{RSH} + \begin{array}{c} \text{CH}_3 \\ | \\ -\text{CH}_2-\text{C}\cdot \\ | \\ \text{COOCH}_3 \end{array} \longrightarrow \text{RS}\cdot + \begin{array}{c} \text{CH}_3 \\ | \\ -\text{CH}_2-\text{C}-\text{H} \\ | \\ \text{COOCH}_3 \end{array}$$

$$\text{RS}\cdot + \begin{array}{c} \text{CH}_3 \\ | \\ \text{CH}_2=\text{C} \\ | \\ \text{COOCH}_3 \end{array} \longrightarrow \begin{array}{c} \text{CH}_3 \\ | \\ \text{RS}-\text{CH}_2-\text{C}\cdot \\ | \\ \text{COOCH}_3 \end{array}$$

Other important variables affecting molecular weight and molecular weight distribution are the decomposition rate of the initiator and the reactivity of the resulting free radicals. If one wished to change the initiator and not change the molecular weight, one would have to adjust the temperature to equalize the rates of decomposition of the initiators, but this change would probably alter the relative rates of propagation, termination, and chain transfer leading perhaps to significant differences in the polymer.

To achieve a low $\overline{M}_w/\overline{M}_n$ it is desirable to keep the concentrations of reactants as constant as possible throughout the polymerization. It is undesirable to simply charge all of the monomers, solvents, and initiators into a reactor and heat the mass to start the reaction. This undesirable procedure is sometimes used on small scale reactions in the laboratory but is seldom, if ever, used in production. At best, it yields a high $\overline{M}_w/\overline{M}_n$; at worst, the reactions may run out of control because they are highly exothermic. Instead, one charges some of the solvent into the reactor, heats it to reaction temperature, and then adds monomer, solvent, and initiator solutions to the reactor at rates such that the monomer and initiator concentrations are kept as constant as possible. Adding monomer at a rate that maintains a constant temperature leads to maintaining a fairly constant monomer concentration. The appropriate rate of addition of initiator can be calculated from a knowledge of the rate of its decomposition at the temperature being used. Keeping the solvent concentration constant is more complex since we have to bear in mind that, as the polymerization proceeds, polymer is accumulating and its presence obviously reduces the concentrations of the other components. In a sense, the polymer becomes a part of the "solvent" for the polymerization. One attempts to add solvent at a decreasing rate so that other concentrations stay as constant as possible. Obviously, perfect control is not possible, but careful attention to the details makes an important difference in $\overline{M}_w/\overline{M}_n$ of the polymer produced.

Copolymerization of mixtures of vinyl monomers complicates the situation. The rates of reaction involved in the various addition reactions depend on the structures of the monomers. If the rate constants for all the possible reactions were the same, the monomers would react randomly and the average composition of molecules of substantial length would all be the same. However, the rate constants are not equal. If polymerization were carried out by putting all the reactants in a flask and heating, the first molecules formed would contain more than proportional amounts of the most reactive monomer. Toward the end of the reaction, the last molecules formed would have an excess of the least reactive monomer, usually an undesirable situation. Equations have been developed to predict the results with different monomers. Reference [1] discusses copolymerization in detail. In actual practice, however, the problem is less complex. Reactions are not run in bulk, but rather, as mentioned above, monomers, solvents, and initiators are added gradually to the reaction mixture. If the addition is carefully controlled so that the rate of addition equals the rate of polymerization, copolymers having reasonably uniform average composition corresponding to the feed ratio will be obtained with most monomers used in coatings polymers. This procedure, sometimes called monomer-starved polymerization, results in polymerization under conditions where the concentration of monomer is low and fairly constant.

Acrylic polymers for automotive lacquers must be designed to permit formulations that will provide long-term resistance to photodegradation. Furthermore, the coatings must have an appropriate T_g to resist physical changes under relatively extreme weather conditions. For example, the temperature of the top coat on a car parked in the summer sun can reach, or even exceed, 65°C. If a sudden thundershower comes up, the coating surface will become dimpled if the T_g is too low. On the other hand, if the T_g of the coating is too high, cracking may occur on a winter morning when a car is driven from a heated garage out into the cold. Such cracking problems are especially likely if the molecular weight of the polymer is low since the strength of the film is then lower (see Chapter 24 for a discussion of mechanical properties of films).

Poly(methyl methacrylate) has the greatest resistance to photodegration (or photoxidative degradation) of any acrylic polymer but the T_g (105°C) is too high to resist cracking. Early acrylic lacquers used PMMA with butyl benzyl phthalate as a plasticizer to lower the T_g. However, evaporation of the phthalate plasticizer when the coating was baked in automobile assembly plants and slowly in service proved problematic. Therefore monomeric phthalates were at least partly replaced with oligomeric polyester plasticizers in OEM lacquers (see Chapter 33). Oligomeric plasticizers are less efficient in reducing T_g; acrylic copolymers, such as methyl methacrylate–n-butyl acrylate, with lower T_gs than PMMA were adopted to avoid the necessity of excessive amounts of plasticizer.

As briefly discussed in Chapter 2, polymers of acrylic esters have lower T_gs than polymers of the corresponding methacrylate esters. Also, as the length of the alkyl ester carbon chains increases, T_g decreases. Branched alkyl esters, on the other hand, have higher T_gs than their straight-chain isomers. Table 4.1 gives the T_g of high molecular weight homopolymers of a series of acrylate and methacrylate esters and other monomers commonly used as comonomers. These values are from the chapter on glass transition in the second edition of the *Polymer Handbook* [2]; many values can be found in the literature but in this chapter the attempt was made to select the most comparable data.

Copolymers have T_g values intermediate between those of the homopolymers. It is common to use the so-called "Fox equation" to estimate the T_g of copolymers, where w_1, w_2, w_3, and so on, are the weight fractions of the various monomers in the copolymer and T_{g1}, T_{g2}, T_{g3}, are the T_gs (in kelvin) of their high molecular

Table 4.1. Glass Transition Temperatures (°C) for the Homopolymers of Methacrylic and Acrylic Acid and their Esters[a]

Monomer	Methacrylate	Acrylate
Free acid	228	106
Methyl	105	10
Ethyl	65	−24
n-Butyl	20	−54
IsoButyl	53	−24
t-Butyl	114	74
2-Ethylhexyl	−10	−50
n-Octyl	−45	−65
2-Hydroxyethyl	70	
2-Hydroxypropyl	76	

Other Monomers Used in Acrylic Copolymers

Styrene	100
Vinyl acetate	32
Vinyl chloride	81
Vinylidene chloride	−18

[a]Data selected from Ref. [2]. When multiple data are given, the numbers were averaged.

weight homopolymers. The calculations provide an estimate of the T_g of the high molecular weight copolymer.

$$\frac{1}{T_{g(\text{copolymer})}} = \frac{w_1}{T_{g1}} + \frac{w_2}{T_{g2}} + \frac{w_3}{T_{g3}} + \cdots$$

Somewhat better approximations can be calculated using a different mixing equation, also devised by Fox.

$$T_{g(\text{copolymer})} = v_1 T_{g1} + v_2 T_{g2} + v_3 T_{g3} + \cdots$$

Here v_1, v_2, v_3, and so on, are the volume fractions from the various monomers in the copolymer. This equation is not as widely used because of the inconvenience of using densities of the various homopolymers in order to calculate volume fractions. Other equations for estimating the T_g of copolymers have been proposed, but their discussion is beyond the scope of this book.

The most appropriate T_g for the polymer varies according to end use and to the other components of the coating. For baking automotive lacquers, polymers with T_gs of about 70°C are typically used. Choice of monomers is not controlled by T_g considerations alone. Since MMA imparts high resistance to photodegradation, its content is usually maximized. Conversely, esters having readily abstractable hydrogen atoms, such as isobutyl and 2-ethylhexyl esters are questionable for many applications both because exterior durability would diminish and a greater extent of branching would occur during polymerization. Styrene is commonly used along with MMA to lower cost. Exterior durability of styrene containing copolymers is not well understood, requiring field testing of each combination to assure adequate weather resistance. Specialized comonomers are sometimes incorporated to impart special properties. For example, adhesion of a lacquer top coat to primers is commonly enhanced by using a small amount of an amine functional acrylate as a comonomer. 2-(Dimethylamino)ethyl and 2-(aziridinyl)ethyl methacrylates, **1** and **2**, respectively, are two such monomers.

A commercially important modification was the use of a blend of two acrylic polymers having T_gs differing by 20 to 40°C [3]. Although the solutions were miscible, the acrylic polymers were incompatible. When the solvent evaporated during film formation, submicroscopic phase separation occurred. Such lacquers resisted water spotting at higher temperatures than lacquers formulated with a single acrylate copolymer with otherwise similar properties.

4.1.2. Nonaqueous Dispersion Acrylic Polymers

The high levels of solvent required for solution lacquers led to the development of alternative types of acrylic coatings. One approach that was extensively used for automotive top coats for several years was nonaqueous dispersions (NAD) of thermoplastic acrylic polymers. In NADs most of the polymer is not in solution but rather is present in a dispersed phase. At the same molecular weight and viscosity, the solids can be higher with dispersed than with dissolved polymer. This results because the viscosity of dispersions depends primarily on the volume fraction of the internal phase rather than on the molecular weight of the polymer. Molecular weight controls the viscosity of dissolved polymers (see the Mooney equation in Section 2.3).

Nonaqueous dispersion polymers are usually graft copolymers in which a hydrocarbon-soluble segment, such as poly(isooctyl acrylate) is grafted to a hydrocarbon-insoluble segment, such as PMMA or vice versa. If properly designed such a copolymer can form a stable dispersion of aggregates of polymer molecules in certain mixtures of organic solvents. The key is to chose the copolymer and the solvents so that one segment of the copolymer is soluble and the other is not. Then the soluble segments of the copolymer will orient to the periphery of dispersed polymer aggregates and the nonsoluble segments will orient toward the center of the aggregates. The aggregates are stabilized by entropic repulsion imparted by soluble segments at the particle surface (see Section 2.5). Actual NAD polymers contain ungrafted segments of both types; the ungrafted soluble material dissolves in the medium and the ungrafted insoluble material resides in the core or the polymer aggregate particles.

Nonaqueous dispersion copolymers can be prepared by starting with a *precursor* resin, which is soluble in hydrocarbon solvent and graft polymerizing an insoluble polymer to the precursor. An example of a precursor is a low molecular weight polymer of lauryl acrylate. This polymer is readily soluble in hydrocarbon solvents and has a large number of abstractable hydrogen atoms. Benzoyl peroxide is used as initiator for graft polymerization of methyl methacrylate chains onto the poly(lauryl acrylate). Many other precursors have been studied. Other processes have proven useful in some cases (see Ref. [4] for a detailed discussion of NAD polymerization).

It is possible to make NAD polymers that would provide quite high solids, on the order of 50 NVV. In practice, for automotive top coats, useful application solids are much lower. It is necessary to use a resin with a T_g no less than 65°C, yet to have polymer particles that will coalesce rapidly to a film that will flow well after application and during baking. This can only be achieved if there is some solvent dissolved in the polymer particles to lower the T_g during film formation. The lower T_g permits more rapid coalescence and also gives a less viscous coalesced film that can level better in the baking oven before the last of the solvent evaporates out of the film. However, the solvent dissolved in the polymer particles increases their volume. The higher volume fraction internal phase, as we have seen in Section 2.3, leads to a more viscous coating. There is also some fraction of the polymer that is soluble in the solvent mixture; this increases the external phase viscosity and the bulk viscosity.

In the early days of dispersion lacquers, the solids were only 15–17 NVV. This was higher than the 10–12 NVV of the solution lacquers but still gave very high VOC emissions. Over two stages, formulators were able to develop dispersion lacquers that gave acceptable leveling with NVV of the order of 27 for low pigment content lacquers and somewhat higher for higher pigment content colors. Since prospects for further significant increase in solids were poor, this type of automotive top coat has been abandoned in the United States.

Cross-linkable NADs will be discussed in Chapter 7. While their commercial use has also declined, they remain an active research subject.

4.2. HALOGENATED THERMOPLASTIC POLYMERS

Halogenated polymers are another important group of thermoplastic polymers for coatings applications. An important characteristic of such polymers is their tendency to provide films with lower moisture permeability than films from most other polymers. Therefore, they are widely used in coatings where corrosion protection is a major requirement. The halogenated polymers used on the largest scale in thermoplastic coatings are copolymers of vinyl chloride, copolymers of vinylidene fluoride, chlorinated rubber, and chlorinated ethylene vinyl acetate copolymers.

4.2.1. Solution Grade Thermoplastic Vinyl Chloride Copolymers

Solvent-soluble homopolymers of vinyl chloride have not been found useful in the coatings field; the T_g (81°C) is higher than appropriate, solvent selection is limited, and adhesion tends to be poor. These shortcomings can be overcome by using copolymers. An example, which has been widely used in container coatings, is a terpolymer of vinyl chloride, vinyl acetate, and maleic acid with a weight ratio of 86:13:1 and a mole ratio of 81:17:1. The $\overline{M}_w$ is on the order of 75,000. The vinyl acetate reduces the T_g and broadens the range of useful solvents. The maleic acid promotes adhesion. Other solution vinyl resins are made using hydroxy-functional comonomers. Examples of applications for such resins are beverage can linings and top coats for corrosion protection systems.

Poly(vinyl chloride) (PVC) and vinyl chloride copolymers require stabilization to prevent thermal and photochemical degradation. The polymers undergo dehydrochlorination, which is an autocatalytic chain reaction, to yield conjugated polyenes. As the number of conjugated double bonds increases, the polymer discolors progressively, finally absorbing all wavelengths of light and becoming, therefore, black. The resultant highly unsaturated polymer undergoes autocatalytic oxidation resulting in a high degree of cross-linking and hence embrittlement. It has been proposed that at least one major weak point (defect) results from addition of a vinyl chloride monomer in a head-to-head fashion (Eq. 4.1) to the growing polymer chain followed by chain transfer of Cl to monomer (Eq. 4.2). The resulting allylic chloride is susceptible to dehydrochlorination, which generates a new allylic chloride (Eq. 4.3), as shown in Scheme 4.1 [5]. The continuing dehydrochlorination is apparently autocatalyzed by CHl.

A wide variety of stabilizing agents has been found, including organotin esters such as dibutyl tin dilaurate; barium, cadmium, and strontium soaps; maleates, especially dibutyl tin maleate; and oxirane compounds (see Chapter 25 for further discussion).

While there are other soluble chlorinated vinyl polymers, such as copolymers of vinylidene chloride also made by chain-growth polymerization, their use in the coatings field has been small. Use of solution vinyl copolymers has declined in recent years, primarily because of the very low solids at which they can be applied, about 10 to 12 NVV.

Scheme 4.1

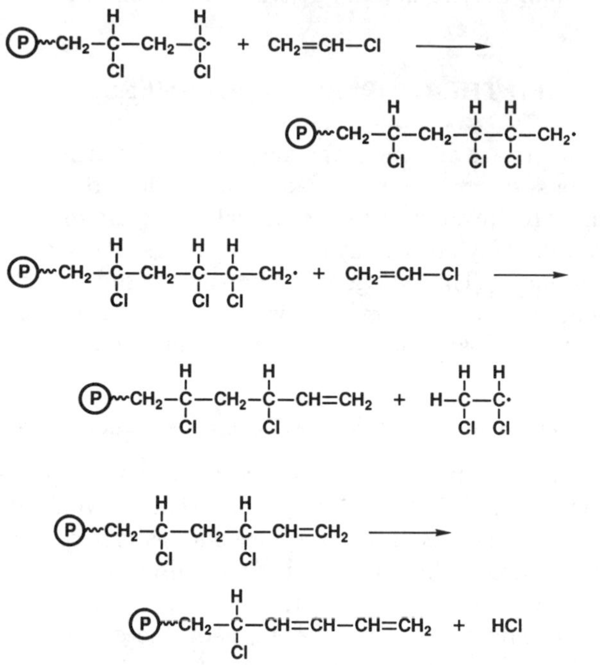

$$\text{P}\sim CH_2-\underset{\underset{Cl}{|}}{\overset{\overset{H}{|}}{C}}-CH_2-\underset{\underset{Cl}{|}}{\overset{\overset{H}{|}}{C}}\cdot \quad + \quad CH_2\!=\!CH\!-\!Cl \quad \longrightarrow$$

$$\text{P}\sim CH_2-\underset{\underset{Cl}{|}}{\overset{\overset{H}{|}}{C}}-CH_2-\underset{\underset{Cl}{|}}{\overset{\overset{H}{|}}{C}}-\underset{\underset{Cl}{|}}{\overset{\overset{H}{|}}{C}}-CH_2\cdot \qquad (4.1)$$

$$\text{P}\sim CH_2-\underset{\underset{Cl}{|}}{\overset{\overset{H}{|}}{C}}-CH_2-\underset{\underset{Cl}{|}}{\overset{\overset{H}{|}}{C}}-\underset{\underset{Cl}{|}}{\overset{\overset{H}{|}}{C}}-CH_2\cdot \quad + \quad CH_2\!=\!CH\!-\!Cl \quad \longrightarrow$$

$$\text{P}\sim CH_2-\underset{\underset{Cl}{|}}{\overset{\overset{H}{|}}{C}}-CH_2-\underset{\underset{Cl}{|}}{\overset{\overset{H}{|}}{C}}-CH\!=\!CH_2 \quad + \quad H-\underset{\underset{Cl}{|}}{\overset{\overset{H}{|}}{C}}-\underset{\underset{Cl}{|}}{\overset{\overset{H}{|}}{C}}\cdot \qquad (4.2)$$

$$\text{P}\sim CH_2-\underset{\underset{Cl}{|}}{\overset{\overset{H}{|}}{C}}-CH_2-\underset{\underset{Cl}{|}}{\overset{\overset{H}{|}}{C}}-CH\!=\!CH_2 \quad \longrightarrow$$

$$\text{P}\sim CH_2-\underset{\underset{Cl}{|}}{\overset{\overset{H}{|}}{C}}-CH\!=\!CH\!-\!CH\!=\!CH_2 \quad + \quad HCl \qquad (4.3)$$

4.2.2. Poly(vinyl chloride) Dispersion Polymers

An important class of vinyl polymers where high molecular weights can be used without necessitating very low solids for application are dispersion grade vinyl chloride copolymers. These copolymers are manufactured as particles with an average diameter of a few micrometers. The particles are prepared by *suspension polymerization*. In this process, a solution of an initiator in the monomer mixture is added with vigorous stirring to a hot dilute solution of poly(vinyl alcohol) (PVA) in water. In this manner, the monomers are dispersed into small droplets, inside of which the polymerization is initiated and completed, resulting in the formation of the particles. The PVA stabilizes the suspension by minimizing coalescence of the particles when the temperature of the polymerizing system is above the T_g. Since PVA has many easily abstractable hydrogen atoms, it is probable that graft side chains of PVC are grown on the PVA molecules to some extent. After the polymerization is complete, the polymer particles are separated by filtration, washed, and dried.

Note that a monomer-soluble initiator is used in suspension polymerization. This is in contrast to the water-soluble initiators used in emulsion polymerization, which will be discussed in Chapter 5. The molecular weight of polymers made by suspension polymerization depends primarily on initiator structure and the combination of reaction temperature and initiator concentration in the monomer. Commonly, $\overline{M}_w$ is on the order of 100,000.

In the plastics industry, dispersion grade PVC is used in "plastisols," Plastisols are based on polymer particles dispersed in a plasticizer. Since the T_g of the polymer is well above room temperature, and also because the polymer is partially crystalline, the polymer does not dissolve in the plasticizer at room temperature at an appreciable rate. When the plastisol is heated above the polymer T_g and above the melting point of the crystalline regions in the polymer particles, the polymer dissolves in the plasticizer and the particles coalesce to a molten mass. When cooled the product is a plastic material consisting of a thermodynamically stable, homogeneous solution of PVC and plasticizer. The viscosity of plastisols is generally too high for application as coatings. However, the viscosity can be lowered by addition of solvent. Solvents must be chosen that dissolve the plasticizer without significantly swelling the polymer particles. Application viscosities can generally be achieved with 20% solvent or less. These systems are properly called "organosols," but they are also frequently called plastisols even though there is volatile solvent present.

In spite of many attempts, it has not been possible to make analogous plastisols from acrylic polymers. Presumably, the partial crystallinity of the PVC permits formulating stable dispersions in plasticizer. Acrylic polymers with similar T_gs swell rapidly in plasticizers and solvents at room temperature resulting in lower solids as in NAD lacquers.

4.2.3. Fluorinated Polymers

Polytetrafluoroethylene (PTFE) has the greatest degree of exterior durability and heat resistance of any polymer, which has found use in coatings. However, PTFE is insoluble in solvents and its fusion temperature is so high that its coatings uses are limited to those applications where the substrate can withstand very high temperatures [6]. For example, aqueous dispersions of PTFE are used for coating the interiors of chemical processing equipment and cooking utensils [7]. After application the polymer particles are sintered together at temperatures as high as 425°C. In common with all highly fluorinated compounds, PTFE has such a low surface tension that it is not wet by either oils or water hence it provides a "nonstick" cooking surface. Further discussion of coatings made with PTFE are presented in Chapter 36.

Since its fusion temperature is somewhat lower, poly(vinylidene fluoride) (PVDF) can be used in additional applications. Poly(vinylidene fluoride) is used in coil coatings as a plastisol like dispersion in a solution of acrylic resin [6,8]. The fusion temperature is reported to be 245°C. The exterior durability of coatings with such a binder is outstanding. Copolymers of PVDF are finding use as powder coatings.

Polyethers derived from perfluorinated oxiranes have low T_gs; oligomers with $\overline{M}_n$ of a few thousand are relatively low viscosity oils [9]. The oligomers are durable and have low surface tensions but are high in cost. They are not used to form solid films but rather are used on porous surfaces primarily to reduce surface tension. For example, they are used as lubricants on magnetic coatings on rigid computer disks. Stone, concrete, and brick surfaces can be treated to protect the surfaces against water penetration. There has been particular interest in their use to protect old buildings, monuments, and statues against deterioration [9].

4.2.4. Chlorinated Rubber and Ethylene Vinyl Acetate Copolymers

Chlorinated rubber is used in top coats for heavy duty corrosion protection paints because of the barrier properties resulting from its low moisture permeability.

Chlorinated rubber is compatible with some alkyd resins and is used in substantial volumes to impart fast drying qualities to alkyd-based traffic paints. Like PVC, chlorinated rubber tends to dehydrochlorinate and requires stabilization. Stabilizers similar to those used with PVC are used.

Chlorinated rubber is prepared from natural rubber, which has been milled to reduce its molecular weight. A solution of the milled rubber in CCl_4 is reacted with chlorine or with some other chlorinating reactant. The reactions are complex. Addition to the double bonds, substitution, and cyclization reactions all occur. In order to eliminate most of the double bonds from the rubber, the final product contains 65–68 wt% chlorine. Excess free chlorine and HCl are removed by washing; the polymer is precipitated by addition of alcohol, filtered, and dried. Various molecular weight grades are manufactured. As would be expected, polymer strength increases with molecular weight, but the viscosity of solutions also increases. The formulator makes an appropriate compromise between coatings solids and film properties for a particular application.

Recent environmental regulations in the United States have increased the cost of producing chlorinated rubber to prohibitive levels and production has, at least temporarily, been discontinued. Imported chlorinated rubber is being used. Chlorinated ethylene–vinyl acetate copolymers have been developed, which can be used to replace chlorinated rubber in at least some applications [10]. Since the initial resin does not contain double bonds, the chlorine content need not be as high as is the case with chlorinated rubber; grades with 52, 55, and 58% chlorine are available. The higher the chlorine content, the better the barrier properties but the lower the flexibility and impact resistance. The performance is said to be comparable to chlorinated rubber with the advantage of better paint storage stability.

4.3. CELLULOSE DERIVATIVES SOLUBLE IN ORGANIC SOLVENTS

Cellulose is a naturally occurring polymer of repeating anhydroglucose units; it is insoluble in water and in all organic solvents. However, derivatives of cellulose can be made by reactions with the hydroxy groups. Some derivatives are soluble in organic solvents, others are soluble in water. Water-soluble cellulose derivatives are used to modify the flow properties of latex paints and are discussed in Chapter 35.

Anhydroglucose Unit

While many organic solvent-soluble cellulose derivatives have been prepared, those of most importance in coatings are nitrocellulose and cellulose acetobutyrate. Ethyl cellulose, cellulose acetate, and cellulose propionate (the ethyl ether, acetate, and propionate esters of cellulose, respectively) have been used to some extent.

Table 4.2. Composition of Nitrocellulose Derivatives

Type	Nitrogen (%)	ONO$_2$ Groups per Anhydroglucose Unit
SS (spirit soluble)	10.7–11.2	1.9–2.0
RS (regular solubility)	11.8–12.3	2.15–2.25
Explosive	12.3–13.5	2.25–2.5

4.3.1. Nitrocellulose

Nitrocellulose (NC) is another example of poor nomenclature since it is not really a nitro compound. It is a nitrate ester derivative and hence should be called cellulose nitrate. There are three types of nitrocellulose manufactured with differing extents of esterification. Their compositions are given in Table 4.2. An isomeric dinitroanhydroglucose unit of nitrocellulose is

Dinitroanhydroglucose Unit

The SS type is alcohol-soluble and is widely used in flexographic printing inks but has few, if any, uses in coatings. The RS type is soluble in esters and ketones and is the type generally used in coatings. The explosive type is mentioned in Table 4.2 to emphasize the hazards of nitrocelluloses. When RS nitrocellulose is used in coatings it is not explosive but is extremely flammable. To reduce the handling hazard, nitrocellulose is shipped wet with a nonsolvent. In most cases for RS types, alcohols are used and the nitrocellulose is shipped at 70 NVW with the balance being ethyl or isopropyl alcohol. While the RS type nitrocellulose is not soluble in the alcohol, it is soluble in mixtures of ketones and esters with alcohols and hydrocarbons (see Chapter 14).

Nitrocellulose is made by reacting cellulose with nitric acid. Small amounts of water are present and some hydrolysis of the cellulose occurs, reducing molecular weight. Several molecular weight grades are available. As is generally the case, the higher the molecular weight, the greater the polymer strength, and the lower the solids of coatings that can be made. Viscosity of nitrocellulose solutions has, for many years, been tested with a falling ball viscometer in which the time for a ball to fall a specified distance through a nitrocellulose solution of standard concentration is measured. The various molecular weight grades are designated as $\frac{1}{4}$, $\frac{1}{2}$, 1 s, and so forth. The times are proportional to the time in the falling ball test. The larger the number of seconds, the higher the molecular weight.

Nitrocellulose polymers were used in lacquer formulations for automotive top coats before acrylic lacquers were developed. They continued to be used for some years in refinish lacquers for repairing automobiles because they polished more easily and gave a deeper gloss appearance. By now nitrocellulose lacquers have been displaced by acrylic lacquers, except for show cars, because acrylics are far more durable. Nitrocellulose lacquer primers continue to be used to some degree in refinishing automobiles.

The principal use of NC lacquers now is in wood finishing (see Chapter 34). These lacquers have relatively low solids but continue to be used to a significant, if decreasing, extent because they enhance the appearance of wood grain more than any other coating. Other desirable characteristics of nitrocellulose lacquers for wood finishing are its very rapid drying, which permits handling and shipment soon after finishing, and its easy damage repair, since the films are thermoplastic. Increasingly stringent VOC emission regulations can be expected to force reductions in the use of NC substantially during the 1990s.

4.3.2. Cellulose Acetobutyrate

A variety of organic esters of cellulose are used in fibers, plastics, and, to a degree, coatings. Although cellulose acetate has been used to a small extent in lacquers as a replacement for nitrocellulose, the principal ester derivatives used in coatings have been mixed acetate, butyrate esters (CABs). CAB lacquers are used to some extent; they exhibit lighter color and better color retention with reduced risk in handling relative to NC lacquers. The principal use, however, has been as a component in acrylic automotive coatings. CABs help control flow and particularly help facilitate orientation of aluminum pigment flakes parallel to the surface of the wet film immediately after application. Orientation of the flakes in the dry film enhances the appearance of metallic finishes (see Ref. [11] and Chapter 33).

Several types of CAB are available with different ratios of acetate/butyrate/ unesterfied hydroxyl groups and with differing molecular weights. The ratio affects the solubility and compatibility of the polymer. Generally speaking, as one would expect, the larger the ratio of butyrate ester the larger the amount of aromatic hydrocarbon that can be tolerated in a solvent mixture. Also, in general, as the butyrate content increases the T_g tends to be lower. Compatibility with acrylic resins can be very dependent on the composition of the CAB and the acrylic polymer. An example of a CAB grade used in acrylic lacquers has an average of 2.2 acetate groups, 0.6 butyrate groups, and 0.2 unreacted hydroxyl groups per anhydroglucose unit. This corresponds to about 30% acetoxy content, 17% butyryloxy content, and 1.5% hydroxyl content, based on the total weight of the polymer.

GENERAL REFERENCES

B. B. Kine and R. W. Nowak, "Acrylic and Methacrylic Ester Polymers," *Encyclopedia of Polymer Science and Engineering*, 2nd ed., Vol. 1, Wiley, New York, 1986, pp. 234–299.

G. W. Odian, *Principles of Polymerization*, 3rd ed., Wiley-Interscience, New York, 1991.

REFERENCES

1. G. W. Odian, *Principles of Polymerization*, 3rd ed., Wiley-Interscience, New York, 1991, pp. 198–334 and 452–531.

2. W. A. Lee and R. A. Rutherford, "The Glass Transition Temperatures of Polymers," in J. Branderup and E. H. Immergut, Eds., *Polymer Handbook*, 2 ed., Wiley, New York, 1975, pp. III-139–III-192.

3. W. S. Zimmt, U.S. Patent 3,823,205 (1974).

4. K. E. J. Barrett, Ed., *Dispersion Polymerization in Organic Media*, Wiley, London, 1975.

5. W. H. Starnes, Jr., *Pure Appl. Chem.*, **57**, 1001 (1985). G. Georgiev, L. Christiv, and T. Gancheva, *J. Macromol. Sci. Chem.*, **A27 (8)**, 987 (1990).

6. S. Munekata, *Prog. Org. Coat.*, **16**, 113 (1988).

7. K. Batzar, *Proc. XVth Internat. Conf. Org. Sci. Technol.*, Athens, Greece, 13 (1989).

8. J. E. Gaske, *Coil Coatings*, Federation of Societies for Coatings Technology, Blue Bell, PA, 1987.

9. G. Moggi, *Proc. XVIth Internat. Conf. Org. Sci. Technol.*, Athens, Greece, 251 (1990).

10. Anonymous, *Hypalon CP Chlorinated Polyolefins*, Technical Bulletin, Du Pont Company, Wilmington, DE, (1989).

11. B. N. McBane, *Automotive Coatings*, Federation of Societies for Coatings Technology, Blue Bell, PA, 1987.

CHAPTER V

Latexes

A "latex" is a dispersion of polymer particles in water. Most latexes are made by a free radical initiated chain-growth polymerization process in which the liquid monomers are emulsified in water at the start; hence the process is called "emulsion polymerization." Emulsion polymerization begins with a dispersion of liquid droplets in an aqueous medium, that is, an emulsion, and ends with a dispersion of solid polymer particles in the aqueous medium, that is, a latex. Consequently, it is not surprising that latex paints are sometimes misnamed "emulsion paints," but this terminology is best avoided since it can cause confusion with systems that are true emulsions. The process, however, is properly called emulsion polymerization.

Molecular weights of polymers prepared by emulsion polymerization are generally high; $\overline{M}_w$s of 1,000,000 or higher are common. However, unlike solution polymers, the molecular weight of the polymer in latex particles does not affect the viscosity of the latex. Instead, latex viscosity is governed by the viscosity of the medium in which the polymer particles are dispersed (the continuous phase) and by the volume fraction and packing factor of the particles (the discontinuous phase) as described by the Mooney equation discussed in Sections 2.3 and 19.5. This makes it possible to formulate latex coatings at much higher solids than would be possible with polymer solutions of such high molecular weight. Furthermore, in most cases the continuous phase is mainly water, so VOC is low.

Aqueous dispersions of polymers are also prepared by other methods, for example, polyurethane dispersions in water, prepared by step-growth polymerization. Frequently, these systems are just called aqueous dispersions rather than latexes. They are discussed briefly in Chapter 12.

A large majority of architectural coatings use latexes as the principal vehicle; small, but growing, parts of the OEM product and special-purpose coatings markets are latex based. Most latex paints form films by coalescence at ambient temperatures (see Section 3.3.1). The limitations of most present-day latex-based paints are poor leveling after application, inability to achieve high gloss, lower hardness and solvent resistance than enamels applied from solution, as well as greater sensitivity and higher permeability to water, On the other hand, the durability of the films is far superior to films formed from drying oil and alkyd paints, and VOC emissions are generally lower.

5.1. EMULSION POLYMERIZATION

Emulsion polymerization, is carried out in water using monomer(s), surfactant(s), and a water-soluble initiator. Many of the same monomers described in Chapter 4 are used, and the polymerization reactions are broadly similar at the molecular level. However, the physical circumstances of polymerization are different, affecting the polymerization chemistry. The process is easily varied, and the properties of the latex can be strongly affected by changes in the physical conditions of the polymerization. Many early laboratory studies were carried out by a small scale "batch process" in which all of the ingredients were put into a sealed bottle, which was then shaken in a temperature controlled water bath. This process is often called the "pop-bottle" process. Current textbook hypotheses describing the mechanism of emulsion polymerization are frequently based largely on such studies. However, batch processes cannot be used commercially because the heat evolved by the exothermic polymerization process would be uncontrollable in a large vessel. Instead, commercial latexes are produced by a *semicontinuous batch process* using an apparatus such as that shown schematically in Figure 5.1. Initially, a "seed latex" is made by a batch process from a small charge of monomer(s), initiator, and surfactant(s), and the main part of the monomer(s) and other ingredients are added subsequently at a rate that permits the heat of reaction to be removed.

Since the products obtained from the two processes, batch and semicontinuous emulsion polymerization, are often very different, conclusions derived from studies on the batch process are often misleading when applied to the semicontinuous process and vice versa. As this distinction is not universally appreciated, the literature on emulsion polymerization must be read with caution. Much of our discussion will be related to semicontinuous processes, the method by which commercial latexes are produced. However, we begin by discussing the batch polymerization process in order to help the reader visualize the interactions of the many variables involved. Then the discussion is extended to the semicontinuous process. First, however, the raw materials used in both processes are discussed.

5.1.1. Raw Materials for Emulsion Polymerization

Monomers. Many monomers have been converted to latexes by emulsion polymerization. The main requirements are that the monomers can undergo free radical chain polymerization and that they do not react with water; monomers with slight but limited solubility in water are generally required, although water miscible comonomers can be used in small amounts. The two major classes of latexes used in coatings are based on acrylic esters and on vinyl esters. The distinction between these classes is not clear-cut; there are many commercial copolymer latexes of acrylic esters with vinyl acetate. Nevertheless, latexes are classified on the basis of the predominant monomer type.

The earliest latexes used in coatings were copolymers of styrene with butadiene; this type is seldom used today in architectural coatings, but it is used in large volume in coatings for paper. Paper coatings are usually not classified as part of the coatings market and are, therefore, outside the scope of this book.

Most of the acrylic monomers described in Chapter 4 are well suited to emulsion polymerization and copolymerization; ethyl, butyl, and 2-ethylhexyl acrylate, as

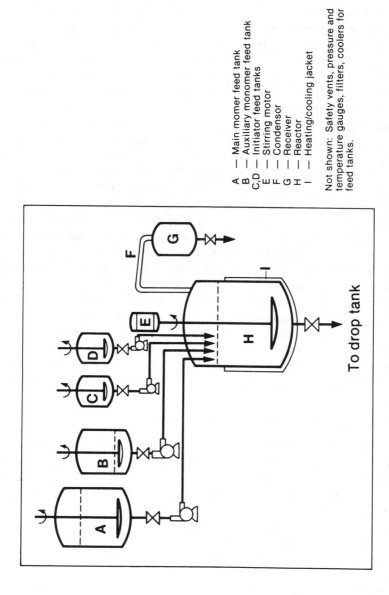

Figure 5.1. Schematic diagram of a semicontinuous batch process production unit. (Adapted from Ref. [1], with permission.)

To drop tank

A — Main momer feed tank
B — Auxiliary monomer feed tank
C,D — Initiator feed tanks
E — Stirring motor
F — Condensor
G — Receiver
H — Reactor
I — Heating/cooling jacket

Not shown: Safety vents, pressure and temperature gauges, filters, coolers for feed tanks.

well as methyl and butyl methacrylate are commonly used in latexes for coatings. Styrene can be readily copolymerized with these acrylic monomers. In the United States, vinyl acetate (VAc) is the predominant vinyl ester monomer. In Europe the vinyl ester of a branched, ten-carbon carboxylic acid (mixed isomers) is used as a comonomer with VAc. Vinyl acetate can be copolymerized with ethylene under pressure to produce low cost latexes widely used in adhesives.

$$H_2C=CH-O-\overset{\overset{\displaystyle O}{\|}}{C}-CH_3$$

Vinyl Acetate

$$H_2C=CH-O-\overset{\overset{\displaystyle O}{\|}}{C}-\overset{\overset{\displaystyle R_1}{|}}{\underset{\underset{\displaystyle R_3}{|}}{C}}-R_2$$

Vinyl Ester of C_{10} Acid
(R_1, R_2, and R_3 are Alkyl
Groups Totaling to C_8H_{19})

As a very broad generalization, the U.S. market uses acrylic latexes for exterior applications as well as for applications requiring excellent hydrolytic stability, whereas vinyl latexes are used for interior flat paints. Details are described in Sections 5.2 and 5.3.

The Fox equation, described in Section 4.1, can be used to estimate roughly the T_g values of latex copolymers for cases when copolymerization is relatively uniform. However, latex copolymers are often nonuniform. As we will see, it is possible to make latex copolymers in which the inside (core) of each particle has a different composition than the outside (shell). Nonuniform latex copolymers may have two or more T_gs or a broad single T_g; in such cases the Fox equation is of limited use.

Water-soluble monomers such as acrylic acid, acrylamide, and 2-hydroxyethyl acrylate (and the corresponding methacrylate derivatives) cannot be homopolymerized to latexes; they form water-soluble polymers rather than polymer particles. However, modest amounts of such monomers can be copolymerized with less soluble monomers by emulsion polymerization. Acrylic and methacrylic acid have been used, for example, to prepare core–shell latexes having copolymers rich in COOH groups near the surface of each particle. Such latexes have the sometimes useful property of undergoing changes in viscosity as the pH is changed, as is explained in Section 5.2.

Hydroxyethyl methacrylate and acrylate are used as comonomers in making latex polymers that can be thermoset with melamine and urea formaldehyde cross-linkers. Use of acrylic and methacrylic acid as comonomers leads to formation of latex polymers with COOH substituents, which can be cross-linked with zinc salts and with epoxy resins.

Another important group of monomers is tertiary amine functional acrylates and methacrylates, for example, 2-(dimethylamino)ethyl methacrylate. Incorporation of 1 or 2 wt % of an amine bearing monomer imparts superior adhesion to both old paint surfaces and metal surfaces. The difference is particularly marked when the dry latex paint surface is wetted with water.

The utilization of vinylidene chloride as a comonomer in latexes is of increasing importance. Vinylidene chloride–acrylic ester copolymer latexes give films with much lower water permeability than other acrylic latexes, a desirable characteristic for coatings for metals.

Initiators. The principal initiators used in emulsion polymerization are soluble in water. The most common initiators are persulfate salts, especially ammonium persulfate (more correctly, ammonium peroxydisulfate). The persulfate ion cleaves thermally to produce sulfate ion radicals in the water phase (Eq. 5.1) that can initiate free radical polymerization, as shown for ethyl acrylate (Eq. 5.2). Note that the terminal group is a sulfate half-ester ion.

$$^-O_3S-O-O-SO_3^- \longrightarrow 2 \ ^-O_3S-O\cdot \tag{5.1}$$

$$^-O_3S-O\cdot \ + \ CH_2=\overset{\overset{\displaystyle H}{|}}{\underset{\underset{\displaystyle O}{\overset{\displaystyle \|}{C}}-O-CH_2-CH_3}{C}} \longrightarrow \ ^-O_3S-O-CH_2-\overset{\overset{\displaystyle H}{|}}{\underset{\underset{\displaystyle O}{\overset{\displaystyle \|}{C}}-O-CH_2-CH_3}{C}}\cdot \tag{5.2}$$

The sulfate ion radical can also abstract hydrogen from water leading to the formation of an acid sulfate ion and a hydroxy free radical. In this case initiation leaves a terminal hydroxy group. While this reaction is disfavored by the strength of the O—H bond toward homolytic cleavage, it has been shown to occur to an appreciable extent under conditions where monomer concentrations are low.

$$^-O_3S-O\cdot \ + \ H_2O \longrightarrow HSO_4^- \ + \ HO\cdot$$

The half-life of persulfate is such that a 0.01 M solution of $(NH_4)_2S_2O_8$ at pH 10 produces 8.4×10^{12} radicals per mL per s at 50°C and 2.5×10^{15} radicals per mL per s at 90°C. For rapid polymerization at lower temperatures, radical production can be accelerated by catalysts. For example, mixtures of ferrous, thiosulfate $S_2O_3^{2-}$, and persulfate salts react faster than persulfate alone. Since catalysis involves a sequence of oxidation and reduction reactions, processes using this combination are called "redox" emulsion polymerization. The reactions involved are complex; the following are said to be representative:

$$^-O_3S-O-O-SO_3^- \ + \ Fe^{2+} \longrightarrow \ ^-O_3S-O\cdot \ + \ SO_4^{2-} + \ Fe^{3+}$$

$$S_2O_8^{2-} + \ 2 \ S_2O_3^{2-} \longrightarrow \ S_4O_6^{2-} + \ 2 \ ^-O_3S-O\cdot$$

$$Fe^{3+} \ + \ 2S_2O_3^{2-} \longrightarrow \ Fe^{2+} \ + \ S_4O_6^{2-}$$

Using redox systems, polymerization can be initiated at room temperature; the exothermic polymerization reaction warms the reaction mixture to the desired temperature (often 50–80°C). Cooling is required to prevent overheating.

Other water soluble initiators are available as alternatives to persulfate and redox initiators. For example, water soluble azo initiators are used commercially, perhaps in the expectation that the initiator residues will have minimal adverse effect on exterior durability (see Chapters 2 and 4).

It is common to add a second, less hydrophylic initiator in the late stages of the semicontinuous process to facilitate conversion of monomer to high levels, preferably greater than 99%. An example of such an initiator is *t*-butyl hydroperoxide, which is only slightly soluble in water and presumably tends to migrate into the latex particles. It can be speculated that an initiator that is more soluble in the polymer particles than in water would be more effective than ammonium persulfate in the late stages of reaction when most of the unreacted monomer is dissolved in the polymer particles.

Surfactants. Surfactants (see Section 2.3.) are a very important component in emulsion polymerizations. Anionic and nonionic surfactants are both used. Typical anionic surfactants are sodium stearate,

$$H_3CCH_2CH_2CH_2CH_2CH_2CH_2CH_2CH_2CH_2CH_2CH_2CH_2CH_2CH_2CH_2CH_2COO^- \ Na^+$$

sodium dodecylbenzene sulfonate,

$$H_3CCH_2CH_2CH_2CH_2CH_2CH_2CH_2CH_2CH_2CH_2CH_2\text{—}\bigcirc\text{—}SO_3^- \ Na^+$$

and sodium lauryl sulfate (often incorrectly called sodium lauryl sulfonate).

$$H_3CCH_2CH_2CH_2CH_2CH_2CH_2CH_2CH_2CH_2CH_2CH_2OSO_3^- \ Na^+$$

A typical nonionic surfactant is ethoxylated nonylphenol,

$$H_3CCH_2CH_2CH_2CH_2CH_2CH_2CH_2CH_2\text{—}\bigcirc\text{—}O(CH_2CH_2O)_{\bar{n}}H$$
$$\bar{n} = 20 \text{ to } 40$$

In latexes, surfactants play two crucial roles, which are to facilitate emulsion polymerization and to stabilize the resulting polymer particles. In general, anionic surfactants are used at the levels of 1 to 4 wt%, they cost less than nonionic surfactants, and they reduce particle size more efficiently on a weight basis. On the other hand, nonionics, used at levels of 2 to 6 wt%, are more effective in stabilizing the latex against coagulation during freeze–thaw cycling, make it less sensitive to coagulation by salts (especially salts of polyvalent cations), and less sensitive to changes in pH. Mixtures of anionics and nonionics are often used in efforts to combine the best features of both. In general terms, anionic surfactants stabilize primarily by charge repulsion and nonionic surfactants stabilize primarily by entropic repulsion (see Section 2.5 for a discussion of the stabilization of dispersions). More explicit generalizations about the use of surfactants in emulsion polymerization are treacherous because there will be exceptions to almost any statement that can be made.

Additional entropic stabilization can be achieved by including in the polymerization recipe a small amount of a water-soluble polymer, which can form graft copolymers readily. A polymer commonly used for this purpose is poly(vinyl alcohol) (PVA), which has many easily abstractable hydrogen atoms. During polymerization, initiator radical ions can abstract a hydrogen from PVA to form a free radical on a PVA chain. A graft of the monomers being polymerized will now grow on the PVA chain. More than one such graft could grow on a single PVA molecule. In any case, as the graft chain becomes longer, it will become more hydrophobic and will associate with other polymer molecules in particles, thereby carrying the PVA part of the molecule to the particle surface. The PVA part remains hydrophylic and will associate with water to form an entropic stabilizing layer analogous to that formed by nonionic surfactants. Such layers can be thicker than the layer from adsorbed surfactant molecules and hence may be more effective in promoting stability. When used in this way, PVA is called a "protective colloid." Use of PVA or other protective colloids may reduce the water sensitivity of films made from latex paint because less water-soluble surfactant is needed.

Hydroxyethyl cellulose (HEC) can also be used as a protective colloid in latex preparation. It has been shown that grafting occurs and that the resulting latex is thixotropic and more shear thinning than the nonthixotropic latex obtained when no water-soluble polymer is present during the emulsion polymerization [2]. It has also been shown that monomer–HEC grafting, while improving latex stability, favors large particle size, broad particle size distribution, and low film clarity [3].

Commonly, both anionic and nonionic surfactants are used in an emulsion polymerization recipe, together with PVA or some other water-soluble polymer as a protective colloid. The anionic surfactant provides charges at the surface for charge repulsion and can be used in lower concentration than nonionic surfactant, as discussed in Section 5.1.2. The nonionic surfactant and the protective colloid provide stabilization by entropic repulsion, which has the distinct advantage of reducing sensitivity to multivalent cations that may be introduced into a latex paint formulation from the water or from other ingredients such as pigments contaminated with soluble salts. In semicontinuous batch polymerization processes it is possible, and sometimes desirable, to use anionic surfactants in the initial batch charge and to add nonionic surfactants in the later additions.

The two types of surfactants impart different rheological characteristics to the latex. Anionic surfactants lead to formation of essentially rigid particles with relatively broad particle size distribution. Such latexes can have low viscosity at relatively high solids, that is, in the range 45–50 NVV. On the other hand, the thicker swollen layer on the surface of entropically stabilized particles and the narrower distribution of particle sizes generally obtained using nonionic surfactants lead to lower polymer solids, or higher viscosity at a given solid, than when stabilization is accomplished with anionic surfactants alone. Furthermore, the surface layer of an entropically stabilized latex particle is not rigid and hence can be distorted when shear stress is applied, imparting shear thinning characteristics to the fluid.

All surfactants impart some water sensitivity to films from latex paints. This sensitivity is reflected, for example, in a tendency of house paints to water spot if rained on soon after application and in the somewhat limited corrosion resistance when most latex coatings are used on steel. Selection of surfactant combinations that minimize such shortcomings is an ongoing subject of research. For example, so-called "soap-free" latexes have been developed in which emulsion polymerization is effected without use of conventional surfactants. Such current approaches may reduce the water sensitivity associated with surfactants; however, they are unlikely to eliminate this problem because the processes are designed to produce an unconventional surfactant in situ (e.g., by copolymerization of acrylic acid).

Water and other additives. Deionized water is used in emulsion polymerization and is usually used in latex paint production. Where it is not, variations in water quality can cause a variety of problems, especially when anionic surfactants are used. Other ingredients sometimes used in emulsion polymerization processes include buffers, which protect sensitive monomers from hydrolysis and sensitive surfactants from deactivation; and thickeners, which control viscosity.

5.1.2. Batch Process

This process is carried out in a small reactor loaded with water, surfactant(s), monomer(s), and initiator(s). The reactor is sealed, mounted on a shaker, and immersed in a thermostatically controlled water bath where it is agitated. Let us imagine an idealized situation in which the mixture has been heated to reaction temperature but polymerization has not yet begun; in other words, we will look at a snapshot of the system before reaction has started.

The situation is illustrated in Figure 5.2a. Most of the monomer is emulsified in droplets whose size depend on the nature and amount of the surfactant and the agitation level. Some surfactant is oriented on the surface of the droplets, and a small amount dissolves in the water. However, many surfactants are only slightly soluble; when a certain concentration (the critical micelle concentration, CMC) is exceeded, surplus surfactant forms *micelles*, which are submicroscopic, thermodynamically stable clusters containing, typically, 30–100 surfactant molecules. In micelles the hydrophobic parts of each molecule orient toward the center and the hydrophylic parts orient outward in contact with water. Typical concentrations of anionic surfactant (~2 wt%) exceed the CMC significantly, so much of the surfactant resides in micelles. The CMC of different surfactants varies over a wide range from about 0.01% to about 1%.

The initiator is dissolved in the water; in reality it would begin reacting immediately, but in this idealized picture it has not yet begun to react. While the monomers used in emulsion polymerization to produce latexes for coatings are not miscible with water, most are at least slightly soluble, about 0.5 to 5 wt%; thus a small fraction of the monomer leaves the droplets and dissolves in water. Furthermore, some fraction of the monomer enters the micelles, where it is attracted to the interior hydrophobic part of the surfactant. A dynamic equilibrium is established in which monomer molecules move rapidly from the droplets to the solution to the micelles and back. Surfactant molecules are involved in a similar dynamic equilibrium.

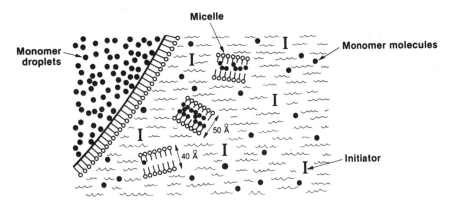

Figure 5.2. Schematic, idealized representations of emulsion polymerization (a) before polymerization has begun and (b) during polymerization. (Adapted from H. Gerrens, *Ber. Bunsen. Phys. Chem.*, **67(8)**, 741 (1963), with permission.)

Approximate populations of the various species involved in the idealized system of Figure 5.2a are indicated in Table 5.1. The values in Table 5.1 were estimated for a system with the following characteristics: (1) 2 wt% of a surfactant of $M = 300$, a density of 1, a CMC of 0.1 wt%, and assuming that the average micelle has 50 molecules of surfactant; (2) 2 wt% of initiator of $M = 228$ and a density of 2; (3) 45 wt% of monomer of $M = 100$, 1 wt% solubility in water, and a density of 0.93; (4) emulsion droplets assumed to have an average diameter of 5 μm; and (5) polymer particles assumed to have an average diameter of 0.5 m and a density of 1.2. At the beginning of the process a very large number of micelles coexists with a much smaller number of much larger monomer droplets. During emulsion polymerization both the monomer droplets and the micelles are consumed and are replaced with an intermediate number of stabilized polymer particles. The number of particles formed, of the order of 10^{15}/mL, may be 10^3 times greater than the original number of droplets but only a small fraction of the original number of micelles. These values are only rough approximations for several reasons; no account is taken of the amount of surfactant on the surface of the emulsion droplets or of the monomer dissolved in the inside of the micelles in the early stage. Furthermore, due to shrinkage resulting from polymerization, product has higher density than the initial mixture. For the same reasons and because monomers, surfactants, and conditions differ widely, the ranges in Table 5.1 should be considered only as estimates for the sake of illustration.

When polymerization begins, the initiator reacts to generate free radicals dissolved in water, typically at a rate such that about 10^{13} to 10^{15} free radicals are generated per second per milliliter. The free radicals quickly react with monomer molecules to initiate polymerization. Due to the relative concentrations of monomer in the various phases, free radicals are most likely to encounter a monomer molecule dissolved in water, and initiation will primarily occur there. To a lesser extent, the free radicals will encounter a micelle and react with a monomer molecule there. However, due to the relatively low concentration of droplets and the insolubility of the initiating radical ion in monomer very little initiation occurs in the monomer droplets.

If initiation occurs in a micelle, propagation occurs there. If initiation occurs in the aqueous phase, propagation may continue for several steps there. However, as the molecular weight of the growing polymer chain increases its solubility in water decreases, and one of two things will happen while its molecular weight is still in the oligomer range. The growing chain may enter a micelle where propa-

Table 5.1. Approximate Concentrations of Species in an Emulsion Polymerization Mixture

Component	Concentration (per mL)
Dissolved monomer molecules	3×10^{19}
Monomer droplets	2×10^{12}
Dissolved surfactant molecules	10^{18}
Micelles	8×10^{17}
Dissolved initiator molecules	5×10^{19}

gation continues. Alternatively, it may acquire surfactant from the solution. Either way, a surfactant-stabilized polymer particle containing a single free radical is formed. Such particles rapidly imbibe monomer, and propagation continues within the particle.

Termination (by combination or by disproportionation as described in Chapter 4) occurs when a second free radical enters the particle. Arrival of a second radical increases the concentration of free radicals within the particle to several orders of magnitude higher than in solution polymerization, and termination is almost instantaneous. After termination the surfactant stabilized polymer particle is temporarily inactive because it contains no free radicals.

Monomer rapidly migrates to the inactive polymer particle. A third free radical (IMMMMM·) enters the monomer-swollen particle and again initiates polymerization. Chain growth continues until a fourth free radical enters and causes termination. This process is repeated over and over, leading to formation of additional polymer molecules in the particle. Dissolution of monomer from the monomer droplets replenishes the monomer in solution.

With many monomers of interest in coatings a steady state is reached in which at a given instant about half the particles contain one free radical and the other half contain none. Very few contain more than one because of the speed of termination. Concentrations of unreacted monomers may build to over 50 vol% while the particle is inactive. When a free radical enters an inactive particle, monomer is rapidly consumed; since much of the polymerization occurs in particles containing a high concentration of polymer, conditions favor chain transfer to polymer.

As the polymer particles grow they become starved for surfactant. Their surfactant level is replenished by adsorbing more surfactant molecules from the solution. The result is growth of the particles that acquired a polymer molecule early on and disappearance of micelles that did not. Before consumption of monomer is complete, virtually all micelles that did not acquire a polymer molecule will have disappeared and the corresponding surfactant molecules will have moved over to the growing polymer particles. At this point in the process the number of latex particles is fixed. It follows that under these circumstances the number of particles per unit volume will depend on the concentration of micelles (a function of the amount of surfactant present and its critical micelle concentration) and on the rate of initiation (a function of the type of initiator, its concentration, and the temperature), but not to any significant degree on the amount of monomer. Early studies led Smith and Ewart to propose that, when other factors are constant, the number of polymer particles N is related to the concentration of initiator [I], and of surfactant [S], by Eq. 5.3.

$$N = K[I]^{0.4}[S]^{0.6} \qquad (5.3)$$

Polymer particle growth continues until unreacted monomer or initiator is used up; enough initiator is used to assure high conversion of monomer. While the total amount of monomer has little bearing on the number of polymer particles it affects their size. After the early stages of the process, all polymerization occurs within a fixed number of polymer particles; consequently the more monomer present, the bigger the particles can grow.

5.1.3. Semicontinuous Batch Process

As indicated earlier, the semicontinuous process must be used in large-scale production to avoid overheating and runaway reactions. This process has the additional advantage of being much more versatile, as it permits addition of the various ingredients at different rates and during different stages of the process. Thus an almost unlimited variety of latexes with distinctly different properties can be made from a relatively short list of monomers and surfactants. The apparatus is illustrated in Figure 5.1.

The raw materials used in this process are similar to those used in the batch process. Monomers and initiators are added in proportions and at rates such that rapid polymerization occurs. The monomer is sometimes emulsified before addition, and the emulsion is added at such a rate that the monomer droplet concentration in the reactor is low. When the process is performed in this way, it is said to be run under monomer-starved conditions. When all ingredients are emulsified and added together the monomer and initiator concentrations remain essentially constant. As with the batch process, micelles are consumed early in the process. After they have been consumed there is no mechanism for starting new particles, and further polymerization occurs within existing particles. As polymerization continues, the size of the latex particles increases.

When emulsion polymerization has reached the stage at which the micelles have been used up, the intermediate product at that stage is a latex with very small particle size. As indicated earlier, such a material is sometimes called a *seed latex*; its characteristics will effect the subsequent course of the process. In many cases the seed latex is simply a transient stage in the process, but in many others the reaction conditions and monomer feed are changed after the seed latex has formed. Sometimes it is advantageous to produce a large batch of seed latex and to subdivide it for subsequent use.

As with batch polymerization, particle size of the latex is controlled by the number of seed latex particles in proportion to the amount of monomer added during the process. Increasing the number of seed particles will, other factors being equal, yield a latex of smaller average particle size. Substantial variations in particle size are possible, and they have significant affects on the rheological and film properties of latex paints.

Particle morphology. The semicontinuous process can be carried out under monomer starved conditions by adding monomer at a rate no faster than the rate at which it can react. Under these conditions it is possible to make latexes having "core-shell" morphology by changing from one monomer feed composition to another part way through the process. The composition near the center of each particle in a core-shell latex is different from the composition near the surface, and many core-shell latexes display two different Tgs after coalescence. The mechanism by which core-shell particles form remains a subject of continuing research.

Microgels. Inclusion in an emulsion polymerization monomer mixture of a few percent of a monomer having two reactive sites, such as divinyl benzene or a diacrylate, will lead to formation of latexes in which the polymer in the particles is cross-linked, called "microgels." Such latexes are of little use as primary film formers, but microgel particles are used as rheology control additives in automotive topcoats.

Molecular weight The $\overline{M}_w$s obtained by both batch and semicontinuous processes are generally very high; $\overline{M}_w$s over 1,000,000 are common. In many cases there is a broad distribution of molecular weights. High $\overline{M}_w$s result from several factors. First, it is thought that during polymerization the fraction of polymer particles containing more than one free radical at a given instant is extremely low. As a result the rate of termination reactions within the particle is reduced, and a single growing chain may consume virtually all the monomer present in the particle before it terminates. Second, viscosity inside the particles is very high, reducing the mobility of the polymer chains. It is then easier for more mobile monomer molecules to add to the growing chain and increase the M, than for two less mobile growing chains to react with each other and terminate growth. Third, if the structure of the polymer has hydrogen atoms that have even a slight tendency to be abstracted, chain transfer to the polymer will be significant. The environment within the polymer particle strongly favors chain transfer to the polymer because the concentration of free radicals is very low (often one per particle) while the concentration of the polymer approaches 100%. Chain transfer to the polymer increases $\overline{M}_w$ and $\overline{M}_w/\overline{M}_n$ because it is statistically probable that when growth of short chains is ended by chain transfer the new chain will grow as a branch on a polymer molecule that already has a high molecular weight.

In one study comparing homopolymerization and copolymerization of vinyl acetate and butyl acrylate both by batch and semicontinuous processes, it was found that $\overline{M}_w/\overline{M}_n$ for batch homopolymers and copolymers was broad, ranging from 15 to 21, but the semicontinuous process generally gave even broader distributions, with $\overline{M}_w/\overline{M}_n$ ranging from 9 to 175 [4]. The semicontinuous process commonly afforded bimodal distributions with considerable amounts of relatively low molecular weight polymer. Semicontinuous poly(vinyl acetate) homopolymer had the broadest distribution; GPC indicated substantial fractions of material with M below 660 and above 1,700,000. Presumably, the low monomer concentration in the semicontinuous process especially favors chain transfer to polymer.

In practice, molecular weight of latex polymers is seldom determined. In contrast to solution polymers, viscosity of latexes is independent of the molecular weight. Furthermore, in solution coatings, one tends to be working with lower molecular weight polymers in the region where film properties are highly molecular weight dependent. On the other hand, it seems likely that, even if the molecular weight of a latex is low in comparison to other latexes, it is still above the level where there are major effects on film properties.

Particle size and shape. The particle size of a latex from a semicontinuous batch process is determined by the number of particles relative to the amount of monomer. The number of particles is governed primarily by the amount and type of surfactant(s) and initiator(s). Anionic surfactants tend to produce more particles than nonionics when used on an equal weight basis, and hence lead to smaller particles.

Particle size distribution is an important variable in formulation of latex paints. A broad distribution is often advantageous because the packing factor is strongly affected by particle size distribution. The volume fraction of latex particles that can be present at a standard viscosity will be highest with broad particle size distribution. In general, particle size distribution is broader with anionics and more nearly approaches monodisperse particle size with nonionics. The effect of changing the ratio of nonionic (nonylphenol ethoxylate with 40 mol of ethylene oxide) to anionic (sodium dodecylbenzenesulfonate) on particle size and particle size distribution with a constant amount of total surfactant in the polymerization of a vinyl acetate–dibutyl maleate copolymer has been studied [4]. The lowest ratio gave the smallest particle size and the broadest distribution of particle sizes.

Until fairly recently, latexes with essentially spherical particles have been the rule. During the 1980s, latexes were commercialized in which groups of a few particles are fused together to form nonspherical lobed particles [5]. At equal concentration, the viscosity of these latexes is significantly higher than the viscosity of comparable latexes with spherical particles. Expressed in terms of the Mooney equation, the shape factor for these particles is greater than 2.5 and the packing factor is smaller than for a spherical particle latex. As we discuss in Chapter 35 on latex paints, these lobed particle latexes require modified formulating approaches. One reason is that lobed particle latexes have higher viscosity than spherical particle latexes at a given V_i, and another is that they also exhibit more pronounced shear thinning rheology.

5.2. ACRYLIC LATEXES

Acrylic latexes are widely used for exterior paints because of their resistance to photodegradation. They are also much more resistant to hydrolysis and saponification than are vinyl acetate latexes. These properties are critical in exterior paints; they make acrylic latex paints particularly useful for alkaline substrates such as masonry and galvanized metal and for other applications where there is exposure to high humidity. Acrylic and styrene–acrylic latexes are being used increasingly for industrial maintenance coatings. Acrylic latexes are finding increasing interest in OEM automotive applications, and their use may grow in response to the need to reduce VOC emissions. For example, latexes are being incorporated into waterborne basecoats for OEM automotive basecoat–clear coat applications.

A critical decision in designing a latex is monomer selection. An important consideration is to select a monomer combination that will give a copolymer with the appropriate T_g. The effect of monomer selection on other film properties is also critical. Cost is an important consideration—the paint business is highly price competitive; particularly for lower-priced paints, adequate film properties must be attained using low cost monomers.

The T_g must be low enough to permit coalescence of the latex at the lowest anticipated application temperature, yet high enough to assure adequate film hardness and toughness. Coalescence is aided by coalescing solvents, as explained in Section 3.3.1. T_gs in the vicinity of 5 to 15°C are common for exterior house paints, which may be applied at temperatures as low as 5°C. The minimum film forming temperature (MFT) of latexes is measured on a metal block with graduated temperatures, observing the lowest temperature at which the latex coalesces. This temperature is related to T_g but is influenced by other factors such as particle size, phase separation within the latex particles, and plasticizing effects of water and surfactants. The MFTs tend to be somewhat lower than T_gs. T_gs and MFTs in the desired range are usually attained by copolymerizing monomers whose homopolymers have much higher and much lower T_gs than the target values.

As a high T_g comonomer, methyl methacrylate (MMA, homopolymer T_g = 105°C) imparts excellent exterior durability and hydrolytic stability at moderate cost. Styrene is often substituted partly or completely for MMA because it costs less and provides a similar T_g effect (homopolymer T_g = 100°C). Such products are sometimes called acrylic–styrene latexes and sometimes simply acrylic latexes. It is clear that styrene imparts excellent hydrolytic stability, but it may adversely affect exterior durability. It is known that styrene homopolymer degrades relatively rapidly outdoors. However, in a copolymer, some MMA can be replaced by styrene without measurably decreasing exterior durability. The amount of styrene that can be substituted apparently varies depending on the other comonomers and probably upon other variables, such as process conditions. For example, processes that would tend to produce blocks of homopolystyrene are suspect. It is safest to test each composition thoroughly outdoors before use. Other properties may be affected. For example, it cannot be assumed that the mechanical properties of the copolymer will be equal when styrene is substituted for MMA even though its T_g may be about the same. In this regard polystyrene has a significantly higher brittle-ductile transition temperature than poly(MMA) (see Chapter 24), even though polystyrene has a slightly lower T_g.

Acrylic esters are generally used as the low T_g monomer(s), and the choice is strongly affected by cost considerations. If the cost of the common acrylic esters is intermediate between that of MMA and of styrene, it is more economical to use ethyl acrylate (EA; homopolymer T_g = −24°C) than butyl acrylate (BA; homopolymer T_g = −54°C) with MMA in order to get as high a proportion of EA in the composition as possible. However, with styrene–acrylic latexes, BA may be more economical when styrene is less costly than EA or BA. Presumably the different acrylic esters impart different film properties at a given T_g, but published data is inadequate to evaluate the effects. An example of laboratory semicontinuous batch process for preparing a 40:59:1 MMA/EA/methacrylic acid copolymer latex is provided in Table 5.2. Reference [6] provides many more examples.

Note from Table 5.2 that the measured MFT of the latex at 9°C, is somewhat lower than the T_g value estimated by the Fox equation for a 40:60 MMA/EA copolymer, 17°C. Also note that changing to a 50:50 MMA/EA ratio increases MFT to 21°C while the Fox equation predicts a T_g value of 28°C. These figures give a general picture for how T_g and MFT relate to composition. The former latex is suitable for an architectural paint.

Acrylic acid (AA), or the somewhat less water-soluble methacrylic acid (MAA), is commonly used in preparation of acrylic latexes, usually at 1 to 2 wt% of the monomer charge. The carboxylic acid functionality incorporated with these monomers enhances mechanical stability of latex paint, reducing the amount of surfactant needed. Furthermore, when such latexes are neutralized with ammonia, the viscosity increases substantially, and the latex becomes shear thinning. The effects of these acid monomers on stability and viscosity are maximized when they are incorporated in the last part of the monomer feed and when the polymerization medium is acidic [1,7]. Viscosity of a latex made in this way is very dependent on pH. As ammonia is added, there is little change in viscosity until the pH reaches about 7; then viscosity increases steeply until the pH reaches 9 or 10; at still higher pH the viscosity decreases. The pH of acrylic latex paints is usually adjusted to about 9, where there is a substantial viscosity effect; not only is the viscosity high,

Table 5.2. Laboratory Procedure for Preparation of a MMA/EA/MAA Copolymer Latex

Materials

1 L	Deionized water
96 g	Triton X-200 (sodium salt of an alkylaryl polyether sulfate ester) (Union Carbide Co.)
320 g	Methyl methacrylate (10 ppm MEHQ stabilizer) (MEHQ is monomethylether of hydroquinone)
480 g	Ethyl acrylate (15 ppm MEHQ stabilizer)
8 g	Methacrylic acid (100 ppm MEHQ stabilizer)
1.6 g	Ammonium persulfate

Procedure

Prepare an emulsion of all reactants in 800 mL of water. Place 200 mL of the emulsion and 200 mL of water in a 3-L flask fitted with an inert gas inlet tube, a thermometer, a stirrer, an additional funnel, and a reflux condenser. Heat in a 92°C water bath while stirring until the internal temperature reaches 82°C. The mixture will begin to reflux, and the temperature will rise to about 90°C within a few minutes indicating rapid polymerization. When refluxing subsides, add the remaining emulsion continuously over a period of 1.5 h. Heat to maintain refluxing; internal temperature will be 88–94°C. After monomer addition is complete, heat to 97°C to complete conversion of monomer. Cool and strain.

Properties

NVW 42.9 (calculated 43.1); pH 2.7; viscosity (Brookfield), 11.5 mPa·s; MFT 9°C; film hardness (Tukon), 1.2 KHN. A 50:50 MMA/EA copolymer latex (no MAA), prepared similarly, had similar properties except that MFT was 21°C and film hardness was 6.2 KHN.

Source: See Ref. 6.

the paint exhibits shear thinning, which is sometimes a desirable application characteristic (see Chapter 35).

Hoy [1] describes a latex that shows this effect, composed of MMA/EA/BA/AA (40:52:6:2 wt/wt), in which the acrylic acid is added late in the process. The T_g of this copolymer was reported to be 15°C. Neutralization with ammonia to pH 9 causes the diameter of the particles to expand to about 1.8 times that at pH 7. The corresponding increase in volume is almost sixfold. The expansion results from association of water with the highly polar salt groups in the polymer near the particle surface. The presence of the expanded layer at pH 9 substantially increases viscosity at low shear rates. However, because this layer can be distorted by shear forces, the viscosity increase at high shear rates is less marked.

In the same paper, Hoy [1] describes effects of changing monomer addition schedules during semicontinuous batch polymerization. In one experiment an MMA/BA ratio of 40:60 was used, with 2 wt% of AA added late in the process. When the 40:60 ratio was maintained throughout monomer addition, the product, called a "uniform feed" latex, had a T_g of +20°C.

A second "staged feed" latex of the same overall composition was made by a process designed to yield core-shell particles [1]. In the first one-half of the monomer addition the MMA/BA ratio was 70:30 and in the second one-half it was 10:90. The resulting latex showed two T_g values, one at +60°C and the other at −10°C. Because the shell of this latex has a low T_g, its MFT is substantially lower than that of the uniform feed latex. However, coalesced films are cloudy. Apparently, the high T_g cores remain uncoalesced; the composition of the cores and the shells is sufficiently different that the polymers are not miscible with one another. Because the refractive index of the two polymers is different, light passing through the film is scattered. Since the T_g of the continuous phase is low, the blocking resistance of a film of staged feed latex is inferior to a film of uniform feed latex.

A third latex with the same monomer composition was made by a process called "linear power feed addition" of the monomers [1], using the apparatus sketched in Figure 5.3. There are two stirred tanks for monomer mixtures. The "near tank" was loaded with one-half of the monomer charge with an MMA/BA ratio of 70:30, and the far tank was loaded with the other one-half with a 10:90 ratio. During polymerization, monomer was pumped from the near tank into the reactor in the usual way. At the same time, monomer was being pumped at the same rate from

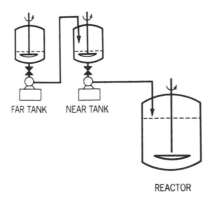

Figure 5.3. Schematic representation of apparatus for semicontinuous latex polymerization using a linear power feed addition. (Adapted from Ref. [1], with permission.)

the far tank to the near tank. Again, 2 wt% of AA was added in a late stage of the process. With this procedure, the monomer feed starts at 70:30 MMA/BA and ends at 10:90, but it changes continuously throughout the process. Each latex particle presumably contains high T_g polymer at its center and low T_g polymer (with COOH groups) at its surface. It differs from the usual core-shell latex in that there is a linear gradient of composition and T_g from the center to the surface. Films cast from this latex have equal clarity and the same measured T_g, about $+20°C$, as those cast from the uniform feed latex. However, the glass transition is much broader for the linear power feed latex than for the uniform feed latex. Very importantly, the linear power feed latex has a lower MFT than the uniform feed latex, but it has a higher blocking resistance temperature.

Exterior house paints for large surfaces are usually formulated at low gloss in the United States; the formulations are heavily pigmented. Block resistance requirements are modest, and the pigment helps improve this property; for this reason T_gs of about 5–15°C are adequate. Latex paint films are more permeable to water vapor than films from oil or alkyd paints, an advantage because it reduces blistering of paints applied to wood surfaces (but a disadvantage for coatings applied to metal).

Applications such as gloss trim and door paints, as well as interior trim paints (kitchen cabinets, windowsills, etc.), have much different requirements. To be glossy, they must be formulated at lower pigment levels, yet much greater block resistance is required. This situation presents a challenge to designers of latexes. The problems have been at least partly overcome as described by Mercurio [8]. A latex T_g of about 55°C is reported to be required to achieve adequate block resistance. Coalescence of acrylic latexes with this T_g can be attained by using a small particle size latex with substantial amounts of carefully selected coalescing solvent(s). The coalescing solvent increases VOC to near the current regulatory limit, 250 g VOC/L of paint as applied, excluding water. Commercial products based on these principles are available; details of how they are made are proprietary.

The above discussion illustrates a significant point. Many refinements of process and composition can be used to improve the characteristics of latexes without substantially increasing raw material cost and, possibly, reducing it. While a simple acrylic latex made as described in Table 5.2 could be used to make a house paint, such a latex would probably not be competitive in today's sophisticated market. Commercial production involves many process nuances that are held confidential by the latex producing companies. It is difficult to replicate commercial latexes even though the monomer composition can be precisely analyzed.

For latexes to be used over metal it is desirable to have low moisture vapor permeability. Acrylic latexes give films with quite high permeability. One approach to reducing water permeability is to use vinylidene chloride as a comonomer. Since the reactivities of vinylidene chloride and acrylic ester monomers are quite different, the polymerization must be carried out under monomer starved conditions to achieve reasonably uniform copolymerization. Reference [9] provides details of procedures for the copolymerization and characterizations of several vinylidene chloride–acrylate latexes. While the moisture vapor permeability is much reduced, there can be a serious problem of stabilizing the chlorinated copolymer against photodegradation.

5.3. VINYL ESTER LATEXES

Vinyl acetate (VAc) is less expensive than (meth)acrylic monomers. However, VAc latexes are inferior to acrylics in both photochemical stability and resistance to hydrolysis. When an acetate group on a PVAc chain hydrolyzes, the liberated acetic acid may catalyze hydrolysis of more acetate groups, and the hydroxy group may exert a neighboring group (anchimeric) effect that promotes hydrolysis of adjacent acetate groups. Hence PVAc latexes find their principal use in interior coatings that do not have to withstand high humidity exposure or frequent wetting. One such use, flat wall paints, is the largest volume type of paint sold in the U.S. market.

The T_g of a PVAc homopolymer is 32°C, too high for film formation under ambient conditions. Thus one must either formulate PVAc with plasticizer or copolymerize VAc with a monomer that will reduce T_g. The most widely used comonomer for this purpose is probably n-butyl acrylate (BA), although other comonomers, such as di-n-butyl maleate have been used.

Copolymerization of VAc with BA poses potential problems because the rate of reaction of a free radical on a terminal BA group with another molecule of BA is much greater than its rate of reaction with VAc; furthermore, the rate of reaction of a free radical on a terminal VAc group is much greater with BA than with VAc. Thus a 50:50 mixture of BA and VAc will produce a polymer very rich in BA at the outset of polymerization and very rich in VAc in the late stages. This problem is overcome by use of the semicontinuous batch process; when the monomer mixture is added at a rate equal to the polymerization rate, that is, under monomer-starved conditions, a relatively uniform copolymer will form with a composition essentially equal to the ratio of monomers being fed.

Such copolymers are made on a large scale commercially. An extensive study of the copolymerization of VAc with BA by both batch and semicontinuous processes and extensive characterization of the copolymers has been reported [5,10] and is partly discussed in Section 5.1. Since the reaction rates are relatively low, the rate of monomer addition must be quite slow. If the initial charge is high in the least reactive monomer, the addition can be faster while still obtaining reasonably uniform copolymerization. Mathematical simulations can be used to predict the effects of different addition modes [11].

An essential component of the recipe for polymerization of VAc is a buffer such as sodium bicarbonate to maintain a pH near 7. It is essential because VAc monomer hydrolyzes at appreciable rates under either acidic or basic conditions. This hydrolysis is irreversible because the products are acetaldehyde and acetic acid. At pH less than 7 hydrolysis is autocatalytic because acetic acid catalyzes further reaction.

$$CH_2{=}CH{-}O{-}\overset{\overset{\displaystyle O}{\|}}{C}{-}CH_3 \ + \ H_2O \ \longrightarrow \ CH_3{-}\overset{\overset{\displaystyle O}{\|}}{C}{-}OH \ + \ CH_3{-}\overset{\overset{\displaystyle O}{\|}}{C}{-}H$$

Furthermore, acetaldehyde is oxidized by persulfate, which consumes initiator persulfate and generates still more acetic acid to further catalyze hydrolysis.

Copolymer latexes of VAc and (meth)acrylic monomers have been designed for dual-purpose use in both exterior and interior flat paints. The objective is to reduce cost in two ways—lower raw material cost for exterior paints and lower inventory and storage costs by basing all production on a single resin. This approach has been commercially successful, although the latexes undoubtedly represent a compromise between the best resins for exterior and interior use.

Copolymers of VAc with ethylene, produced by emulsion polymerization under pressure, are widely used in adhesives and are being investigated as vehicles for coatings; in this case the intent is to reduce monomer cost below the level of VAc, although the gain is partly offset by the cost of pressure processing.

There are differences in coatings latex technology in Europe and United States. In Europe, a common monomer is the a vinyl ester of C_{10} branched acid (see Section 5.1). Copolymer latexes of this monomer with VAc are apparently suitable for both interior and exterior use, perhaps aided by the less stringent exterior exposure conditions in much of Europe.

GENERAL REFERENCES

Anonymous, *Emulsion Polymerization of Acrylic Monomers*, Technical Bulletin CM-104 A/cf, Rohm & Haas C., Independence Mall West, Philadelphia.

D. C. Blackley, Ed., *Science and Technology of Polymer Colloids*, NATO ASI Series, M. Nijhoff, Brussels, 1983.

G. Odian, *Principles of Polymerization*, 3rd ed., Wiley-Interscience, New York, 1991, pp. 335–355.

I. Piirma, Ed., *Emulsion Polymerization*, Academic, New York, 1982.

J. W. Vanderhoff, *J. Polym. Sci. Polym. Symp.*, **72**, 161–198 (1985).

REFERENCES

1. K. L. Hoy, *J. Coat. Technol.*, **51** (651), 27 (1979).

2. B. D. Nguyen and A. Rudin, *J. Coat. Technol.*, **58** (736), 33 (1986).

3. D. R. Craig, *J. Coat. Technol.*, **61** (779), 49 (1989).

4. M. S. El-Asser, T. Makgawinata, and J. W. Vanderhoff, *J. Polym. Sci. Polym. Chem. Ed.*, **21**, 2363 (1983).

5. C-S. Chou, A. Kowalski, J. M. Rokowski, and E. J. Schaller, *J. Coat. Technol.*, **59** (755), 93 (1987).

6. Anonymous, *Emulsion Polymerization of Acrylic Monomers*, Tech. Bull. CM-104 A/cf, Rohm & Haas, Philadelphia,

7. B. Emelie, C. Pichot, and J. Guillot, *Makromol. Chem.*, **189**, 1879 (1988).

8. A. Mercurio, K. Kronberger, and J. Friel, *J. Oil Col. Chem. Assoc.*, **65**, 227 (1982).

9. H. R. Friedli and C. M. Keillor, *J. Coat. Technol.*, **59** (748), 65 (1987).

10. S. C. Misra, C. Pichot, M. S. El-Asser, and J. W. Vanderhoff, *J. Polym. Sci. Polym. Chem. Ed.*, **21**, 2383 (1983).

11. G. Arzamendi and J. M. Asua, *J. Appl. Polym. Sci.*, **38**, 2019 (1989).

CHAPTER VI

Amino Resins

Amino resins are by far the predominant cross-linking agents for thermosetting coatings. The amino resins most commonly used in coatings are derived from melamine, that is, 2,4,6-triamino-1,3,5- triazine. The related aminotriazine, ben-zoguanamine, as well as urea, glycoluril, and copolymers of (meth)acrylamide are also utilized.

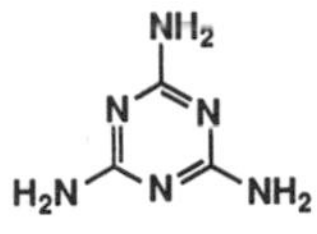

Melamine

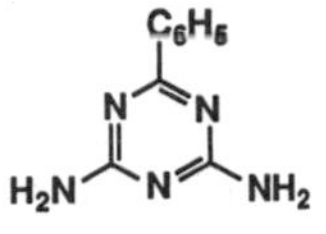

Benzoguanamine

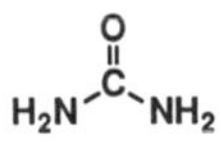

Urea

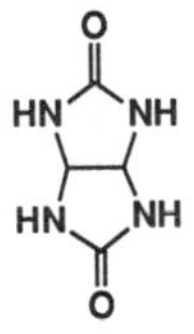

Glycoluril

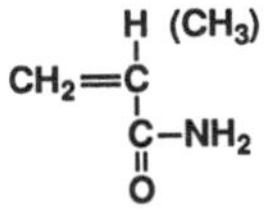

(Meth)Acrylamide

The —NH$_2$ (or >NH) groups of these precursors have unshared electron pairs, but they are very weak bases, much weaker than aliphatic amines. To emphasize the chemical differences these groups are properly called amidine (aminotriazines), urea, or amide groups, not amine groups. The resins are made by reacting the precursors with formaldehyde and subsequently with alcohols to yield ethers of general structure 1. Such ethers are activated toward nucleophilic substitution by the neighboring N and are thereby much more reactive than common aliphatic ethers.

It is this enhanced reactivity that make these resins useful as cross-linkers. When the nucleophile is the alcohol functionality of a polyol (POH) transetherification can occur, as shown in Eq. 6.1, resulting ultimately in formation of a cross-linked polymer network. The reaction is generally catalyzed by acid. Carboxylic acids and carboxamides also react with the activated ethers, as shown in Eqs. 6.2 and 6.3, respectively. The corresponding polyfunctional resins are utilized in cross-linking reactions.

$$>N-CH_2-OR + \text{(P)}\sim OH \rightleftharpoons$$

$$>N-CH_2-O\sim\text{(P)} + R-OH \qquad (6.1)$$

$$\cdot 1 + \text{(P)}\sim\overset{\overset{O}{\|}}{C}-OH \rightleftharpoons$$

$$>N-CH_2-O-\overset{\overset{O}{\|}}{C}\sim\text{(P)} + R-OH \qquad (6.2)$$

$$1 + \text{(P)}\sim\overset{\overset{O}{\|}}{C}-NH_2 \rightleftharpoons$$

$$>N-CH_2-NH-\overset{\overset{O}{\|}}{C}\sim\text{(P)} + R-OH \qquad (6.3)$$

$$\text{(P)}\sim \quad \text{a polymer}$$

Since the precursors contain —NH$_2$ (or >NH) groups of amidine, urea, or amide types, rather than of the more strongly basic amine type, the term amino resins is a misnomer and is misleading. Nevertheless, this terminology is so firmly entrenched in the polymer literature that we feel compelled to use it. We begin with melamine resins, followed by briefer coverage of other types.

6.1. SYNTHESIS OF MELAMINE–FORMALDEHYDE RESINS

The first reaction in the synthesis of melamine–formaldehyde (MF) resins is "methylolation," the reaction of melamine with formaldehyde. It is generally carried out under basic conditions, utilizing aqueous sodium carbonate. With excess formaldehyde, the reaction can be driven to form hexamethylolmelamine, **2** (R = H). A mixture of partially methylolated derivatives, including species such as the symmetrical trimethylolmelamine, **3** (R = H), is formed with less than the stoichiometric 6 mol of formaldehyde per mol of melamine.

2   **3**

The second reaction in the synthesis of MF resins, "etherification," is the acid-catalyzed reaction of methylolmelamines with an alcohol, most commonly methyl, butyl, or mixtures of methyl and butyl alcohols. Complete etherification of methylolmelamines **2** and **3** (R = H) yields MF resins **2** and **3** (R = alkyl). Many commercial MF resins are partially etherified.

In addition to the *monomeric* species **2** and **3**, all commercial melamine resins contain oligomeric species in which triazine rings are linked by methylene ether bridges ($>NCH_2OCH_2N<$), as we discuss in Section 6.1.3.

6.1.1. The Methylolation Reaction

A probable mechanism for the base catalyzed reaction of melamine with formaldehyde is outlined in Scheme 6.1, where $-NH_2$ represents the melamine amino groups. The first step involves nucleophilic attack by the amino group on the electrophilic C of formaldehyde, which is facilitated by removal of a proton from N by the base B^-. This is followed by proton transfer from the resulting B—H to the negative oxygen atom, which yields the methylolated product and regenerates the basic catalyst. It should be noted that both steps are reversible.

Scheme 6.1

Main Reaction:

$$-NH_2 \ + \ H_2C{=}O \ + \ B^- \ \rightleftharpoons \ -NH_2{-}CH_2{-}O^- \ + \ B{-}H$$

$$-NH_2{-}CH_2{-}O^- \ + \ B{-}H \ \rightleftharpoons \ -NH_2{-}CH_2{-}OH \ + \ B^-$$

Side Reaction:

$$-NH_2{-}CH_2{-}O^- \ + \ H_2C{=}O \ \rightleftharpoons \ -NH_2{-}CH_2{-}O{-}CH_2{-}O^-$$

$$-NH_2{-}CH_2{-}O{-}CH_2{-}O^- \ + \ B{-}H \ \rightleftharpoons \ -NH_2{-}CH_2{-}O{-}CH_2{-}OH \ + \ B^-$$

Studies of the kinetics of the reaction of formaldehyde with melamine indicate that the presence of one methylol group on N deactivates the group for a second reaction by a factor of 0.6. On the other hand, there seems to be little effect of substitution on one amino group on the reactivity of the other amino groups. These kinetic factors tend to favor formation of the symmetrical trimethylolmelamine, **3** (R = H). However, the kinetic preference is not strong enough to overcome the thermodynamic tendency to produce mixtures of products. Accordingly, at equilibrium, the reaction of 6 mol of formaldehyde with 1 mol of melamine yields a mixture of products including all levels of methylolation and free formaldehyde [1].

A significant side reaction is methylolation on oxygen, also shown in Scheme 6.1.

6.1.2. The Etherification Reaction

Following methylolation, the base catalyst is neutralized with acid and the appropriate alcohol (or alcohol mixture) is introduced together with an acid catalyst. The ensuing reversible reaction leads to formation of the ether of the methylol; that is, formation of an alkoxymethyl group. Nitric acid is commonly used as a catalyst because nitrate salt byproducts are relatively easily removed.

There is substantial disagreement in the literature about the mechanism of the etherification reaction, as well as the closely related transetherification reaction shown in Eq. 6.1. The contending mechanisms for these substitution reactions are S_N1 and S_N2, both of which are outlined in Scheme 6.2, where R = H and R'OH = methyl or butyl alcohol. Note that the conjugate base (A^-) is omitted from the intermediate steps.

Scheme 6.2

S_N1 Mechanism

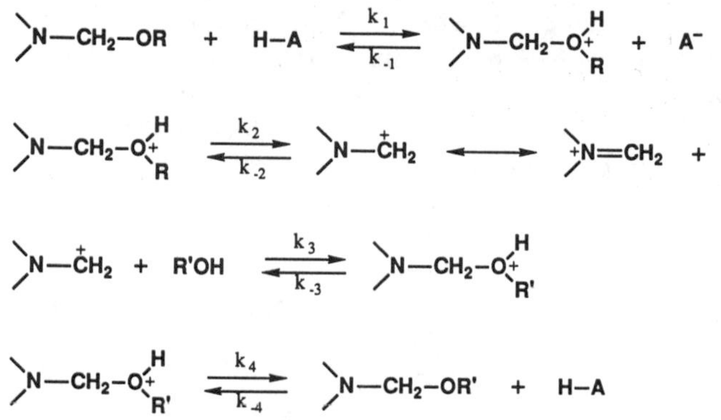

$$\text{>N—CH}_2\text{—OR} \ + \ \text{H—A} \ \underset{k_{-1}}{\overset{k_1}{\rightleftharpoons}} \ \text{>N—CH}_2\text{—O}^+\!\!\!<^H_R \ + \ A^-$$

$$\text{>N—CH}_2\text{—O}^+\!\!\!<^H_R \ \underset{k_{-2}}{\overset{k_2}{\rightleftharpoons}} \ \text{>N—}\overset{+}{\text{C}}\text{H}_2 \ \longleftrightarrow \ {}^+\text{N}=\text{CH}_2 \ + \ \text{ROH}$$

$$\text{>N—}\overset{+}{\text{C}}\text{H}_2 \ + \ \text{R'OH} \ \underset{k_{-3}}{\overset{k_3}{\rightleftharpoons}} \ \text{>N—CH}_2\text{—O}^+\!\!\!<^H_{R'}$$

$$\text{>N—CH}_2\text{—O}^+\!\!\!<^H_{R'} \ \underset{k_{-4}}{\overset{k_4}{\rightleftharpoons}} \ \text{>N—CH}_2\text{—OR'} \ + \ \text{H—A}$$

(6.4)

S_N2 Mechanism

$$\text{>N—CH}_2\text{—OR} \ + \ \text{H—A} \ \rightleftharpoons \ \text{>N—CH}_2\text{—O}^+\!\!\!<^H_R \ + \ A^-$$

$$\text{>N—CH}_2\text{—O}^+\!\!\!<^H_R \ + \ \text{R'OH} \ \rightleftharpoons \ \text{>N—CH}_2\text{—O}^+\!\!\!<^H_{R'} \ + \ \text{ROH}$$

$$\text{>N—CH}_2\text{—O}^+\!\!\!<^H_{R'} \ + \ A^- \ \rightleftharpoons \ \text{>N—CH}_2\text{—OR'} \ + \ \text{H—A}$$

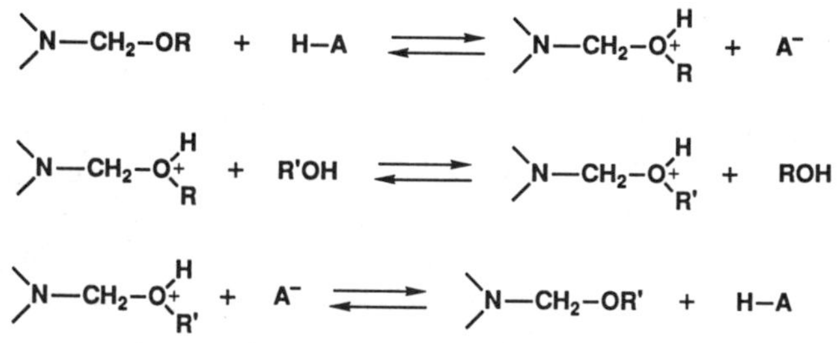

(6.5)

The distinguishing feature is whether the protonated methylol group ionizes to an intermediate, resonance-stabilized carbocation, Eq. 6.4, before reacting with the alcohol, characteristic of the S_N1 mechanism, or reacts directly with the alcohol (Eq. 6.5) characteristic of the S_N2 mechanism.

The controversy about mechanisms in MF resin synthesis extends to the mechanisms of the transetherification reaction involved in cross-linking MF resins with polyols, where R = methyl or butyl and R'OH is the polyol. The experimental evidence bearing upon this controversy has been acquired, primarily, in studies of the transetherification reaction in cross-linking coatings and will be discussed in Section 6.3.1.2. Based on the available evidence, we favor the S_N1 mechanism.

An alternative reaction pathway is available for etherification of methylol groups on singly substituted N, that is, an N that also bears an H atom. This mechanism is outlined in Scheme 6.3, where R = H and R'OH = methyl or butyl alcohol.

Scheme 6.3

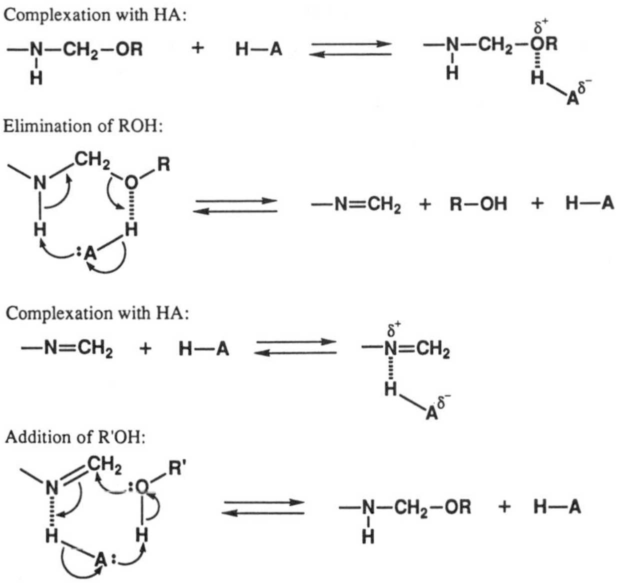

Complexation with HA:

$$—N—CH_2—OR \quad + \quad H—A \quad \rightleftharpoons \quad —\overset{\delta^+}{N}—CH_2—\overset{\delta^+}{O}R$$

Elimination of ROH:

$$\rightleftharpoons \quad —N{=}CH_2 \; + \; R—OH \; + \; H—A \qquad (6.6)$$

Complexation with HA:

$$—N{=}CH_2 \quad + \quad H—A \quad \rightleftharpoons \quad —\overset{\delta^+}{N}{=}CH_2$$

Addition of R'OH:

$$\rightleftharpoons \quad —N—CH_2—OR \; + \; H—A$$

The distinguishing feature compared with the mechanisms in Scheme 6.2 is formation of the neutral imine intermediate, Eq. 6.6, which is possible owing to the presence of the NH group. An important consequence of imine formation is that the elimination of water is catalyzed efficiently by relatively weak acids, such as carboxylic acids (RCOOH). Complexation of the methylolmelamine by the acid (A—H) is sufficient to yield the imine, either in a concerted push–pull mechanism, as shown in Scheme 6.3, or step-wise. Complexation of the reactive imine by the acid is also sufficient to effect the subsequent addition of alcohol, resulting in the overall substitution of water by alcohol. This reaction is also reversible.

Since the catalysis in Scheme 6.3 involves complexation rather than protonation by the acid, the rate is dependent on the nature of H—A as well as on its concentration. Indeed, both the more positive H end and the more negative A end of H—A participate by facilitating breaking of both the C—O and the N—H bond, respectively. By convention, this is called *general acid catalysis*. The mechanism in Scheme 6.3 applies as well to cross-linking of polyols by MF resins with N—H groups by transetherification, where R = methyl or butyl and R'OH represents the polyol.

In contrast, elimination of water from methylolmelamines lacking N—H groups, either by the S_N1 or S_N2 mechanism (Scheme 6.2) requires protonation of the methylol group, necessitating the utilization of a strong, ionizing acid, such as nitric, sulfuric, or a sulfonic acid. Such catalysis, requiring protonation, is called *specific acid catalysis*, signifying that the rate is dependent only on the concentration of protons (H^+) and independent of the conjugate base (A^-)

6.1.3. Self-Condensation Reactions

Self-condensation refers to those reactions that lead to the formation of bridges between triazine rings, thereby resulting in dimers, trimers, and oligomers. The nature and degree of such reactions depend on process factors including pH, ratio of reactants, reaction temperature, rate of removal of water, and probably others. Apparently, the predominant self-condensation reaction is the formation of methylene ether bridges between melamine rings, as evidenced by ^{1}H and ^{13}C NMR [2], and represented by the general structure **4** for a dimer.

$$\text{\textbackslash N—CH}_2\text{—O—CH}_2\text{—N}/$$

4

Self-condensation by methylene ether bridge formation between melamines may occur during etherification by base or acid catalyzed reaction of methylolmelamines with melamine imines, as shown in Eq. 6.7. Melamine imines can be formed by acid or base catalyzed elimination of water from methylolmelamines.

$$\text{N—CH}_2\text{—OH} \ + \ \text{—N}{=}\text{CH}_2 \ \rightleftharpoons \ \text{N—CH}_2\text{—O–CH}_2\text{—NH—} \qquad (6.7)$$

Alternatively, methylene ether bridges may form from condensation of two methylolmelamines (during acid catalyzed etherification), as shown in Eq. 6.8, in accordance with one or the other of the mechanisms in Schemes 6.2 and 6.3.

$$2 \ \text{N—CH}_2\text{—OH} \ \rightleftharpoons \ \text{N—CH}_2\text{—O–CH}_2\text{—N} \ + \ \text{H}_2\text{O} \qquad (6.8)$$

Self-condensation reactions 6.7 and 6.8 are presumably reversible.

The literature frequently shows self-condensation reactions to form methylene bridge derivatives of the general structure **5** as side reactions during etherification (e.g., see Ref. [3]). However, attempts to detect such side products from etherification by spectroscopic methods, including NMR analysis, have been unsuccessful [2]. It is important to note that the NMR results cannot be attributed to the inability to detect the methylene bridge derivatives, since they have been detected by solid state ^{13}C NMR in MF–polyol cross-linked films [2].

$$\text{N—CH}_2\text{—N}$$

5

Formation of methylene bridges has been demonstrated in studies of the acid-catalyzed reaction of a model compound for Class I MF resins, N,N-dimethoxymethyl-N',N', N'', N''- tetramethylmelamine **6** and water [4]. In the presence of *para*-toluenesulfonic acid (pTSA) and water at elevated temperatures, a cyclic trimer **7**, with methylene groups connecting the three nitrogen atoms is formed. Recent studies demonstrated that the formation of the trimer is a reversible reaction [5]. When trimer **7** is heated with methyl alcohol and formaldehyde at 130°C, Compound **6** is regenerated.

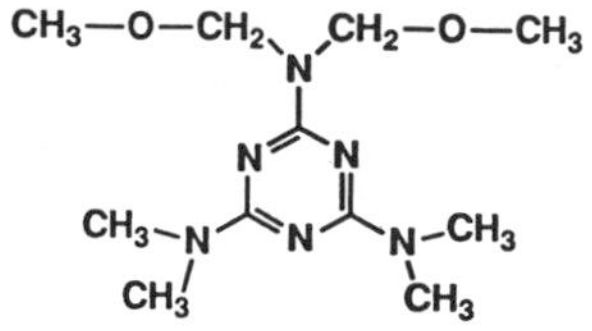

6

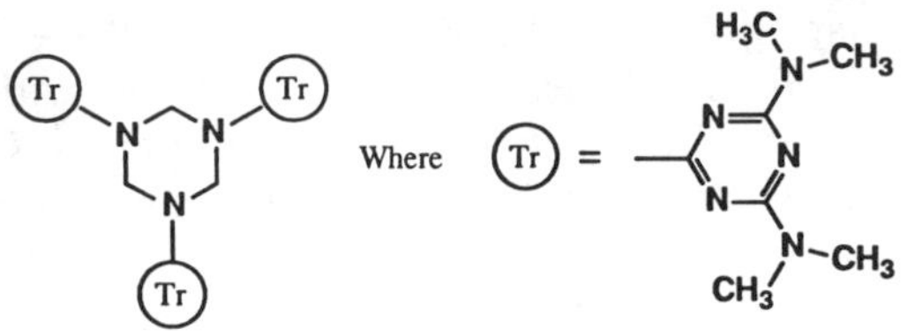

Where (Tr) =

7

6.2. TYPES OF MF RESINS

A large variety of MF resins is made commercially with differences in functional groups as well as degree of polymerization (P). To facilitate discussion, MF resins are classified into two broad groups, shown in Table 6.1. The two classes differ by the relative proportion of reactants utilized in their synthesis, the resulting characteristic functionality, and selected characteristics of the resins. The reader should bear in mind, however, that a variety of MF resins is produced commercially within each class including methyl and butyl ethers with different ranges of P, as well as mixed-ether resins, commonly mixtures of methyl with *n*-butyl, isobutyl, or 2-ethylhexyl ethers.

Commercial MF resins were first introduced in 1936, and they quickly gained popularity because of their ability to improve the properties of baked alkyd coatings. From the 1930s through the 1950s, the predominant commercial melamine resins were the so-called "polymeric, butylated" Class II types (see Table 6.1). These resins had enough bridging so that P was three or more, and the alcohol was usually *n*-butyl or isobutyl alcohol. Such resins were economical, they were readily miscible in alkyd formulations, and they imparted very wide latitude; that is, the coatings did not require exacting control of formulation to produce acceptable application characteristics and film properties.

Monomeric, methylated Class I MF resins were developed in the 1940s. The term *monomeric* is somewhat optimistic as shown below. Class I resins proved more compatible than butylated resins with certain coreactant resins, and in some cases they provided tougher films. Introduction of water-borne and high solids

Table 6.1. Classes of Melamine Resins

Class	Synthesis	Characteristic Functionality[a]	Selected Characteristics
I	Excess HCHO Excess alcohol (ROH)	—N(CH$_2$OR)$_2$	Highest functionality Strong acid catalysis Lowest P
II	Deficient formaldehyde Usually excess alcohol (ROH)	—NHCH$_2$OR	Lower functionality Weak acid catalysis Usually higher P Self-condense readily

[a]Most commonly R = —CH$_3$, —*n*-C$_4$H$_9$, *iso*-C$_4$H$_9$, branched —C$_8$H$_{17}$.

coatings during the 1970s accelerated the shift toward monomeric resins. Methylated Class I resins are more miscible with the water–solvent–resin blends used in water-borne coatings, and monomeric Class I resins impart lower viscosity to high solids coatings. These generalities are not universal—some Class II resins are used in low-solvent coatings.

One commercial *monomeric* methylated Class I MF resin has been reported to contain 62% monomers (i.e., one triazine ring per molecule), 23% dimers, together with 15% trimers and higher oligomers [6]. The equivalent weight of pure hexamethoxymethylmelamine would be 65 g/eq, but the equivalent weight of such a resin is closer to 80 g eq^{-1}, resulting from the presence of dimers, trimers, and so forth. (In estimating equivalent weight it is necessary to make an assumption as to whether or not the $>NCH_2OCH_2N<$ bridging groups react to form cross-links. Here it is assumed that they do not.) Chromatographic analysis of a different but similar resin of this type is shown in Figure 6.1. The chromatograms show that at least 30 different chemical species are present in significant amounts [7]. While the predominant groups are $—N(CH_2OCH_3)_2$, various species containing incompletely etherified groups $[—N(CH_2OCH_3)(CH_2OH)]$ and incompletely formylated groups $[—NHCH_2OCH_3]$ are present. These *monomeric* methylated MF resins are frequently called *HMMM resins*, an acronym for hexamethoxymethylmelamine, but the terminology "high HMMM resin" is more appropriate.

Other Class I resins are available with different degrees of polymerization (P) and with different extents of formylation and etherification. In addition to the methylated derivatives, butylated as well as mixed methyl/*n*-butyl, methyl/isobutyl, methyl/ethyl, and methyl/2-ethylhexyl derivatives are also available.

Class I resins, which generally have lower Ps, exhibit lower viscosity than Class II resins at the same percent solids. At a given P, butyl ethers exhibit lower viscosities, attributable to lower T_g values, than the corresponding methyl ethers. Other considerations, which are discussed in the following section on the use of MF resins in coatings, bring out possible advantages of using mixed methyl–butyl ether resins.

Broader ranges of Class II resins are available because, in addition to all the variables affecting Class I resins, the ratio of formaldehyde/melamine and of alcohol/formaldehyde can be varied through a wider range. It is harder to suppress bridge-forming reactions during synthesis of Class II than with Class I resins, and very low Ps are almost unattainable. As discussed in the previous section, methylene ether bridges can form both under acidic and alkaline conditions.

The predominant reactive group present in most Class II resins is $—NHCH_2OR$. They also contain, $—NH_2$, $>NCH_2OH$, as well as $—N(CH_2OR)_2$ groups. In order to minimize viscosity and maximize reactivity, resin producers strive to reduce P and maximize the amount of symmetrical trimethoxymethylmelamine **3** (R = Me), known as TMMM. This is possible to a limited degree because methylolation of $—NH_2$ groups, yielding $—NHCH_2OH$, is kinetically favored over methylolation of $—NHCH_2OH$, yielding $—N(CH_2OH)_2$. Owing to advances in HPLC analysis techniques, it has been possible to follow the effect of small changes in process conditions on composition. Continuing refinement of processes has made possible commercial production of resins with as high as 50% TMMM. As with Class I resins, a variety of alcohols is used in making different resin grades.

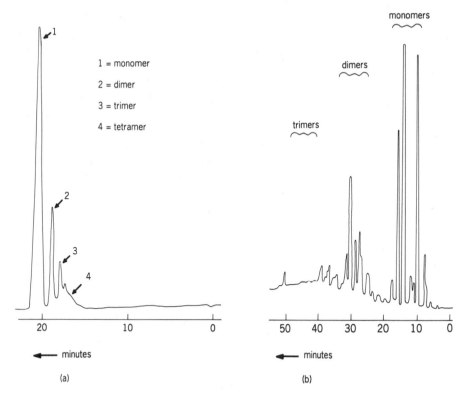

Figure 6.1. High-performance liquid chromatograms (HPLC) of a typical Class I, high HMMM resin. (*a*) An SEC chromatogram and (*b*) a gradient HPLC chromatogram. (From Ref. [7], with permission.)

A few Class II resins are produced using processes that are deficient in both formaldehyde and alcohol. Such resins have substantial levels of —NHCH$_2$OH groups, imparting high reactivity and water miscibility. Coatings formulated with such resins are expected to suffer from poor package stability and relatively high levels of free formaldehyde.

6.3. REACTIONS OF MF RESINS IN COATINGS APPLICATIONS

Melamine–formaldehyde resins are almost never used by themselves in coatings. Rather, they are used to cross-link coreactant resins having hydroxyl, carboxylic acid, and/or amide groups. Acrylic (Chapter 7), polyester (Chapter 8), alkyd (Chapter 10), epoxy (Chapter 11), and polyurethane (Chapter 12) resins are the most important classes of coreactant resins, and each will be discussed in the appropriate chapters.

As of 1991 there are two broad concerns regarding the use of MF resins in coatings. The first is that most (all?) amino resin cross-linked coatings give off small amounts of formaldehyde vapor during the cross-linking process, posing a potential hazard to workers who are exposed daily. Some resins are substantially better than others in this respect. Formaldehyde exposure levels are strictly regulated in the United States, and current technology appears adequate to satisfy these regulations. While the regulations already appear conservative, it is probable that allowable exposure levels will be further reduced. Intensive research on ways to reduce formaldehyde emissions is underway. Besides improving ventilation, approaches include changing the resins, changing the formulations, and introducing formaldehyde scavengers.

A second concern is specific to automobile top coats, most of which are cross-linked with MF resins in the United States. Top coats are more or less vulnerable to an "acid etching," appearance of small blemishes on the coating surface on exposure to acid rain. Again, this problem is under intensive investigation. Study of the information presented in the this chapter may suggest to the reader ways to address these concerns.

6.3.1. Polyol Coreactants

Polyols are by far the most common resins cross-linked by MF resins. The hydroxyl groups of polyols react by either transetherification with the activated alkoxymethyl groups or by etherification of methylol groups of MF resins to form new ether cross-links. Strong acid catalysts, such as sulfonic acids, are used for Class I resins, and weak acid catalysts, such as carboxylic acids, are used for Class II resins. The reactions are reversible but are driven towards cross-linking by volatilization of the monofunctional alcohol or water produced, as shown in Eq. 6.9, where R = alkyl or H and $\circledP\!\!\sim$ OH is the polyol.

$$\begin{array}{l} \text{>N—CH}_2\text{—OR} \ + \ \circledP\!\!\sim\!\text{OH} \ \xrightarrow{\text{H—A}} \\[12pt] \text{>N—CH}_2\text{—O}\!\!\sim\!\circledP \ + \ \text{R—OH} \end{array}$$

(6.9)

Rates of these cross-linking reactions depend on the structure of the polyol and the MF resin, on the type and amount of catalyst, and, of course, on temperature. With regard to the polyol, primary alcohols usually react faster than secondary alcohols, an exception being highly hindered primary alcohols such as neopentyl alcohols. These reactivities are in accord with predictions based on steric effects, that is, the less sterically hindered alcohols react faster. However, evidence has been presented that the reactivity of primary and secondary alcohols with a high HMMM resin is the same; whereas the reverse reaction is faster with secondary alcohols [8]. This is a valid alternative explanation for the apparent higher reactivity of primary alcohols, since the curing reactions occur under reversible conditions. Furthermore, this proposal is not unreasonable within the context of the S_N1 mechanism for transetherification (Scheme 6.2). Unfortunately, the utilization of an over-simplified rate expression in these studies raises some question about the conclusion. Specifically, rate constants were determined from a rate expression that was inappropriate for the equivalent amounts of alcohol and HMMM resin utilized in the reported studies.

The rate at which MF resins cross-link polyols increases with increasing volatility of the alcohol in the order n-butyl < ethyl < methyl. These rates are probably influenced by the rate of diffusion of the alcohol from the reactive site and evaporation from the film.

Factors in addition to cure response should, however, be considered in selecting the alcohol for ether formation in the synthesis of MF resins. For example, viscosity at a given percent solids content can be reduced by substituting Class I *mixed ether resins* for HMMM types, an important consideration in formulating high solids coatings. The alcohols in mixed ether resins are generally 1:6–1:3 *n*-butyl or isobutyl alcohol, the balance being methyl alcohol. Only modest reductions of VOC are attained because mixed ether resins contain a higher weight fraction of volatile byproducts; butyl groups represent a higher weight fraction of resins than do methyl groups. Perhaps a greater advantage of mixed ether resins is that they impart lower surface tension to coatings than high HMMM types. As we discuss in Chapters 23 and 26, high surface tension is often associated with development of film defects such as crawling, cratering, leveling, and poor intercoat adhesion [9].

As one might anticipate, the side reactions that occur during the acid-catalyzed etherification step in MF resin synthesis tend to occur as well during acid-catalyzed cross-linking with polyols. The self-condensation reactions of MF resins include formation of methylene ether bridges of the general structure **4** and, very probably, methylene bridges of the general structure **5**. These self-condensation reactions of MF resins result in cross-linking as do reactions of MF resins with polyol (or other coreactant), which have been termed cocondensation [3]. It is important to recognize that both self- and cocondensation reactions contribute to the structure of the resulting cross-linked polymer network and to its film properties.

With strong acid catalysis, the apparent rate at which Class I resins react with most polyols (cocondensation) is substantially faster than self-condensation reactions. However, with Class II resins the apparent rates of co- and self-condensation are similar. An important factor here is the presence of substantial amounts of —NHCH$_2$OR and —NCH$_2$OH groups in Class II resins; both groups are expected to promote self-condensation reactions, as illustrated in Eqs. (6.7) and (6.8). The methylol groups can, of course, react analogously to a polyol; and the —NHCH$_2$OR groups can readily eliminate alcohol to yield reactive melamine imine (—N=CH$_2$) groups, which can react to form methylene ether bridges and methylene bridges (by reaction with —NH groups).

6.3.1.1. *Catalysis of MF–Polyol Reactions*

With strong acid catalysts, commonly aryl sulfonic acids, usually in the range of 0.5 to 1.5 wt% of MF resin, the cocondensation reactions of MF Class I resins with polyols occur readily (within 10–30 min) at about 110–130°C, providing that the alcohol groups are not severely sterically hindered. Of course, cure time and/or temperature may be reduced by increasing the catalyst concentration. However, it is inevitable that the storage (or package) stability is also reduced by this approach, since the reaction at ambient temperatures is also catalyzed by acid.

Another danger in increasing acid catalyst concentration to reduce cure time and/or temperature relates to the durability of the cured coating. In addition to catalyzing the transetherification (and etherification) reactions of MF resins with polyols, acid also catalyzes hydrolysis of cross-links in the cured coatings. The course of the hydrolysis reaction can be followed by reference to Schemes 6.2 and 6.3, where —N—CH$_2$—OR represents a cross-linked polymer (derived from self- or cocondensation) and R'OH represents water. The hydrolysis reaction is seen to break the cross-linked bonds and generate methylol groups, which are known, at least in part, to eliminate formaldehyde (see Eq. 6.10) [3].

$$\text{>N—CH}_2\text{—OH} \quad \rightleftharpoons \quad \text{>NH} \quad + \quad \text{H}_2\text{C=O} \qquad (6.10)$$

Clearly, higher acid concentration utilized for cross-linking results in higher acid residues in the cured coating and faster rates of hydrolytic degradation. In this regard, it has been determined that sunlight enhances the rate of acid hydrolysis, which reduces exterior durability of MF–polyol top coats [3].

In any large parking lot, automobiles can be seen on which part of the paint is in excellent condition and part is badly deteriorated. The deteriorated areas were possibly recoated after the car was assembled, necessitated by damage of the finished coating during assembly or in shipment. This so-called repair coating is catalyzed with relatively high acid concentration (>2 wt%) in order to effect cure at relatively low temperature (around 70°C). The low temperature is necessary when the repair is done after temperature-sensitive plastic and rubber components have been installed in the automobile.

Similar principles apply to automobile production when no repair is needed. From the standpoint of exterior durability, it is highly desirable to bake at relatively high temperatures and/or for relatively long times so levels of acid catalyst can be minimized. This assertion applies both to long-term durability and to the resistance of relatively new autos to spotting by acid rain (acid etching). However, other important considerations, such as energy savings, production rates, and the increasing use of plastic body components motivate low temperature, short bakes. The optimum compromise involves many variables, some of which vary with time. Thus, exact baking conditions vary from one automobile assembly plant to another and with time in a given plant.

Another important consideration with MF–polyol coatings is storage stability. Catalyst considerations are discussed here; other considerations affecting storage stability were described in Chapter 3 and will be discussed in Section 6.3.1.3. Stability for 6 months is a common requirement of one-package OEM coatings; the liquid coating must retain satisfactory application characteristics and the ability to yield satisfactory film properties during 6 months of storage under conditions prescribed by the user. This level of stability sometimes cannot be attained when strong acid catalysts, such as p-toluene sulfonic acid (commonly abbreviated pTSA or TsOH) are used in the formulation. Such strong acids catalyze cross-linking under ambient conditions at a rate that may increase the viscosity of the liquid coating above the range for successful application in less than 6 months. Consequently, for one-package systems it is common to use acid catalysts that are, by one means or another, deactivated in the liquid coating. Such catalysts are called acid-precursors, latent acids, and blocked acids [10].

Thermally-activated blocked acids are most commonly amine salts or esters of sulfonic acids. The salts are relatively weak acids ($pK_a = 8$–10), which are prepared from the reaction of an amine (R_3N) with a sulfonic acid, as shown in Eq. 6.11 for TsOH ($pK_a = -6$). It is hypothesized that in the presence of weakly basic MF resins, the amine salt takes part in an equilibrium reaction in which the MF resin is protonated (see Eq. 6.12).

$$H_3C\text{—}\langle \rangle\text{—}SO_3H + R_3N \rightleftharpoons$$

$$H_3C\text{—}\langle \rangle\text{—}SO_3^- + R_3\overset{+}{N}\text{—}H \qquad (6.11)$$

$$H_3C\text{—}\langle \rangle\text{—}SO_3^- + R_3\overset{+}{N}\text{—}H + \rangle N\text{—}CH_2\text{—}OR \rightleftharpoons$$

$$\left(\rangle N\text{—}CH_2\text{—}\overset{+}{\underset{H}{O}}R \ ^-O_3S\text{—}\langle \rangle\text{—}CH_3 \right) + R_3N \qquad (6.12)$$

While the equilibrium in Eq. 6.12 favors the structures on the left, it may be shifted to the right by volatilization of the amine, increasing the concentration of protonated MF resin, the initial intermediate in the transetherification reaction (see Scheme 6.2). Generally, the storage stability of the coating with a blocked catalyst approaches the stability of the uncatalyzed coating. In some cases the cure rate approaches that of the coating catalyzed with the free sulfonic acid, while in others the curing rate is somewhat reduced [11].

Certain esters of sulfonic acids liberate the free acids when heated either by alcoholysis with the polyols or by decomposition reactions. An example of a tosylate ester, which has proven useful as a catalyst, is the oxime ester shown in Eq. 6.13. When heated this substance decomposes in a complex reaction to liberate TsOH. Unlike amine salts, tosylate esters do not increase conductivity of the coatings, an important consideration for coatings that will be applied by electrostatic spray.

$$\underset{H_5C_6}{\overset{H_5C_6CO}{>}}C\text{=}N\overset{OTs}{\diagup} + \text{(P)}\text{—}OH \longrightarrow TsOH + byproducts \qquad (6.13)$$

In all types of MF–polyol coatings the acid strength of the medium can be no stronger than that of protonated melamine resin, which exerts a leveling effect on acid strength much the same as water does. In water, the strongest acid is, of course, the hydronium ion. Apparently, pTSA (TsOH) is completely ionized by MF resins, which explains why even stronger acids, such as hexafluorophosphoric acid, exhibit similar catalytic activity. On the other hand, pTSA is more effective for Class I MF resins than weaker acids, such a butylphosphoric and carboxylic acids. With pTSA, the rate of cocondensation of an HMMM resin and acrylic polyol is reported to be proportional to the square root of the acid concentration [2].

In water-borne MF–polyol coatings, the amine utilized to neutralize the solubilizing carboxylic acid groups in the polyol (see Section 7.3 for a discussion) also neutralizes the sulfonic acid. Commonly utilized hydroxy-functional amines, such as N,N-(dimethyleamino)ethanol (DMAE), may also participate in transetherification of MF resins and/or transesterification of ester groups in the polyol during storage, which may retard or prevent their volatilization during cure, causing loss of coating cure response [12].

While the acid strength of a strong acid catalyst makes little difference in cure rates, the choice of acid (or blocked acid) catalyst can make important differences in the film properties of cured coatings. The most widely used catalyst has been pTSA. The grade of pTSA is important, since some grades contain significant quantities of sulfuric acid, which can lead to pronounced yellowing of coating films during baking. It has been found that water resistance of films can be improved by use of a more hydrophobic sulfonic acid, such a dinonylnaphthalene disulfonic acid (DNNDSA), which is reported to give films that are less likely to blister when exposed to high humidity conditions [13]. Dinonylnaphthalene disulfonic acid is particularly effective in coatings that are applied directly to metal which, with pTSA, are especially prone to blistering on humidity exposure.

Other factors involved in the selection of acid catalysts can be appreciated by comparing another commercial catalyst, *p*-dodecylbenzenesulfonic acid (DDBSA), to DNNDSA. *Para*-dodecylbenzenesulfonic acid provides roughly similar catalytic activity and film properties as DNNDSA when used in coatings that are applied over a prime coat. However, when a DDBSA catalyzed coating is used directly on steel, the adhesion of the coating is likely to be poor. It seems probable that the sulfonic acid group on DDBSA is strongly adsorbed on the steel surface, leading to the surface becoming covered with dodecyl groups. The low surface tension of the long hydrocarbon dodecyl groups on DDBSA may cause dewetting by the rest of the coating or formation of a weak boundary layer that reduces adhesion. The better adhesion imparted by DNNDSA might be related to the presence of two sulfonic acid groups.

The activity of acid (or blocked acid) catalysts can be affected by pigmentation of the coating. Specifically, some grades of titanium dioxide pigments lead to loss of catalyst activity with storage time of the coating. This phenomenon is apparently related to the composition of surface treatments on the TiO_2. Silicon dioxide treated TiO_2 is preferable to aluminum oxide treated TiO_2 in this respect, probably owing to the basicity of alumina, which tends to neutralize the acid. When alumina treated TiO_2 is required for other reasons, higher concentration of acid catalyst must be used to counteract this effect.

Cocondensation of MF Class I resins with polyols can be catalyzed by weak acids, such as carboxylic acids; however, elevated cure temperatures, generally greater than 140°C, are required [14]. In the case of Class II resins, carboxylic acids, are more effective in catalyzing cocondensation in accordance with the mechanism provided in Scheme 6.3. Since many polyol resins contain some carboxylic acid groups, which are present to promote adhesion and facilitate pigment dispersion, no added catalyst may be needed for cross-linking with Class II resins.

Under baking conditions, methylol groups can undergo both cocondensation with polyols and self-condensation with other MF resins. Furthermore, formaldehyde may be eliminated as shown in Eq. 6.10, increasing the population of —NH groups, characteristic of Class II resins. The relative rates and extents of these reactions are yet unclear from the literature. At the high temperatures used in coil coating ovens (air temperatures as high as 375°C), strong acids are most effective with Class II resins.

6.3.1.2. Kinetics and Mechanism of MF–Polyol Cocondensation

Many studies have been directed at elucidating the mechanism(s) of the reactions between MF resins and polyols. Until recently, most of these studies were based on following the time necessary to reach some degree of hardness in the film. However, both cocondensation and self-condensation reactions contribute to hardness [6]. Therefore, one must view with caution almost all conclusions from such studies that are applied specifically to the cocondensation reaction between hydroxyl groups and MF resins.

Based on the time and temperature required to achieve hard films, MF Class I and II resins have been classified as high and low cure temperature resins, respectively. This conclusion is usually explained by a statement that the —NHCH$_2$OR group reacts more rapidly with hydroxyl groups than does the —N(CH$_2$OR)$_2$ group. However, direct evidence for this statement is in short supply. It is well established that Class II resins self-condense more rapidly than Class I resins, but there is no published evidence, of which we are aware, that the rate constant for cocondensation with Class II resins is higher than with Class I resins. It may be higher, but evidence for this conclusion is lacking.

It is desirable to develop methods for measuring the extent of cocondensation directly. Infrared (IR and FTIR, Fourier transfer infrared) spectroscopy has been used to follow changes in functional group concentrations as a function of time and temperature [2]; particular care must be taken in interpreting the results because of band overlap in the hydroxyl region. The rate of evolution of volatile reaction products, including formaldehyde and methanol, has been followed by gas chromatography [3,14] and by thermogravimetric analysis [11]. Major contributions have been made by application of dynamic mechanical analysis to MF–polyol cross-linking coatings [6,15]. A promising, relatively new technique is oscillating plate rheometry [16].

One of the misconceptions resulting from using hardness results to study kinetics was that the cocondensation reaction of Class I resins was thought to be much slower than it has now been found to be. Disappearance of polyol hydroxyl groups, as followed by FTIR, showed that the reaction was substantially more rapid than development of film properties [14,16,17]. Possibly, the relatively slow development of optimal film properties reflects the longer time required for self-condensation reactions, which may be necessary to achieve the optimal cross-link density.

The results of studies of reaction rates of the model compound, N,N-dimethoxymethyl-N',N',N'',N''-tetramethylmelamine (**6**), with alcohols are also explicable within the context that film properties continue to change after evolution of methanol from cocondensation is essentially complete [4,5]. Such model compounds are potentially very useful because their relative simplicity may permit monitoring both the disappearance of reactants and the appearance of products in the same experiment. It must, of course, be borne in mind that the simpler the model system, the less applicable the results will be to the more complicated system of interest.

A widely believed but often mistaken conclusion is that steric hindrance limits the number of active ether groups on high HMMM resins that can participate in cross-linking of coatings films to three or at the most four of the six groups [1]. This conclusion was based on the need to use an excess over the stoichiometric amount of MF Class I resin, relative to polyol, to achieve desired film properties at baking schedules such as 30 min at 120°C, utilizing around 0.3% pTSA. However, Hill et al. [6] provided strong evidence in support of an essentially complete reaction of —NCH$_2$OCH$_3$ groups with certain polyols, showing that all —NCH$_2$OCH$_3$ groups of high HMMM resins can cocondense if there are enough hydroxyl groups present for them to react with. His conclusions were based on studies of changes in mechanical properties, we discuss in Chapter 24. The results of Hill et al. further substantiate the hypothesis that the excess MF resin used in most commercial formulations is needed not to complete the cocondensation cross-linking, but to permit some degree of self-condensation cross-linking that may be necessary to achieve the desired film properties.

As noted in Section 6.1.2, the mechanism of cocondensation of Class I MF resins with polyols remains controversial with experimental evidence continually being interpreted in support of (or against) the S_N1 or S_N2 mechanisms, provided in Scheme 6.2. Based on studies with the model analog **6** and monofunctional alcohols, it was proposed that both S_N1 and S_N2 mechanisms occur [4]. Subsequently, Meijer [8] reported a study on the reactivity of a high HMMM resin with chiral monofunctional alcohols. He concluded that the S_N1 mechanism dominates and also reinterpreted the earlier results of Ref. [4] in favor of the S_N1 mechanism. Unfortunately, these conclusions remain in question owing to the utilization of an oversimplified rate expression.

Using an HMMM–acrylic composition, and following the cocondensation reaction by FTIR, Bauer [2] provided support for the S_N1 mechanism. Furthermore, they elaborated on the importance of the role of methyl alcohol in interpreting the kinetic order of the reaction and the rate dependence of polyol. A key point is that the rate of cocondensation by the S_N1 mechanism is expected to be dependent on the concentration of polyol if the methyl alcohol, formed from HMMM, competes with the polyol for the intermediate carbocation (refer to Scheme 6.2). It can be conjectured that water, when present, also competes with the polyol for this carbocation. Under most real-life conditions it is probable that concentrations of methyl alcohol and water are significant in the early stages of coating cure and decline in the later stages. If so, the rate of cocondensation will only become independent of polyol concentration in the late stages of cure if at all, and by then concentration of hydroxyl groups is small.

This valid interpretation of the S_N1 mechanism is an important consideration, since dependence of the rate on the polyol had mistakenly been interpreted as evidence against the S_N1 mechanism. Competition for the intermediate carbocation by methyl alcohol is expected to gain relative importance as the reaction proceeds because the methyl alcohol concentration increases, at least temporarily, whereas the polyol concentration decreases with increasing conversion. Thus, kinetic studies based on hardness or other film properties, which tend to develop only at high conversions, are expected to exhibit a strong dependence on the concentration and nature of the polyol. Such results had generally been interpreted in favor of the S_N2 mechanism, but are equally consistent with the S_N1 mechanism.

Considerable progress has been made toward understanding the complex combinations of reactions that occur during the curing of MF cross-linked coatings, but there is a long way to go before the process will be fully understood. Because all of the principal reactions involved are reversible, the coatings probably never reach an equilibrium state during curing; the system probably remains in a dynamic state and continues to change as long as the coating is kept at baking temperature. The effects of water in coating films during the curing can lead to hydrolysis and loss of formaldehyde [3]. Cyclization reactions may occur such as illustrated by the conversion of model compound **6** to the trimer **7**. The possibility of reactions involving the triazine ring cannot be ruled out. Rates of diffusion of byproduct alcohol and of water remain a matter of conjecture. More research on this subject is clearly warranted by the need to overcome the concerns, described above, of formaldehyde emissions and acid etching.

6.3.1.3. Package Stability Considerations

The limited package stability of coatings containing MF resins can be a difficult problem, especially with Class II resins. Due to the presence of substantial amounts of >NH and >NCH$_2$OH groups, Class II resins can slowly self-condense during storage at room temperature in reactions that are catalyzed by weak acids, such as carboxylic acids. Stability is somewhat improved by addition of small quantities of tertiary amines. Primary or secondary amines, which react with formaldehyde, should not be used with Class II resins. Free formaldehyde is in equilibrium with melamine methylol groups in these resins (see Eq. 6.10). Reaction of the formaldehyde with primary or secondary amines displaces the equilibrium reaction in favor of formaldehyde formation (i.e., demethylolation), which reduces the methylol group functionality of the MF resins.

An important approach toward increasing package stability is to use as much monofunctional alcohol as possible in formulations containing both classes of MF resins. The presence of monofunctional alcohol in the formulation extends the storage ability since its reaction with the MF resin does not lead to cross-linking. On the other hand, cross-linking and viscosity buildup occur when the MF resin self-condenses or reacts with the polyol. Generally, it is desirable to utilize the same alcohol that is used to synthesis the MF resin. If a different alcohol is used, undesirable changes may occur. For example, if n-butyl alcohol is used in the solvent with a methoxymethylmelamine resin, the cure response will gradually become slower as the amount of butoxymethyl increases and methoxymethyl decreases. In this case, it can be speculated that the viscosity might decrease during storage because butoxymethylmelamine is less viscous (it has a lower T_g) than the corresponding methoxymethylmelamine and because of breakage of methylene ether bridges by the excess alcohol.

6.3.1.4. Selection of MF Resins

In using MF resins as cross-linkers for polyols, as well as other coreactants, formulators face a critical decision as to whether to use a Class I or II resin. In general terms, the major advantage of Class II resins is that lower temperature or shorter time schedules can be used for curing. One the other hand, films having greater toughness and impact resistance are generally obtained using Class I resins. Both of these characteristics can be related to the more facile self-condensation of Class II resins, although other factors probably contribute.

No studies on the stoichiometry of cocondensation reactions with Class II resins have been published. The average functionality of Class II resins for cocondensation is almost certainly smaller than that of comparable Class I resins. The difference, however, may not be evident since most formulations contain amounts of MF resin far in excess of the requirement for stoichiometric reaction with the polyol hydroxy groups.

The amounts of the MF resin and catalyst are determined empirically in the laboratory and are optimized for a specified baking time and temperature. Many formulations are designed to achieve optimum properties when cocondensation is nearly complete and self-condensation is partly complete. If the conditions actually experienced in the ultimate use of the coating differ from those for which the formulation was designed, it is predictable that the extent of self-condensation and perhaps also cocondensation will differ from the optimum levels. Accordingly, coating properties such as hardness, adhesion, exterior durability, and impact resistance may be affected.

This potential problem is much more serious with high solids coatings where the hydroxyl equivalent weight and average functionality of the polyol is substantially less in comparison with low solids, higher MW polyols. These characteristics result in greater sensitivity of cross-linked density of high solids coatings to cure variables, including time and temperature. These considerations have been addressed in terms of a "cure window," corresponding to the range of cure times and temperatures that provide films of acceptable properties [18]. The cure window was shown to be significantly smaller for high solids MF–polyol coatings, especially when Class II resins were used.

6.3.2. Other Coreactant Resins

Carboxylic acid-functional resins will react with MF resins to form the corresponding ester derivatives (see Eq. 6.2). Although the reaction is apparently slower with carboxylic acids than with hydroxyl groups, carboxylic acid-functional resins can be, and are, cross-linked with MF Class I resins. Carboxylic acid groups are also present in substantial levels in water-borne hydroxy-functional coreactant resins to enhance dispersibility, as well as in solvent-borne polyols, at low levels, to enhance the stability of pigment dispersions. When significant levels of carboxylic acid groups are present, it seems desirable to cure under conditions that will ensure that they will react, because residual carboxylic acid groups are expected to enhance the water sensitivity of the cured films.

Primary amide (Eq. 6.3), urea, and urethane groups also react with Class I MF resins. While the reaction of MF resins with amides is important in specific applications, cross-linking of amides is not widely used. Coreactant resins of mixed hydroxyl and urethane functionality are being used more widely. As this usage increases, it will become very desirable to determine the relative rates of the forward and reverse reactions of urethane and hydroxyl groups with MF resins.

6.4. OTHER AMINO RESINS

Although MF resins are the dominant amino cross-linking agents for thermosetting coatings, other amino resins are also used. The structures of starting materials for important amino resins are provided in the beginning of this chapter.

6.4.1. Benzoguanamine–Formaldehyde Resins

Using benzoguanamine as a starting material, a range of resins analogous to MF resins can be prepared. As with melamine, benzoguanamine may be methoxy-methylated and/or butoxymethylated to various extents. The average functionality is, of course, lower because there are only two —NH$_2$ groups per molecule. Etherified benzoguanamine–formaldehyde (BF) resins yield cross-linked films with significantly greater resistance to alkali and to alkaline detergents, such as sodium tripolyphosphate, compared with MF resins. On the other hand, the exterior durability of BF-based coatings is poorer than MF-based coatings. Thus, BF resins are used for applications where resistance to alkali detergents is more important than exterior durability. Coatings for washing machines and dishwashers are examples.

Reduced exterior durability of BF-based coatings most probably reflects lower photostability arising from the presence of the phenyl group on the triazine ring of benzoguanamine, although supporting evidence for this reasonable hypothesis does not appear available. Aside from the hydrophobicity of the phenyl groups, a convincing explanation for enhanced alkali resistance is not intuitively forthcoming.

6.4.2. Urea–Formaldehyde Resins

Urea also reacts with formaldehyde to form methylol derivatives. The first and second formaldehyde add easily; tri- and tetramethylol derivatives of urea have never been isolated. Methylolated ureas are etherified by reaction with alcohols, analogously to methylolated melamines. Self-condensation reactions also occur under both acidic and basic conditions. A wide variety of etherified urea–formaldehyde (UF) resins are manufactured using different ratios of formaldehyde to urea and different alcohols for etherification.

In general, UF resins are the most economical amino resins; they are also the most reactive. With sufficient acid catalyst, coatings formulated with UF resins can cure at ambient or mildly elevated temperatures. However, such coatings lack satisfactory exterior durability for most outdoor applications, probably because the cross-links are relatively reactive and have poor resistance to hydrolysis. MF or BF resins are preferred even in coatings for indoor use when good hydrolytic resistance is required, as is usually the case for coatings for steel. UF resins are widely used in coatings for temperature sensitive substrates such as wood furniture, paneling, and cabinetry. In such applications, low temperature bake is essential, and corrosion resistance (related to hydrolytic stability of the cross-links) is unimportant.

6.4.3. Glycoluril–Formaldehyde Resins

Glycoluril reacts readily with formaldehyde to give tetramethylolglycoluril (TMGU) [19]. In the range of pH 4 to 8, aqueous solutions of TMGU show an equilibrium level of about 3.6 methylol groups per glycoluril together with 0.4 parts free formaldehyde. Above pH 8.5, demethylolation is more favored. For example at pH 11, there are approximately equal parts of methylol groups and free formaldehyde. Below pH 3, reversible self-condensation occurs to form glycoluril dimers with methylene ether bridges (see Eq. 6.14), where >N—CH$_2$—OH represents methylolated glycoluril groups.

$$\text{>N—CH}_2\text{—OH} \underset{\text{}}{\overset{\text{pH < 4}}{\rightleftharpoons}} \text{>N—CH}_2\text{—O—CH}_2\text{—N<} \ + \ H_2O \qquad (6.14)$$

The behavior of TMGU is unlike the behavior of methylolated melamine and urea resins in this respect. While TMGU does not self-condense to form ether bridges above pH 4, methylolated melamines and ureas self-condense at pH 7.

Tetramethylolglycoluril reacts with alcohols in the presence of a strong acid catalyst to form tetraalkoxymethylglycoluril (GF) resins. Tetramethoxymethylglycoluril is a relatively high melting solid and is used as a cross-linker in powder coatings. In solution coatings, mixed ether (dimethoxymethyldiethoxymethyl)glycoluril and tetrabutoxymethylglycoluril are used, since they are liquids and are more readily handled. The mixed methyl–ethyl ether is water soluble.

Relative to other amino resins, GF resins provide coatings exhibiting significantly greater flexibility at a similar cross-link density. Hence, GF resins are used in applications, such as coil coatings and can coatings, where flexibility is important. GF resins possess the further advantage that significantly less formaldehyde is evolved during cure, compared with MF resins. The higher cost of GF resins limits their use.

6.4.4. (Meth)acrylamide Amino Resins

Acrylic copolymers of N-isobutoxymethylacrylamide can be prepared by two routes. (1) Synthesis of the N-isobutoxymethylacrylamide monomer (from step-wise reaction of acrylamide with formaldehyde and isobutyl alcohol), followed by copolymerization (with other monomers). (2) The copolymerization of acrylamide followed by step-wise reaction with formaldehyde and isobutyl alcohol. Analogous reactions can be carried out with methacrylamide and other alcohols.

Such (meth)acrylamide amino resins are used in coil coatings, where, with proper design, they are weather resistant and more flexible than coatings based on MF–acrylic polyol resins. One can speculate that flexibility is enhanced by the absence of densely cross-linked clusters of self-condensed MF resins with their rigid triazine rings.

REFERENCES

1. J. O. Santer, *Prog. Org. Coat.*, **12**, 309 (1984).
2. D. R. Bauer, *Prog. Org. Coat.*, **14**, 193 (1986).
3. W. J. Blank, *J. Coat. Technol.*, **51** (656) 61 (1979).
4. Z. W. Wicks, Jr., and D. Y. Y. Hsia, *J. Coat. Technol.*, **55** (702), 29 (1983).
5. U. Samaraweera and F. N. Jones, *J. Coat. Technol.*, **64** (804), 69 (1992).
6. L. W. Hill and K. Kozlowski, *J. Coat. Technol.*, **59** (751), 63 (1987).
7. J. H. van Dijk, A. S. van Brakel, W. Dankelman, and C. J. Groenenboom, FATIPEC, XV, II-326 (1980).
8. E. W. Meijer, *J. Polym. Sci., Part A: Polym. Chem.*, **24**, 2199 (1986).
9. N. Albrecht, *Proc. Water-Borne Higher-Solids Coat. Symp.*, New Orleans, LA, (1986), p. 200.
10. S. P. Pappas, B. C. Pappas, X. -Y. Hong, R. Kirchmayr, and G. Berner, *Proc. Water-Borne Higher-Solids Coat. Symp.*, New Orleans, LA, (1988), p. 24.
11. W. J. Mijs, W. J. Muizebelt, and J. B. Reesink, *J. Coat. Technol.*, **55** (697), 45 (1983).
12. Z. W. Wicks, Jr., and G-F. Chen, *J. Coat. Technol.*, **50** (638), 39 (1978).
13. L. J. Calbo, *J. Coat. Technol.*, **52** (660), 75 (1980).

14. M. G. Lazzara, *J. Coat. Technol.*, **56** (710), 19 (1984).

15. L. W. Hill and K. Kozlowski, *Proc. XIIth Intl. Conf. Org. Coat. Sci. Technol.*, July 1986, Athens, Greece, p. 129.

16. T. Yamamoto, T. Nakamichi, and O. Ohe, *J. Coat. Technol.*, **60** (762), 51 (1988).

17. T. Nakamichi, *Prog. Org. Coat.*, **14**, 23 (1986).

18. D. R. Bauer and R. A. Dickie, *J. Coat. Technol.*, **54** (685), 57 (1982).

19. G. G. Parekh, *J. Coat. Technol.*, **51** (658), 101 (1979).

CHAPTER **VII**

Thermosetting Acrylic Resins

As discussed in Chapter 4, thermoplastic acrylic (TPA) polymers have many excellent properties, especially exterior durability, but they require large amounts of solvent to reduce their viscosity low enough for application. Since they are designed to react chemically after application, thermosetting acrylics (TSA) can have lower molecular weights and can be applied at higher solids. The postreaction leads to a polymer network, which is ideally one interconnected molecule with a very high molecular weight and hence potentially good film properties. The cross-linked film is not soluble in solvent while TPA polymers remain soluble in the original solvent. The term thermosetting acrylic resins seems to imply that the resins will cross-link by themselves. While, as we will see, this is possible in some cases, it is not the most common situation. Most TSAs are resins bearing functional groups that can react with a different functional polymer or cross-linker, which is also included in the coating formulation.

To be reactive, the TSA must possess functional groups. Hydroxyl groups are the most widely used; they are cross-linked with either MF resins (see Chapter 6) or with polyisocyanates (see Chapter 12). Carboxylic acid-functional TSAs are designed to be cross-linked with epoxy resins (see Chapter 11) or with 2-hydroxy-alkylamides (see Section 13.4.1). Carboxylic acid groups also react with MF resins but are seldom used as the principal cross-linkable group. Acrylics with amide groups can be used in several ways (see Section 7.1). By using monomers with oxirane substituent groups or isocyanate substitute groups, acrylic copolymers with pendant epoxy and isocyanate groups, respectively, can be prepared. The use of such resins as cross-linkers for carboxy- or hydroxy-functional resins is discussed in Sections 11.1 and 12.3.2, respectively.

7.1 THERMOSETTING ACRYLIC RESINS FOR CONVENTIONAL-SOLIDS COATINGS

Solvent-borne TSAs for conventional-solids coatings were first developed in the 1950s. Their $\overline{M}_n$s are usually 10,000–20,000 with $\overline{M}_w/\overline{M}_n$ of 2.3 to 3.3. Since the early 1970s, solvent-borne TSAs for higher solids coatings have been and continue to be developed, they are discussed in Section 7.2.

Thermosetting acrylics are prepared by free radical initiated chain-growth polymerization under monomer-starved conditions. Procedures are similar to those described in Chapter 4. Polymerization of conventional TSAs is more easily controlled than those of TPAs, since there is not the critical effect of molecular weight on application and film properties. Due primarily to their lower molecular weight, one can use less expensive aromatic hydrocarbon solvents for the polymerization medium. Azo initiators are used predominantly, since they cause relatively few side reactions and minimize formation of end groups, which could reduce weather resistance. Chain transfer to solvent can also result in end groups that lead to reduced exterior durability. It has been suggested that the presence of terminal groups resulting from chain transfer to a ketone solvent such as methyl n-amyl ketone (MAK) may have a particularly deleterious effect on exterior durability [1]. Such terminal groups may not be as deleterious to TPAs since the high molecular weight of TPAs reduces the concentration of chain ends. In the case of the low molecular weight resins discussed in Section 7.2, the deleterious effect is expected to be especially large.

Most of the monomer content in TSAs is the same as that used in TPAs, namely, methyl methacrylate (MMA), styrene (S), butyl acrylate (BA), and so on. Monomers are selected to provide the required T_g and to balance cost and properties based on considerations similar to those for TPAs (Chapter 4). Hydroxyl groups are introduced by using a hydroxy-functional comonomer, commonly, 2-hydroxyethyl methacrylate (HEMA). If a lower T_g reactive monomer is desired, 2-hydroxyethyl acrylate (HEA) can be used but its greater toxic hazard increases handling costs; rather, it is often preferable to reduce T_g by increasing the proportion of low T_g nonreactive monomers such as n-butyl acrylate. Commercial grades of HEMA (and HEA) contain appreciable amounts of diester—ethylene glycol dimethacrylate (EGDMA) in the case of HEMA. Separation of the mono- and diesters is not easy since their boiling points are very similar; even if pure monoester were made, it would be hard to avoid a degree of transesterification to diester during storage and processing. When a relatively small fraction of hydroxy-functional monomer is used, small amounts of diester cause some branching, which does not seem to be harmful. However, larger amounts will lead to higher molecular weight, broader molecular weight distribution, and even gelation.

2-Hydroxypropyl methacrylate (HPMA) can be used as an alternative to HEMA. It is somewhat lower in cost and generally has a lower diester content. It is a mixture of isomers, as shown, in which the secondary alcohol predominates. Since secondary alcohols are less reactive than primary alcohols, TSAs with HPMA substituted for HEMA require higher bake temperatures, longer bake times, or more catalyst to cure to a given cross-link density with MF resins. For example, bake temperatures may have to be increased by 10 to 20°C. Isocyanate cross-linkers also react slower with HPMA-containing TSAs, but there is the accompanying advantage of longer pot life. Use of HPMA in place of HEMA may reduce resistance to photoxidation as a result of the hydrogen atoms on tertiary carbon atoms, which are more easily abstractable by free radicals.

An alternative method for introducing hydroxyl functionality on TSAs is to react a carboxylic acid-functional copolymer made using methacrylic (or acrylic) acid (MAA or AA) with propylene oxide. This procedure has been used commercially for some time. The raw material cost is lower than when HPMA is used but exacting process control is required.

$$\text{(P)}\sim\overset{\overset{\text{O}}{\|}}{\text{C}}\text{-OH} \;+\; \text{H}_2\text{C}\overset{\text{O}}{\diagup}\text{CH-CH}_3 \;\longrightarrow\; \text{(P)}\sim\overset{\overset{\text{O}}{\|}}{\text{C}}\text{-O-CH}_2\text{-}\overset{\overset{\text{OH}}{|}}{\text{CH}}\text{-CH}_3 \;+\; \text{isomer}$$

An example of a hydroxy-functional TSA is an MMA/S/BA/HEMA/MAA copolymer with a weight ratio of 50:15:20:14:1, corresponding to a mole ratio of 54.3:15.6:16.9:11.7:1.5. Typical molecular weights are an $\overline{M}_w$ of about 35,000 and $\overline{M}_n$ of about 15,000 ($\overline{M}_w/\overline{M}_n = 2.3$). With the given ratio of comonomers, the $\overline{P}_w$ is about 320 and the $\overline{P}_n$ about 140. For stoichiometric comparisons, one uses, of course, the number average, $\overline{M}_n$. This resin would have a hydroxy equivalent weight of slightly over 900 with a number average functionality $\overline{f}_n$ of about 16 hydroxyl groups per polymer molecule. The small amount of carboxylic acid functional monomer (MMA) is introduced to reduce the probability of pigment flocculation in the liquid coating.

Countless variations can be prepared. The nonfunctional monomers (MMA, S, and BA, in this case) are selected on a basis of contribution to T_g of the final film, exterior durability requirements, and cost. This monomer combination provides good exterior durability with a relatively high T_g at a moderate cost as is appropriate for an automotive top coat, for example. This resin also has a relatively high level of hydroxy-functional monomer, appropriate when relatively high cross-link density (XLD) films are desired for such applications as automotive top coats. For applications requiring more flexible films, such as coil coating or exterior can coatings, one would use a lower HEMA content. By adjusting T_g and functionality, TSAs can be designed for a wide range of end uses.

The composition of hydroxy-functional TSAs also varies with the cross-linkers. A TSA resin appropriate for a TSA–MF coating would not necessarily be appropriate to use with a particular polyisocyanate cross-linker. Depending on the polyisocyanate used with a TSA, the T_g of the final cross-linked films could be higher or lower than would be obtained cross-linking the same TSA with an MF resin. Thermoset resins and cross-linkers must be selected or designed for use as a system. Thermosetting acrylic resins for polyisocyanate cross-linking tend to have lower T_gs and lower $\overline{f}_n$s than TSAs for cross-linking with MF resins in order to compensate for the effect of the intermolecular hydrogen bonding between urethane groups in the cross-linked films.

Carboxylic acid-functional acrylic resins used in thermoset systems together with epoxy resins are made using an amount of acrylic or methacrylic acid required to provide the desired XLD. Close quality control of acrylic acid is required since it can dimerize.

Acrylics with amide groups can be used in several ways. Acrylamide copolymers can be cross-linked with MF resins by reaction with the amide group; the curing temperature required is higher than for hydroxy-functional resins. Alkoxymethyl derivatives of acrylamide copolymers can be made by reacting the amide groups with formaldehyde followed by etherification, analogous to the preparation of amino resins as described in Chapter 6. These resins can be used as cross-linkers for hydroxy-functional TSAs. Alternatively, one can make alkoxymethyl derivatives of monomeric acrylamides for use as comonomers. For example, N-(isobutoxymethyl)methacrylamide has been used as a comonomer along with hydroxy-functional comonomers to make copolymers that are "self-cross-linking [2]."

Acrylamide–HEMA copolymers with nonreactive comonomers can also be reacted with formaldehyde and then with an alcohol to make a self-cross-linking resin. While self-cross-linking resins are intriguing to resin chemists, it is often more desirable to use two separate resins; each with one of the two functional groups. The formulator has much greater latitude with the two resin system. Furthermore, storage stability is likely to be better since the starting time for storage is when the resins are mixed rather than when the self-cross-linking resin is manufactured.

The solids content of conventional TSA enamels (30–40 NVW and 25–35 NVV) is higher than that of TPA lacquers but they still have relatively high VOC emissions. Solids for automotive TSA enamels, especially metallic enamels, were in the lower end of the range due to their low pigmentation. Thermosetting NAD acrylics were developed, which permitted solids of metallic enamels around 35 NVV. These NADs are prepared in the same general fashion as described in Section 4.1.2 except, of course, that a comonomer such as HEMA is used to provide the sites for cross-linking [3]. The NAD–TSAs were used on a large scale for some years but solvent emissions were still too high and new approaches were needed.

7.2. THERMOSETTING ACRYLICS FOR HIGH SOLIDS SOLVENT-BORNE COATINGS

Still further increase in solids became necessary to meet lower VOC emission requirements. During the 1970s major efforts were launched to make high solids solution acrylic resins. It may seem that the problem is simple; why not just reduce $\overline{M}_n$ from 15,000 to 1500? Actually, the problem is very complex; as molecular weight goes down, greater care must be exercised in carrying out the polymerization. The amount of non- or monofunctional resin must be kept to a very low fraction. Due to the larger number of chain ends, the effect of chain transfer to solvents, such as ketones, which introduces photoreactive end groups, is potentially more serious. More care is needed in establishing the stoichiometric ratio of cross-linker and, as noted in Section 6.3.1.4, the cure window becomes narrower. As we will see later, commonly, problems controlling film defects during application become greater as the solids content becomes higher. It is particularly difficult to avoid sagging of spray applied baking coatings.

The composition and process procedures for making high solids acrylics in commercial use are proprietary information. A TSA resin made for studying MF cross-linking of fairly high solids coatings has the following composition: S/MMA/BA/HEA; weight ratio 15:15:40:30; $\overline{M}_w$, 5200; $\overline{P}_w$, 54; $\overline{M}_n$, 2300; $\overline{P}_n$, 20; $\overline{P}_w/\overline{P}_n = 2.7$; equivalent weight, 400; $\overline{f}_n = 5.7$ [4]. The resin was prepared at 65 NVW in methyl amyl ketone (MAK), a solvent commonly used in high solids coatings. In comparing this TSA with the conventional TSA described earlier in this chapter, note that $\overline{M}_n$ has been reduced by a factor of 6.5, whereas $\overline{f}_n$ is only reduced by a factor of 2.8. This difference results from the higher content of hydroxy-functional monomer required in the high solids resin. In order to attain about the same XLD in the final film, the low molecular weight resin must undergo more reactions. Although not reported, application solids of coatings formulated with this resin would be around 45 NVV. This solids content corresponds to that used in "high solids" metallic automotive top coats. However, the solids are still too low to meet EPA regulations, which are expected in the 1990s for many end uses.

The free radical polymerization process inherently limits the extent to which molecular weight of a TSA can be reduced. Satisfactory performance requires that substantially all of the molecules have at least two hydroxyl groups, which becomes statistically less probable as molecular weight is reduced. The problem can be illustrated by comparing the conventional TSA ($\overline{M}_n = 15,000$, $\overline{P}_n = 140$) described in Section 7.1 with a potential high solids TSA, having the same monomer ratio with $\overline{M}_n = 1070$ ($\overline{P}_n = 10$), about the level needed to formulate with VOCs around 300 g of solvent per liter of coating. The conventional resin has an $\overline{f}_n$ of 16 hydroxyl groups. While individual molecules have more or less than 16 hydroxyl groups, statistically the number of molecules with fewer than 2 hydroxyl groups is very low. Thus virtually all the molecules are capable of cross-linking. In contrast, $\overline{f}_n$ for the high solids TSA would be only about 1.2 and a very large fraction of the molecules could not be cross-linked. Molecules with no hydroxyl groups would either volatilize or remain in the film as plasticizers, probably to the detriment of film properties. Molecules with one hydroxyl group would terminate cross-linking reactions, leaving "loose ends" in the coating. The theory of elasticity predicts that loose ends will seriously diminish the mechanical properties of a network. Experimental experience confirms that loose ends have a substantial effect on initial film properties [5,6]. For example, each percent by weight of monohydroxy oligomer in a TSA may reduce T_g of the TSA/MF enamel by about 1°C [5].

The problem results from both molecular weight and functional group distribution. Figure 2.2 (Chapter 2) shows the distribution of molecules of differing degrees of polymerization of a series of resins with $\overline{P}_n = 12$ with $\overline{P}_w/\overline{P}_n = 1.07$, 1.5, and 3.0. The theoretical minimum polydispersity that can be obtained by anionic polymerization is 1.07, 1.5 is the theoretical minimum achievable by free radical initiated polymerization, and 3.0 is typical of reasonably well-controlled polymerizations. Note particularly the increasing amounts of low degree of polymerization resin as the polydispersity broadens.

The other factor that is involved is sequence length distribution: sequences of different lengths of nonfunctional monomers separating functional monomers. If these sequence lengths are short compared to the chain length of the low molecular weight parts of the resin, multiple functional groups will be incorporated in the molecules. But, if the sequence lengths of nonfunctional monomers are long in comparison to chain lengths, molecules with one or no functional monomer units can be expected. Statistical methods have been used to calculate the proportions of nonfunctional molecules that would be formed during random copolymerization of monomer mixtures with differing monomer ratios to different molecular weights and molecular weight distributions [7,8]. Due to the assumptions involved, all such calculations are approximate; an example of such results for S/BA/HEA (30:50:20 wt%) copolymers of varying $\overline{P}_n$ is given in Table 7.1 [8].

For $\overline{P}_n = 9.5$ ($\overline{M}_n = 1125$) about 36% of all molecules, corresponding to 13% of the weight, will be nonfunctional. (The mol% and wt% differ because of the molecular weight distribution; low molecular weight molecules are more likely to be nonfunctional than the higher molecular weight ones.) The weight fraction of a monofunctional oligomer is more difficult to calculate but would be expected to exceed that of a nonfunctional oligomer. At the higher HEA content (30%) of the example given earlier with a $\overline{P}_n = 20$, the proportion of monofunctional oligomer molecules would be minimized. But, if one wished to have volume solids of about 70 NVV, one would need to have a $\overline{P}_n$ of about 10 with a narrow distribution of molecular weight. The calculations show that the functional monomer content would have to be very high—so high that the XLD of the cross-linked film would be too high for acceptable film properties.

Major efforts have been invested in attempts to make TSAs for very high solids coatings. $\overline{M}_n$ (and $\overline{M}_w$) can be reduced by use of chain-transfer agents or by use of very high initiator levels. The first approach is readily implemented, but the latter approach is sometimes preferred because of suspected adverse effects of chain-transfer agent residues on weatherability [6]. Close control of variation of temperature and concentrations during polymerization can minimize polydispersity. Initiator choice can be critical. Benzoyl and t-butyl peroxides tend to give wide distributions due to chain transfer to the polymer resulting from hydrogen abstraction. Azo initiators such as AIBN give less branching. t-Amyl peroxides, such as ethyl 3,3-di(t-amylperoxy)butyrate, are reported to give particularly narrow distributions [9]. Azobisisobutyronitrile is generally preferred over benzoyl peroxide since exterior durability of films is superior. No data have been published on the exterior durability of amyl peroxide based resins.

Use of a chain-transfer agent with a functional group is an interesting approach. For example, when 2-mercaptoethanol is used as a chain-transfer agent, initiating free radicals bearing a hydroxyl group are formed. These lead to a high fraction of molecules with a hydroxyl group on one end, sharply reducing the fractions of non- and monofunctional molecules in the resin [10]. Improved film properties are obtained but the strong odor of 2-mercaptoethanol causes handling problems and may leave some residual mercaptan odor in the resin.

Table 7.1. Percentage of Nonfunctional Molecules Statistically Predicted for S/BA/HEA (30:50:20 by wt) Copolymers

$\overline{P}_n$	$\overline{M}_n$	Mol%	Wt%
36.8	4357	15	1.8
19.2	2273	24	5.8
9.5	1125	36	12.8

Source: See Ref. [8].

2-Mercaptopropionic acid had been used as a chain-transfer agent leading to the introduction of carboxylic acid terminal groups on a substantial fraction of the molecules [11]. The copolymers with the narrowest molecular weight distribution ($\overline{M}_w/\overline{M}_n = 1.7$) were made using an equimolar ratio of α-methylstyrene and acrylic acid. Use of α-methylstyrene and acrylic acid as comonomers has the advantage that they tend to react to give alternating monomer addition, thus further minimizing the possibility of nonfunctional molecules even at low molecular weight. The resulting carboxy-functional resin was reacted with the glycidyl ester of a mixed isomer branched-chain C_{10} monoaliphatic carboxylic acid. The resulting resin was hydroxy-functional. Since epoxy groups also react with mercaptan groups, the reactions also served to scavenge any residual mercaptan so there was no residual mercaptan odor in the resin. White pigmented coatings based on the resulting hydroxy-functional resins with $\overline{M}_n = 900$ and MF cross-linkers are reported to give sprayable viscosities of 70 to 74 NVW and good cured film properties [11].

Proprietary acrylic resins for general purpose white coatings with less stringent requirements have been reported with sprayable viscosities at about 54 to 56 NVV (70–72 NVW) [12]. Another commercial resin with an $\overline{M}_n$ of 1300 and $\overline{M}_w/\overline{M}_n = 1.7$ can be formulated with Class I MF resins in white coatings at about 77 NVW [13]. These resins provide hard, chemical resistant films with many good properties, although the films tend to be brittle.

Another approach to increasing solids is to blend acrylic polyols with other low viscosity polyols such as polyesters. Some reduction in VOC can be achieved in this way. The film properties such as exterior durability and chemical resistance from such blends are generally not quite as good as those of straight TSA/MF coatings, but they may be adequate for many industrial coatings.

Free radical polymerization tends to give broad molecular weight distribution and random incorporation of monomers. Other methods of polymerizing acrylics have also been investigated. Haggard and his co-workers [14,15] described alkoxide-initiated anionic polymerization of methacrylate esters to prepare very narrow molecular weight distribution oligomers, $\overline{P}_w/\overline{P}_n < 1.2$. Since functional monomers interfere with anionic polymerization, it is usually necessary to postreact the oligomers to introduce functional groups. Although quantities for commercial development were made available, the resins were subsequently withdrawn, apparently owing to manufacturing difficulties.

A versatile relatively new method of polymerization, which has been called *group-transfer polymerization*, has been developed and applied to the problem [16, 17]. A rapid, reversible complexation of small concentrations of enolate anions with silyl ketene acetals is proposed to explain the living nature of these polymerizations and the role of silyl ketene acetals. The over-all reactions involved are proposed to be

The polymerization permits close control of the structure of acrylic polymers; $\overline{M}_w/\overline{M}_n$ can be reduced to 1.2 or less, distribution of functional groups among molecules can be narrowed, functional groups can be placed at specific locations within each molecule, and multiarmed star polymers can be produced. Limited commercial use of silyl ketene acetal based polymerization started in 1988; large scale use will probably depend on the degree of success in reducing costs of the initiators and of processing. In one case, evidence has been presented indicating that the mechanism of group-transfer polymerization actually is a form of anionic polymerization and that group transfer does not occur [18].

Anionic polymerization as a means to make functional acrylic resins with low molecular weight and polydispersity continues as an active research field. Another recent example is the use of aluminum porphyrin initiators [19].

7.3. WATER-REDUCIBLE THERMOSETTING ACRYLIC RESINS

Another way to reduce VOC emissions from TSA coatings is to make TSA resins that can be diluted with water instead of much of the organic solvent. Such resins are commonly called "water-soluble" resins, but this terminology is misleading since the resins are not soluble in water. Rather, solutions of amine salts of these resins in organic solvents can be diluted with water to form reasonably stable dispersions of polymer aggregates swollen by solvent and water. There are truly water-soluble resins used in latex paints such as water-soluble cellulose derivatives and salts of polyacrylic acid (see Chapter 35). In order to minimize confusion, we have chosen to use the following terminology: "water-borne" is used generally for all coatings with aqueous media, including latex coatings; "water-reducible" is used for coatings based on resins having hydrophilic groups in most or all molecules, which excludes latexes; and the term "water-soluble" is limited specifically to substances that are truly soluble in water.

A typical water-reducible acrylic resin is a copolymer of MMA/BA/HEMA/AA in a weight ratio of 60:22.2:10:7.8 prepared by free radical initiated polymerization using an azo initiator [20]. The polymerization is carried out at high solids (generally 70 NVW or higher) in a water-miscible solvent. Glycol ethers, e.g., 1-(n-propoxy)-2-propanol or 2-butoxyethanol, and butyl alcohols are the most widely used solvents. Except for the solvent and the higher proportion of acrylic acid such resins are similar to conventional solids TSAs described in Section 7.1 with $\overline{M}_w$ and $\overline{M}_n$ of about 35,000 and 15,000, respectively.

The water-reducible acrylic resins typically have acid numbers of 40 to 60. (Acid number is determined by titration and is defined as the number of milligrams of KOH required to neutralize 1 g of resin solids. Equivalent weight equals 56,100/acid number.) Residual monomer is removed by distilling off a small fraction of the solvent and the resin is stored as a concentrated solution. The first step in preparing a coating is to add an amine, such as 2-(dimethylamino)ethanol (DMAE), to neutralize the acid groups. As discussed later, less than the theoretical amount of amine required to neutralize all of the carboxylic acid groups is commonly used. The ratio used is frequently called the extent neutralization (EN). For example, if 75% of the theoretical amount were used, the EN would be 75. Other coating components (pigments, MF resin, and sulfonic acid catalyst) are dispersed or dissolved in this solution and then the coating is diluted with water.

The change in viscosity with dilution of all water-reducible systems is abnormal. An example is given in Figure 7.1 showing log viscosity as a function of concentration for a model resin [21]. The height of the peak in the dilution curve is very dependent on the particular resin and formulation, and the systems are highly shear thinning in the peak region. Another abnormality of water-reducible systems is that the pH of the system is well over 7 (commonly 8.5–9.5) even though less than the theoretical amount of amine necessary to neutralize the carboxylic acid is used.

The morphology of water-reducible acrylic *solutions* hs been studied fairly extensively [21,22]. In Figure 7.1, the water-reducible resin is a 54% solution of butyl methacrylate (BMA)/AA 90:10 in *t*-butyl alcohol, at 75 EN with DMAE. One curve shows the viscosity response when the resin was diluted with further alcohol and the other curve shows the result of water dilution [21]. The solvent dilution

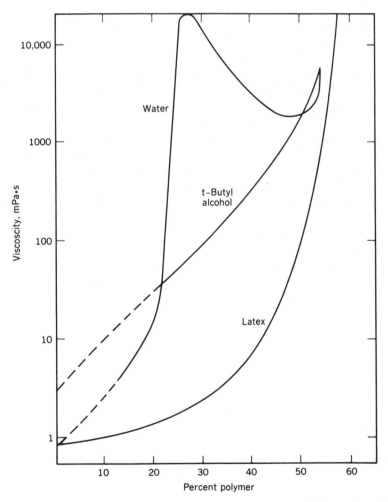

Figure 7.1. Viscosity dependence on concentration for a 10 mol% acrylic acid copolymer, 75 EN with DMAE, dissolved in *t*-butyl alcohol and then diluted with water. Also shown are curves for dilution of the same resin with *t*-butyl alcohol and a typical viscosity–solids latex dilution with water. (From Ref. [21], with permission.)

curve is a fairly straight line typical of the log viscosity versus the concentration relationship of most resin solutions in good solvents. The water dilution curve shows the abnormal response typical of water-reducible resins. In considering the behavior of these systems, it is important to remember that during the dilution with water two kinds of changes are occurring simultaneously: the concentration is being reduced and the ratio of solvent/water is also being reduced.

Notice that in the first stages of dilution, the viscosity drops more rapidly with water than with solvent. It is hypothesized that the viscosity is high before dilution because of association of ion pairs on different molecules. Water associates strongly with ion pairs, separating most of the intermolecular ion pairs and causing rapid reduction in viscosity. However, as dilution with water continues the viscosity levels off and then increases passing through a maximum. Still further dilution causes a very steep drop of viscosity. At an application viscosity of about 0.1 Pa·s, such systems typically have solids in the range of 20 to 30 NVW. The application of solid coatings made with such resins is thus low; however, since a major part of the volatile components is water, the VOC emissions are still low.

The amine salt of the resin is soluble in the t-butyl alcohol and in t-butyl alcohol–water solutions, where the ratio of solvent/water is high. But as additional water is added, the ratio of solvent/water decreases to the point where some of the molecules are no longer soluble in the mixed solvent. These molecules do not precipitate in a separate macrophase, but the nonpolar segments of various molecules associate with each other to form aggregates. The predominantly nonpolar parts of the molecules are in the interior of the aggregates and the highly polar carboxylic acid salt groups are on their periphery. As dilution continues, more and more molecules join in aggregates. Since the solvent is soluble in the resin, some dissolves in the aggregates swelling their volume. Also, water associates with the salt groups and to some degree dissolves in the solvent inside the aggregates, further swelling them. As the aggregates form, the system changes from a solution to a dispersion of aggregates in a continuous phase. As the number and volume of aggregates grows, the volume fraction of internal phase increases and the aggregates get more and more closely packed, leading to the increase in viscosity. At the maximum viscosity, the material is predominantly a dispersion of highly swollen aggregates in water containing some solvent. The high viscosity can lead to difficulty in stirring. On still further dilution, the viscosity drops rapidly. This drop results from two factors. There is a simple dilution effect, that is, the decrease in volume fraction internal phase, even if there were no change in the individual aggregates. However, the drop is even steeper than would result from this effect alone. The balance of the decrease is due to a decrease in the swelling of the aggregates. The partitioning of solvent and water between the aggregates and the continuous phase changes throughout the dilution process, and as more water is added, more solvent moves to the continuous phase, shrinking the swollen aggregates.

Viscosity of the solutions in the organic solvent is Newtonian as it is when only small amounts of water have been added. However, in the mid-range around the peak of the water dilution curve, the dispersions show a high degree of shear thinning. The dilute systems show little or no shear thinning. This behavior can be explained on the basis that, in the shear thinning stage, the swollen aggregate particles (the internal phase) are not rigid. When shear is applied, the particles distort, decreasing the shape factor and increasing the packing factor and, as a result, decreasing the viscosity as a function of increasing shear rate. With further dilution, some organic solvent is *extracted* from the aggregates into the continuous phase, and the aggregates become smaller and less easily distorted. In the more dilute stage, the viscosity is lower so that the shear stress exerted on the aggregates at a given shear rate is less, reducing the probability of distortion. Thus, the flow properties are Newtonian or only slightly shear thinning at application viscosities.

The abnormal pH effect noted earlier can be explained in a way consistent with this picture of the morphology of the system [23]. When one neutralizes a simple carboxylic acid, like acetic acid, with 75% of the theoretical equivalent amount of an amine like DMAE, the pH is about 5.5. However, the situation with the carboxylic acid groups on the polymer chain is different. In making the acrylic resin, the acrylic acid groups are relatively randomly spaced along the polymer chain. In some cases, there will be carboxylic acid groups fairly near each other; in other cases, there will be single acrylic acid groups separated from others on both sides by several hydrophobic ester monomer units. As aggregation occurs during dilution, many of the carboxylic acid groups will be near the surface where all, or almost all, will be neutralized by the water soluble amine. However, geometric factors will require that many of the carboxylic acid groups must be in the interior of the aggregates; those most widely separated from other carboxy groups by hydrophobic monomer residues are presumably most likely to be "hidden" in the interior of the aggregates. 2-(Dimethylamino)ethanol will partition among the continuous (water–solvent) phase, the surface region of the aggregates, and the interior of the aggregates. Since DMAE and DMAE salts are highly soluble in water it is expected that DMAE will concentrate in the first two regions, leaving a fraction of carboxylic acid groups in the interior of aggregates unneutralized. In effect, even when there is only 75% of the amine necessary to neutralize all of the carboxylic acid groups, there is more than enough to neutralize those carboxylic acid groups at or near the surface of the aggregates. Part of the amine is in the continuous phase, resulting in a basic pH reading. Since the interaction is between a weak base and a weak acid, the change of pH with amine addition is very slow. Due to this insensitivity, pH is not appropriate as a quality control specification for such systems.

This description of the morphology of a water-reducible system accounts for its behavior but does not provide direct evidence for the presence of aggregates. Evidence was found by studying the effects of changing solvent/water ratios at a constant resin–amine concentration. A BMA/AA copolymer (mol ratio 84:16) at 75 EN with DMAE was *dissolved* in a series of solvent/water solutions at the same concentration of 21.2% of resin plus DMAE [22]. Viscosity of this series of dispersions goes through a peak at 30:70 solvent/water. The samples were subjected to ultrafiltration. No resin was retained on the filter in the case of samples with a ratio of 80:20 solvent/water up to pure solvent. As the solvent ratio decreased from 80:20 to 30:70, the amount of resin retained increased until at 30:70 essentially all of the resin was retained by the ultrafilter. All of the resin in samples with still less solvent was also retained. The samples were also examined by phase contrast microscopy, which revealed the presence of particles in all those samples where resin was retained by the ultrafilter.

Many variables affect the morphology of these systems. The shape of the viscosity–concentration (dilution) curve can vary substantially from system to system. In some cases, there is a very high viscosity peak—higher than the viscosity of the original undiluted material; in other cases, there may be only a shoulder in the dilution curve rather than a peak.

Dependence of viscosity on molecular weight depends on solvent structure and the ratio of solvent/water in addition to concentration [22]. The effect of molecular weight on viscosity is different at different stages of dilution. The log of the viscosity of the all solvent solution varies with approximately the square root of the molecular weight. The viscosity of systems diluted with water to application viscosity is independent of molecular weight. This is, of course, a major advantage of these systems, which permits application of coatings made with a resin having a molecular weight in the same range as that used in conventional solvent-borne TSA coatings but with VOC contents equivalent to relatively high solids coatings.

Viscosity in the "peak" area of the dilution curve is very dependent on molecular weight. As molecular weight increases, the viscosity of the interior of the aggregate particles is higher so they are more difficult to distort, hence, viscosity of the whole system does not decrease as much at any given shear rate. Very high peak viscosity leads to difficulty in dilution. It is necessary to limit the molecular weight so that the coating can be thoroughly agitated throughout the dilution cycle with the available mixing equipment.

The dilution behavior of TSAs is very dependent on their carboxylic acid content. The effect of varying the mole percent of acrylic acid in a series of BA–AA copolymers from 10 to 50% is shown in Figure 7.2 [21]. With 50% acrylic acid, the salts exhibit a viscosity dependence on dilution behavior approaching that of solution systems. As carboxylic acid content is reduced, the abnormal rheological properties become more pronounced. Notice that the concentration at viscosities near those required for application are highest for the lowest acid content systems. This is an important reason that resins of this type are designed with the lowest carboxylic acid content (acid number 40–60) that will still provide a stable dispersion at application viscosity. Low acid numbers favor formation of relatively unswollen aggregates, keeping viscosity down, but if the acid numbers are too low, the system will separate into macrophases instead of forming a stable microphase dispersion.

The required acid content is lower for resins with increasing hydroxyl group content. While salts of carboxy groups are much more hydrophilic than hydroxyl groups, hydroxyl groups are sufficiently hydrophilic to affect the solubility in water–solvent blends, as indicated by the water solubility of the homopolymer of 2-hydroxyethyl acrylate. Therefore, the minimum required acid content decreases as the content of hydroxyl groups increases.

Melamine–formaldehyde resins are used as cross-linkers for the coatings. They react faster with hydroxyl groups and the resultant ether bonds are more stable to hydrolysis than the ester bonds resulting from cross-linking with carboxylic acids. Thus, while the cross-link density of the final film depends on the total functionality of COOH and OH groups, it is generally desirable to attain as much of the cross-linking as possible through OH groups and to adjust cross-link density by varying the $\bar{f}_n$ of OH. A further advantage of using the lowest possible level of COOH is that less amine is required. Amines are relatively expensive and must be included as part of the VOC emissions.

Another important variable in the use of water-reducible TSAs is the type and amount of amine [21,23,24]. Generally, less than the stoichiometric amount of amine is used. The lower the amine content, the lower the viscosity of the fully

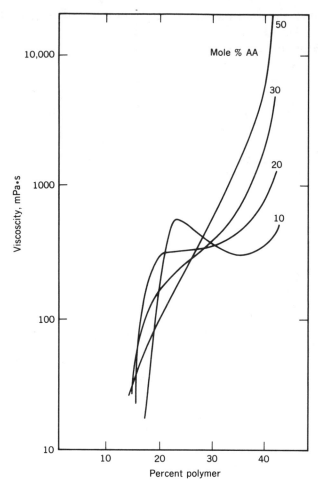

Figure 7.2. Viscosity variations during water dilutions of 42 NVW copolymers having 50, 30, 20, and 10 mol% AA; each at 75 EN with DMAE. (From Ref. [21], with permission.)

diluted systems, that is the higher the solids at a fixed application viscosity. Figure 7.3 shows viscosity response to dilution as a function of extent neutralization of a 90:10 BMA/AA resin with DMAE. For any resin–amine combination there is a minimum amount of amine required to give a stable dispersion at application viscosity, that is, to prevent macrophase separation. In the example shown in Figure 7.3, the 50 EN sample could not be fully diluted without macrophase separation. In a similar experiment using an 80:20 BMA/AA resin, 50 EN with DMAE gave a stable dispersion even when diluted until the viscosity was below 0.1 Pa·s.

Viscosity at application viscosity changes very rapidly with concentration and it is very possible to overshoot a desired reduction. The viscosity is also very sensitive to amine content, so if a coating has been reduced with too much water resulting in too low a viscosity, the viscosity frequently can be brought back up by the addition of small amounts of amine.

Yet another critical variable is the structure of the neutralizing amine. While there may be some effect of base strength (less amine may be required with increasing

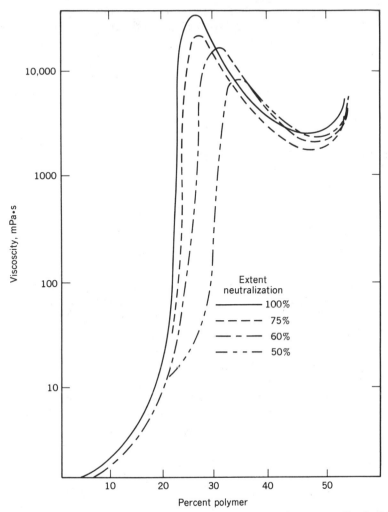

Figure 7.3. Viscosity as a function of weight percent of resin when neutralized with varying levels of DMAE. Resin has 10 mol% AA; dilution started at 54 wt% solids. (From Ref. [23], with permission.)

base strength), the principal variable seems to be the water solubility of the amine. The amount of amine required for stable dispersions increases in the order DMAE < triethyl amine (TEA) < tripropyl amine (TPA). Hydroxy-substituted amines are the most widely used but it has been shown that morpholine derivatives such as N-ethylmorpholine (NEM) are also effective.

It is important in selecting amines to consider not only the effect on dispersion stability but also the effect on package stability and curing of the coating [23]. The amine improves the package stability by minimizing reactions of MF resins during storage. This is especially true since sulfonic acids such as pTSA are used as catalysts. If Class II MF resins are to be used, the amine must be tertiary. If, as is more common, Class I MF resins are used, primary or secondary amines can also be used.

2-(Dimethylamino)ethanol, while widely used, has a disadvantage. It has been shown that transesterification occurs between ester groups on the acrylic resin and the hydroxyl of the amino alcohol, thereby covalently bonding part of the amine to the resin [24]. Since bound amine cannot volatilize during application and baking, it inhibits the acid catalyzed cross-linking reactions. *N*-Ethylmorpholine, which has essentially the same boiling point as DMAE, permits faster curing since it cannot undergo transesterification. 2-Amino-2-methyl-1-propanol (AMP) can be used with Class I MF resins. It provides stable dispersions and, in spite of its substantially higher boiling point, gives coatings that cure more rapidly than even those neutralized with NEM. 2-Amino-2-methyl-1-propanol has a hydroxyl group that can lead to transesterification, but also a primary amine that can undergo aminolysis reactions with ester groups on the resin. By reacting at both sites, AMP can act as a cross-linking agent supplementing cross-linking with the MF resin.

Another factor involved in the choice of amine is the effect on wrinkling [24]. When using TEA as the neutralizing amine, it is common to get low gloss coatings. The low gloss results from the development of a fine wrinkle pattern on the surface during curing of the film, which is caused by partial curing of the surface layer before the interior of the film cures. When the interior of the film does cure it shrinks, causing the immobilized surface of the film to wrinkle. This phenomenon occurs when amine is essentially absent from the surface layer of the film but is still present in the lower layers of the film; the MF resin can cross-link the acrylic at the surface while the reaction is still inhibited in the lower layers.

The probability of wrinkling increases as film thickness increases because there is more likely to be a differential in amine content as film thickness increases. The probability of wrinkling is also affected by amine structure. A series of three amines gave decreasing probability of wrinkling in the order TEA, DMAE, and NEM [24]. Triethyl amine is a relatively strong base (pK_a of conjugate acid = 10.9) and hence diffuses slowly through a film containing COOH groups; it also has a relatively low boiling point (90°C) so that on reaching the surface, it volatilizes rapidly. This combination of factors leads to a differential amine content even with relatively thin films. In contrast, NEM, which is a weaker base (pK_a of conjugate acid = 7.8) and has a higher boiling point (139°C), can diffuse relatively more rapidly through the film but volatilizes more slowly from the surface. These factors result in a more nearly uniform amine gradient and more uniform timing of curing of surface and lower layers of a film. 2-(Dimethylamino)ethanol has an intermediate base strength (pK_a = 9.3) and a relatively high boiling point (134°C), and exhibits intermediate behavior between that of TEA and NEM.

REFERENCES

1. J. L. Gerlock, D. F. Mielewski, D. R. Bauer, and K. R. Carduner, *Macromolecules*, **21**, 1604 (1988).

2. J. Reitberg, J. H. van Roon, and R. van der Linde, *Proc. XVth Intl. Conf. Org. Coat. Sci. Tech.*, Athens, Greece, (1989), p. 209.

3. K. E. J. Barrett, Ed., *Dispersion Polymerization in Organic Media*, Wiley, London, 1975.

4. L. W. Hill and K. Kozlowski, *J. Coat. Technol.*, **59** (751) 63 (1987).

5. T. Nakamichi and K. Shibato, *J. Jpn. Soc. Colour Mater.*, **59**, 592 (1986).

6. H. J. Spinelli, *Org. Coat. Appl. Polym. Sci.*, **47**, 529 (1982).

7. L. W. Hill and Z. W. Wicks, Jr., *Prog. Org. Coat.*, **10**, 55 (1982).

8. K. O'Driscoll, *J. Coat. Technol.*, **57** (705), 57 (1983).

9. V. R. Kamath and J. D Sargent, Jr., *J. Coat. Technol.*, **59** (746), 51 (1987).

10. R. A. Gray, *J. Coat. Technol.*, **57** (728), 83 (1985).

11. R. Buter, *J. Coat. Technol.*, **59** (749), 37 (1987).

12. J. E. Nylund and S. Pruskowski, *Resin Rev.*, **39**, 17 (1989).

13. *Technical Bulletin on Joncryl 500, PSN 0-104-12/83*, S. C. Johnson & Son, Inc., Racine, WI, 1983.

14. S. N. Lewis and R. A. Haggard, U. S. Patent 4,103,093 (1978).

15. R. A. Haggard and S. N. Lewis, *Prog. Org. Coat.*, **12**, 1 (1984).

16. D. Y. Sogah, W. H. Hertler, O. W. Webster, and G. M. Cohen, *Macromolecules*, **20**, 1473 (1987).

17. J. A. Simms and H. J. Spinelli, *J. Coat. Technol.*, **59** (752), 125 (1987).

18. R. P. Quirk and G. P. Bidinger, *Polym. Bull.*, **22**, 63 (1989).

19. Y. Hosokawa, M. Kuroki, T. Aida, and S. Inoue, *Macromolecules*, **24**, 824 (1991).

20. B. C. Watson and Z. W. Wicks, Jr., *J. Coat. Technol.*, **55** (698), 59 (1983).

21. L. W. Hill and B. M. Richards, *J. Coat. Technol.*, **51** (654), 59 (1979).

22. Z. W. Wicks, Jr., E. A. Anderson, and W. J. Culhane, *J. Coat. Technol.*, **54** (688), 57 (1982).

23. L. W. Hill and Z. W. Wicks, Jr., *Prog. Org. Coat.*, **8**, 161 (1980).

24. Z. W. Wicks, Jr., and G. F. Chen, *J. Coat. Technol.*, **50** (638), 39 (1978).

CHAPTER VIII

Polyester Resins

In the fibers and plastics fields the term "polyester" is applied to relatively high molecular weight, partially crystalline, linear, thermoplastic polymeric esters of a short-chain diol and terephthalic acid. In the coatings field, however, the term is used differently. First, it is used to describe a different sort of material—most coatings polyesters have relatively low molecular weights and are amorphous, branched, and ultimately cross-linked. Second, it is applied only to certain polyesters, those prepared from polyols and polybasic acids. Alkyds are also polyesters in the chemical sense of the word, but they are not called polyesters in coatings. As discussed in Chapter 10, alkyds are prepared from polyols and dibasic acids, and, in addition, monobasic acids. The latter are usually derived from vegetable oils. Coatings polyesters are sometimes called "oil-free polyesters" to distinguish them from alkyds.

All polyesters are prepared by step-growth polymerization (see Section 2.2). Often this process involves reaction of a carboxylic acid with an alcohol, called direct esterification. This reaction is acid catalyzed. In the absence of a strong acid catalyst, the rate (k_{ester}) is third order in reactants, as shown in Eq. 8.1, with one carboxyl group reacting with the alcohol and the second catalyzing the reaction. Equation 8.1 disregards the reverse reaction.

$$k_{ester} = k \, [RCOOH]^2 \, [R'OH] \qquad (8.1)$$

Because of the second-order dependence on acid concentration the rate decreases precipitously as the reaction proceeds. For example, polyesterification of diethylene glycol with adipic acid at 166°C without a catalyst is 60% complete in 1 h but requires 27 h to reach 94.5% conversion and would require years to reach 99.8% conversion [1]. This impractical situation can be alleviated somewhat by adding a strong acid or other catalyst so that the reaction becomes second order in reactants and first order in catalyst. Even with catalysts, polyesterifications start readily but must be driven to completion.

The kinetics of ideal step-growth polyesterification can be easily analyzed in terms of p, the fractional extent of reaction; n_p, the number fraction of molecules of differing degrees of polymerization; $\overline{P}_n$, the degree of polymerization; and w_p, the weight fraction of molecules when only difunctional monomers are used [2,3].

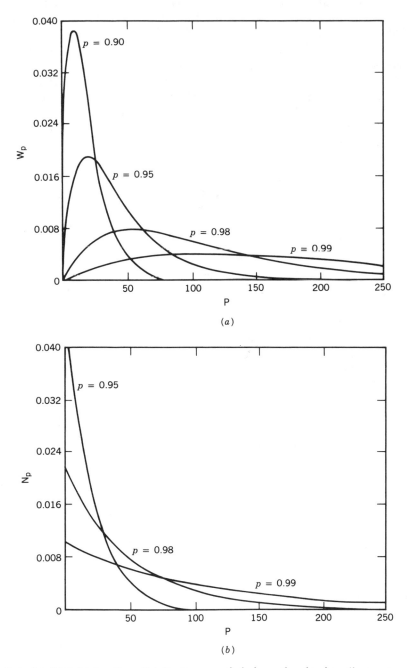

Figure 8.1. (*a*) Weight fraction distribution w_P of chain molecules in a linear step-growth polymer for several extents of reaction p. (*b*) Number or mole fraction distribution n_P. (From Ref. [3], with permission.)

As p increases the degree of polymerization builds up slowly at first—at $p = 0.5$ (corresponding to 50% conversion), $\overline{P}_n$ is only 2. The $\overline{P}_n$ is only 10 at $p = 0.9$ and $p = 0.998$ is required to reach a $\overline{P}_n$ of 500. Thus, with difunctional monomers, high molecular weight can only be attained when the mole ratio of COOH/OH is 1 and when esterification is driven beyond $p = 0.99$. This is difficult because of the decreasing reaction rate at high values of p. Note, as shown in Figure 8.1a, that the number of unreacted monomer molecules remains higher than that of any other single species in the reaction mixture no matter how high p becomes. As shown in Figure 8.1b, $\overline{P}_n$, the peak of the P distribution curve, only reaches substantial levels at high p values. Another noteworthy theoretical conclusion is that the polydispersity $\overline{M}_w/\overline{M}_n$ equals 2 for ideal linear step-growth polymers of high molecular weight.

The relatively low molecular weight, amorphous, branched, cross-linkable polyesters used in coatings are usually made from mixtures of diols, triols, and dibasic acids. Most commonly, excess polyol is used, hence the polyesters are *hydroxy-terminated polyesters*. They are most commonly cross-linked with MF resins or polyisocyanates. Resins with excess carboxylic acid are also made; these *carboxylic acid-terminated polyesters* are cross-linked with epoxy resins, MF resins, or 2-hydroxyalkylamides. Water-reducible polyesters are synthesized with both terminal hydroxyl and carboxylic acid groups; they are usually cross-linked with MF resins.

Branched polyesters are made from monomer mixtures that include one or more monomers having functionality $F > 2$. Kinetic analysis of such polymerizations is complex, but a few generalizations can be pointed out. As the proportion of monomer of $F > 2$ increases, $\overline{M}_n$, $\overline{M}_w/\overline{M}_n$, and the number average functionality $\bar{f}_n$ all increase. The average F of the monomer mixture must be controlled to avoid gelation at high extents of reaction.

In very general terms, thermosetting polyesters give coatings with better adhesion to metal substrates and somewhat lower cost than thermosetting acrylics (TSAs) (see Chapter 7). On the other hand, TSAs give coatings with superior water resistance and exterior durability. The differences result primarily from the presence of the ester linkages in the backbone of polyesters—linkages that impart flexibility but are vulnerable to hydrolysis. Thus, polyesters tend to be used for one-coat coatings on metal, and TSAs tend to be used in applications for which exterior durability and moisture resistance are particularly important, often with primers that provide excellent adhesion of the coating system to the metal substrate. Many exceptions to these broad generalizations can be found.

8.1. HYDROXY-TERMINATED POLYESTERS FOR CONVENTIONAL SOLIDS COATINGS

Most hydroxy-terminated polyesters are made from four raw materials: a diol, a triol, an aromatic dibasic acid (or its anhydride), and an aliphatic dibasic acid. Tribasic acids are also used to some extent. The ratio of moles of dibasic acid to moles of polyol must be less than 1 to give the terminal hydroxyl groups and avoid gelation. Molecular weight is also controlled by this ratio; the smaller the ratio, the lower the molecular weight. The molecular weight distribution $\overline{M}_n$ and $\bar{f}_n$ are all controlled by the diol/triol ratio; $\bar{f}_n$ is especially critical because it affects the potential cross-link density after cure. With the same mole ratio of dibasic acid/

total polyol, and increase in triol component will increase the average number of hydroxyl groups per molecule, decrease the hydroxy equivalent weight, broaden molecular weight distribution, and increase the cross-link density of a fully cross-linked film. The ratio of aromatic to aliphatic dibasic acids is the principal factor controlling the $\overline{T}_g$ of the resin, but differences in polyol structure will also affect $\overline{T}_g$.

The monomers are esterified at 220 to 240°C while removing water. Organotin compounds or titanium orthoesters are frequently used as catalysts. The reaction is continued until a high conversion has been reached. Acid numbers of resins to be cross-linked with MF resins are generally 5–10-mg of KOH per gram of resin solids; those for use with polyisocyanates are commonly esterified until the acid number is less than 2.

Polyesters made from polyfunctional reactants can cross-link during polymerization causing gelation, a very costly occurrence in large scale production. Since the probability of gelation increases with conversion, and high conversions are generally desirable, a relationship between formula composition, extent of reaction, and probability of gelation would be useful. Some progress has been made in predicting gelation in simple systems in which the reactivity of all hydroxyl groups and of all carboxylic acid groups are equal; but even with these "simple" systems, the equations give only fair predictions since they do not take into consideration the formation of cyclic esters. See Refs. [2] or [3] for discussions of such equations. In the case of polyesters used in coatings, the situation is far from simple; the reactivity of the different functional groups commonly varies substantially, and side reactions occur. Therefore, the textbook gelation equations are of little value for polyesters used in coatings. If the mole ratio of dibasic acid/polyol exceeds, or even closely approaches, 1, gelation may occur before esterification is essentially complete. This is seldom a problem in making hydroxy-functional polyesters for coatings unless an error of loading or processing is made, but it can become a critical consideration in the synthesis of alkyds (see Chapter 10).

Recently, Misev [4] published equations that permit the calculation of ratios of polyols and polybasic acids to use in order to make polyesters with a desired $\overline{M}_n$ and $\overline{f}_n$ at a desired extent reaction of the acid groups. The equations are complex and require iterative computer solution. The equations are simplified if the choice of raw materials is limited, as it generally is, to diols, triols, and dibasic acids, and if essentially all the acid groups are to be esterified. In commercial practice, polyesters are cooked until the acid number is reduced to about 5, a level that corresponds to the esterification of about 99% of the carboxylic acid groups. Therefore, the assumption of complete esterification is reasonable.

The simplified equations follow; N_{p2} represents the number of moles of diol; N_{p3}, the number of moles of triol; and N_a, the number of moles of dibasic acid. M_{p2} represents the molecular weight of the diol; M_{p3}, the triol; and M_a, the dibasic acid. If more than one dibasic acid (or other component) is used, then additional terms would be added to the equation for each additional component.

In making hydroxy-terminated polyesters, the number of moles of polyol is one more than that of dibasic acid, as shown in Eq. 8.2.

$$N_{p2} + N_{p3} = N_a + 1 \qquad \text{or} \qquad (8.2)$$

$$N_a = N_{p2} + N_{p3} - 1 \qquad (8.3)$$

The relationship between the average functionality and the moles of reactants is expressed by Eq. 8.4.

$$\bar{f}_n = 2N_{p2} + 3N_{p3} - 2N_a \tag{8.4}$$

Substituting the value of N_a from Eq. 8.3 into Eq. 8.4 gives Eq. 8.5.

$$\bar{f}_n = N_{p3} + 2 \quad \text{or} \tag{8.5}$$

$$N_{p3} = \bar{f}_n - 2 \tag{8.6}$$

When the esterification is carried out 2 mol of water (total molecular weight = 36) are split out for each mole of dibasic acid, resulting in Eq. 8.7 for $\overline{M}_n$. If an anhydride is used, only 1 mol of water ($M = 18$) will be split out for each mole of anhydride and the appropriate adjustment in Eq. 8.7 would be needed.

$$\overline{M}_n = N_{p2}M_{p2} + N_{p3}M_{p3} + N_a(M_a - 36) \tag{8.7}$$

Substituting in Eq. 8.7 for N_a using Eq. 8.3 and substituting for N_{p3} using Eq. 8.6 gives the simplified Eq. 8.8 for $\overline{M}_n$.

$$\overline{M}_n = N_{p2}M_{p2} + (f_n - 2)M_{p3} + (N_{p2} + f_n - 3)(M_a - 36) \tag{8.8}$$

Using these equations, one can calculate how much diol, triol, and dibasic acid should be used in an initial laboratory cook, for a desired $\overline{M}_n$ and $\bar{f}_n$.

A wide range of polyesters has been used commercially. In conventional poly-ester–MF coatings, polyesters with $\overline{M}_n$s of the order of 2000 to 6000, $\overline{M}_w/\overline{M}_n$ of the order of 2.5 to 4, and with $\bar{f}_n$ of 4 to 10 hydroxyl groups per molecule are usual. Table 8.1 gives an example of an initial laboratory formulation calculated for a polyester with an $\overline{M}_n$ of 5000 and an $\bar{f}_n$ of 5 using Eqs. 8.2–8.8. Neopentyl glycol is NPG, trimethylolpropane is TMP, isophthalic acid is IPA, and adipic acid is AA. Structural formulas for these compounds are given in Sections 8.1.1 and 8.1.2.

A formulation such as that in Table 8.1 does not take into consideration all the complexities of polyesterification; thus it can only be expected to provide a starting point for experimentation. A common complication is partial loss of monomer from the reactor by volatilization. The diol is usually the most volatile monomer, so it is generally necessary to use some excess diol. The amount of excess diol depends on the particular reactor and conditions; the efficiency of separating water and diol, the rate of flow of inert gas, the temperature of reaction, and so forth. A formula once established in the laboratory will have to be adjusted when production is scaled up in a pilot plant, again when it is scaled up in a production reactor, and when production is shifted from one reactor to another.

Esterification is a reversible reaction, as shown in Eq. 8.9. The rate of the reaction consequently depends on the rate of removal of the water, particularly near the end of the process. A few percent of a "reflux solvent," such as xylene, is added to accelerate removal of water and to assist in the return of the volatilized diol to the reactor. Polyols can undergo self-condensation to form polyethers, which

Table 8.1. Formulation for a Conventional Polyester Resin

Raw Material	Weight	Moles	Equivalents
NPG	2050	19	38
TMP	270	3	9
IPA	1740	10.5	21
AA	1530	10.5	21

also produces water. Ester groups that form early in the process may hydrolyze or transesterify during the process and reform many times leading to a mixture of kinetic and thermodynamic control of the structure of the final product.

$$\text{RCOOH} + \text{R'OH} \xrightleftharpoons{\text{Catalyst}} \text{RCOOR'} + \text{H}_2\text{O} \tag{8.9}$$

8.1.1. Selection of Polyols

Polyols are selected on the basis of cost, rate of esterification, stability during high temperature processing (minimal decomposition and discoloration), ease of separation from water during processing, viscosity of esters at equal molecular weight and functionality, effect on T_g, rate of cross-linking with MF resins (or other cross-linkers), and the hydrolytic stability of their esters. Obviously, compromises are necessary. The following polyols are representative of the many available.

The most widely used diol is probably NPG and the most widely used triol is TMP. Since both have neopentyl structures, they are relatively difficult to esterify. It is a fair but not always accurate generalization that the slower the rate of esterification, the greater the resistance of the ester to hydrolysis. The hydrolytic stability of the esters of NPG and TMP is significantly better than those from less hindered glycols like ethylene glycol or propylene glycol. Neopentyl glycol (bp 213°C) is volatile at esterification temperatures, so that well-designed fractionating and condensing equipment is needed to permit the removal of water with a minimum loss of NPG during processing at 220 to 240°C. Most diols start decomposing relatively rapidly in the presence of strong acids at temperatures above about 200°C, so strong acid esterification catalysts should be avoided. Organotin compounds and titanium orthoesters are examples of appropriate catalysts.

While the kinetics and mechanisms of esterification and hydrolysis reactions have been extensively studied from a very fundamental point of view, the literature contains little basic data dealing with the polyols of most interest in coatings resins. Turpin [5] proposed that a variant of Newman's "rule of six" may be useful in predicting the relative hydrolytic stability of esters. Newman's rule states that if one numbers the atoms in the structure of an ester starting with the carbonyl oxygen as 1, the larger the number of substituents in the 6 positions, the better the hydrolytic stability. Turpin suggested that substituents in the 7 position might also affect the stability and proposed an empirical "steric factor":

$$\text{Steric factor} = 4 \text{ (No. 6-position atoms)} + \text{(No. 7-position atoms)}$$

For example, for an NPG ester:

3 No. 6 atoms × 4 = 12
9 No. 7 atoms   = 9
Total = 21

The value of 21 for NPG esters is the same for TMP esters, but for ethylene glycol and 1,6-hexanediol esters the steric factors are only 13 and 15, respectively. Turpin's steric factor seems to correlate reasonably well with experience and with the limited published data for esters of acyclic alcohols. Turpin pointed out that another factor, the "anchimeric" or "neighboring group" effect may also be involved. For example, the location of the hydroxyl group in a half ester of ethylene glycol may promote the hydrolysis of its neighboring ester through a cyclic transition state and contribute to the poorer hydrolytic stability of ethylene glycol esters as compared to those of 1,6-hexanediol [5].

Esters of cycloaliphatic diols [e.g., cyclohexanedimethanol (CHDM)], seem more resistant to hydrolysis than would be predicted by Turpin's steric factor. A recent paper reported comparing results of testing films of coatings made with a series of polyesters from several polyols [6]. Coatings based on CHDM polyesters cross-linked with MF resins showed the best test results for properties related to hydrolytic stability.

8.1.2. Selection of Polyacids

Most polyesters contain a mixture of aromatic and aliphatic diacids. The ratio of aromatic to aliphatic acid is the major factor that controls the T_g of the resin. Aromatic acid esters hydrolyze more slowly than aliphatic esters, unless there is an anchimeric effect. Sterically hindered carboxylic acid groups esterify more slowly and their esters are generally more stable to hydrolysis.

Phthalic Anhydride
PA

Isophthalic acid
IPA

Hexahydrophthalic anhydride
HHPA

HOOC COOH

Adipic acid
AA

$(CH_2)_7COOH$
$(CH_2)_7COOH$
$CH_2CH=CH(CH_2)_4CH_3$
$(CH_2)_3CH_3$

Dimer fatty acid
(One isomer of mixture)

HOOC COOH

Azelaic acid
AZA

Isophthalic acid (IPA) is the predominant aromatic acid even though phthalic anhydride (PA) can be processed at lower cost. The preference for IPA is based on superior exterior durability of coatings made with IPA polyesters; this has been attributed to the greater hydrolytic stability of isophthalic esters. It is well established that half esters of (*ortho*)phthalic acid are more easily hydrolyzed than half esters of isophthalic acid in the pH range of 4 to 8 [7]. This difference results from the anchimeric effect of the carboxylic acid group of the phthalic acid half ester in a location that increases the catalytic effect of the acid on the hydrolysis of the ester. Presumably during exterior exposure, resistance to hydrolysis in contact with rain water, which has a pH in the range of 4 to 6, is most important. Under these conditions, IPA polyesters are more resistant than phthalic acid polyesters. On the other hand, both mono- and diesters of IPA hydrolyze more rapidly than those of phthalic acid under alkaline conditions [8].

The processing cost advantage of PA results from its lower melting point (131°C); it can be handled as a molten liquid, it is readily soluble in the reaction mixture, and its reactive anhydride structure rapidly forms monoesters at about 160°C. On the other hand, IPA (mp > 300°C) is more difficult to process because it dissolves slowly in the reaction medium and its lower reactivity slows the processing. These problems are even more acute with terephthalic acid. Due to the required longer time at high temperatures, volatilization losses and acid catalyzed side reactions (e.g., etherification or dehydration of polyols) are more serious problems than with PA.

Adipic acid is probably the most widely used aliphatic dibasic acid. Succinic and glutaric acids give esters with inferior hydrolytic stability presumably due to the anchimeric effect [5]. Longer chain acids such as azelaic and sebacic acids may give somewhat better hydrolytic stability and greater reduction of T_g (on a molar basis) as compared with adipic acid but are more expensive. Dimer acids, derived from dimerization of drying oil fatty acids and, hence, predominantly C_{36} acids, are widely used. Fatty acid dimerization will be discussed in Section 9.3.1. Dimer acids are relatively inexpensive and are available in various grades. The high quality grades of dimer acids normally used in polyesters have very little monobasic or tribasic acid contaminants and have been hydrogenated to saturate any double bonds, which could be a weak point for photodegradation. Dimer acid–isophthalic polyesters impart an excellent balance of properties to MF cross-linked coil coatings, providing high resistance to marring combined with excellent resistance to cracking during fabrication of coated metal.

Cycloaliphatic dibasic acids or anhydrides can be used. Hexahydrophthalic anhydride (HHPA) is said to give exterior durability comparable to IPA. It has been speculated that HHPA half esters may be less susceptible to anchimeric effects than phthalic acid half esters. The esters may isomerize to the E (trans) isomers during processing. 1,4-Cyclohexanedicarboxylic acid is another example; it is usually used as the more easily handled dimethyl ester, which is polymerized by transesterification with polyols.

8.2. POLYESTER RESINS FOR HIGH SOLIDS COATINGS

Low viscosity polyesters are essential for the reduction of VOC emissions. The viscosity of concentrated polyester solutions depends on several variables. Molecular weight and molecular weight distribution are two obviously important factors. The number of functional groups per molecule also affects the viscosity; increasing the number of hydroxyl groups (and to an even greater extent the number of carboxylic acid groups) increases the probability of intermolecular hydrogen bonding, which in turn increases viscosity. The hydrogen-bond effect can be minimized, but not eliminated, by using hydrogen-bond acceptor solvents such as ketones. The T_g of the resin will also affect viscosity at a given concentration. For this reason, reducing the ratio of aromatic/aliphatic dibasic acids and use of acyclic polyols rather than cyclic polyols, such as CHDM [6], give lower viscosities, but there will be a lower limit of T_g below which desired film properties cannot be attained.

There is a distinct advantage of polyesters over acrylics in very high solids coatings. In contrast to the difficulty of making low molecular weight TSAs with at least two hydroxyl groups on essentially all molecules, the synthesis of polyesters with two or more hydroxyl groups on essentially all of the molecules is relatively easy. With the low mole ratios of dibasic acids to polyols (2:3 is typical) required to make low molecular weight resins, the probability of cyclization reactions to form nonfunctional materials is small. Virtually all of the final reactions will result in terminal hydroxyl groups. While the lowest viscosity will be obtained with difunctional resins, such resins also require the most careful formulation and curing to achieve good film properties. If a hydroxyl group is left unreacted on a conventional polyester with an average of five or so hydroxyl groups per molecule, the effect on film properties would probably be relatively small. However, the effect of failing to react all of the hydroxyl groups on a resin with only two hydroxyl groups per molecule would obviously be magnified. A compromise is again required. Commonly, an average $\bar{f}_n$ of between two and three hydroxyl groups per molecule is used.

The major factor controlling molecular weight ($\overline{M}_n$) of a polyester is the mole ratio of dibasic acid/polyol. A ratio of the order of 2:3 is typical. Loss of polyol during the production of the polyester can result in a significantly higher molecular weight than predicted from the starting ratio. Condensers are designed to remove water rapidly, while returning a high fraction of NPG (or other polyol), but the condensers used in coatings resin production are not completely efficient. It is necessary to add some extra glycol to compensate for losses. Since production equipment varies, the final decision on the mole ratio to load must be based on experience in the particular reactor and with the particular formula. The amount of glycol being lost can be estimated by checking the refractive index of the water being removed from the reaction.

A factor that must be considered in designing a high solids polyesters is that the low molecular weight fraction of the resin may be volatile enough to evaporate when a thin film is baked. Such loss has to be counted as part of the VOC emissions together with solvent loss. Belote and Blount [9] studied this volatility both in the absence of cross-linking agent and in its presence. They made a series of model resins from NPG and a 1:1 mol mixture of AA and IPA; the polyacid/polyol ratio was varied from 1:2 to 1:1.15. The presence of the cross-linker (Class I MF resin), as would be expected, reduced the volatile loss. They concluded that an $\overline{M}_n$ of about 800 to 1000 was optimum for achieving lowest VOC with resins having normal molecular weight distributions. At $\overline{M}_n$ around 600, so much of the low molecular weight fraction volatilized that the total VOC was higher than at an $\overline{M}_n$ of 800, *even though the amount of solvent required was lower.*

A broad molecular weight distribution not only increases volatile losses because of the low molecular weight material, but also gives a higher viscosity because of the disproportionate effect on the viscosity of the high molecular weight fraction. Substantial improvement should result from making as narrow molecular weight distribution as possible.

Utilization of 2,2,4-trimethyl 1,3-pentanediol (TMPD) in combination with a two-stage addition of the triol TMP during the synthesis of the polyester, leads to lower viscosity and also to improved film properties in the cured film [10]. Use of TMPD, an unsymmetrical diol, presumably gives narrower molecular weight distribution because of the differential reactivity of the two hydroxyl groups. A drawback of this approach is the need for somewhat higher baking temperatures for coatings, presumably a result of the low reactivity of the hindered terminal hydroxyl groups derived from TMPD.

Three of the resin formulations used in this study are given in Table 8.2. Addition of the TMP in two stages resulted in a substantial reduction in $\overline{M}_n$ from 1500 to

Table 8.2. High Solids Polyester Formulations

Raw Material	Equivalents		
	Resin 1	Resin 2	Resin 3
TMPD	11.96	11.96	12.24
TMP	1.72	0.86	0.88
IPA	4.56	4.56	4.38
AA	4.56	4.56	4.38
Second Stage			
TMP		0.86	0.88
Resin $\overline{M}_n$	1500	1075	1000
Coating NVW	75.6	76.5	77.6

1100 (compare Resins 1 and 2) with a corresponding reduction in viscosity. Further reduction was achieved by adjusting the monomer ratios (Resin 3).

A difunctional polyester with a narrow molecular weight distribution and a reported $\overline{M}_n$ of 425 is commercially available [11]. This resin is presumably manufactured by preparing a polyester with excess CHDM and then distilling off the low molecular weight fraction with a vacuum thin-film evaporator. Low viscosity polyester diols and triols prepared by reacting caprolactone with a diol or a combination of diol and triol are also available [12]. These types of low $\overline{M}_n$ resins seldom provide satisfactory coatings when used by themselves, but they are useful in blends to increase the solids content of somewhat higher molecular weight polyester or acrylic based coatings. For this reason they are often called "reactive diluents."

8.3. CARBOXYLIC ACID-TERMINATED POLYESTER RESINS

While the majority of polyesters used in coatings are hydroxy-terminated, there are also significant uses for those terminated with carboxylic acid groups. Such resins are copolymers of a diol and a molar excess of dibasic acid. A modest amount of a tri- or polyfunctional monomer, most commonly trimellitic anhydride (TMA), is almost always added to increase $\bar{f}_n$. Carboxylic acid functional polyesters can be cross-linked with MF resins, but in most cases they are cross-linked with epoxy resins or with 2-hydroxyalkylamides, as discussed in Sections 11.3.1 and 13.4.1, respectively.

With high ratios of high melting point diacids like IPA, it is usually preferable to use a two-step process. The low solubility and high melting point of IPA prevents clean reaction when it is used in excess. First, one prepares a hydroxy-functional polyester with the IPA and then reacts the terminal hydroxyl groups with the lower melting acid or anhydride. Careful control is needed to assure reproducible products since varying degrees of transesterification occur in the second stage. Final product structure is governed both by kinetic and thermodynamic factors; thus corresponding resin properties are affected by reaction time and other process variables.

8.4. WATER-REDUCIBLE POLYESTER RESINS

Polyesters have also been adapted for use in water-borne coatings with reduced VOC. Such polyesters have both hydroxyls and carboxylic acids as terminal groups. As with other water-reducible resins, acid numbers in the range of 40 to 60 are required to give amine salt solutions in solvent, that can be diluted with water to give reasonably stable dispersions of aggregates of resin molecules swollen with water and solvent. These resins show abnormal viscosity dilution curves similar to those described for water-reducible acrylic resins in Section 7.3.

In theory, one could prepare such water-reducible polyesters by simply stopping the esterification of a combination of diol, triol, AA, and IPA at the desired acid number; in practice it is almost impossible to do so reproducibly. Rather, it is necessary to find combinations of polybasic acids that have sufficiently different reactivity to permit the control of the ratio of unreacted hydroxyl and carboxyl groups. The most widely used method has been to prepare a hydroxy-functional resin and then add enough TMA to esterify a fraction of the hydroxyl groups, generating two carboxyl groups at each site. Table 8.3 gives an example of a formulation for preparing a water-reducible polyester [13]. Advantage is taken of the much higher reactivity of the anhydride groups of the TMA at 180°C, relative to the carboxylic acid groups formed in the reaction.

This approach has been used on a large scale, but it has disadvantages. Most seriously, the ester linkages in the resin are vulnerable to hydrolysis during storage of the formulated coating. The ester group of partially esterified TMA is particularly subject to hydrolysis because of the anchimeric effect of the adjacent carboxylic acid group. Hydrolysis of the TMA partial ester will result in removal of the solubilizing carboxyl groups and, thereby, destabilize the resin dispersion. The film properties may also be adversely affected. In addition, the use of primary alcohols as a solvent, as recommended in Table 8.3, has been found to be undesirable. Primary alcohols have been found to esterify carboxylic acid groups and transesterify ester groups during the thinning at 160°C and, at a slow but appreciable rate, during storage of the resin. This problem can be minimized by use of a secondary alcohol as solvent, along with careful control of process temperatures.

An alternative approach that provides somewhat better, but still limited, hydrolytic stability is to use 2,2-dimethylolpropionic acid as one of the diol components. The carboxylic acid group of this monomer is highly hindered by being located on a tertiary carbon. The resulting differential reactivity makes it possible to esterify the hydroxyl groups while leaving most of the acid groups unreacted. While they are too hindered to esterify readily, the acid groups are accessible to protons and readily form salts.

An unavoidable problem with water-reducible polyesters results from intramolecular reaction of terminal hydroxyl and carboxylic acid groups to form some low molecular weight nonfunctional cyclic polyesters. When coatings are baked, a small amount of the cyclic esters can volatilize out of the coating and gradually accumulate in cool spots in the oven. Eventually, sufficient resin can accumulate to drip down on products going through the ovens, seriously marring the finish. Since the amounts are small, dripping may only start after weeks or months of operation of the coating line.

While water-reducible polyester coatings are used on a reasonable scale commercially, problems of hydrolytic stability and cyclic esters have caused many users

Table 8.3. Water-Reducible Polyester Formulation

Materials	Weight	Mole Ratio	Equivalent Ratio
NPG	685	1.0	1.0
Adipic acid	192	0.2	0.2
TMA	84	0.067	0.1
IPA (85:15 IPA/TPA)	655	0.6	0.6
React at 235°C until the acid number is 16–18, cool 180°C, then add			
TMA	84	0.067	0.1

React at 180°C until the acid number is 40–45, cool to 160°C, and thin to 80% solids with diethylene glycol monobutyl ether.

to favor coatings made from water-reducible acrylic resins, which do not exhibit these problems.

8.5. POLYESTER RESINS FOR POWDER COATINGS

Powder coating polyester resins are brittle solids with a relatively high T_g (50–60°C), so that the powder coating will not sinter (partially fuse) during storage. These requirements are met with terephthalic acid (TPA) and NPG as the principal monomers. Smaller amounts of other monomers are added to increase $\bar{f}_n$ and to reduce T_g to the desired level. Widely used commercial products are amorphous, not crystalline [14]. Both hydroxy- and carboxy-terminated polyesters are widely used. The former are most commonly cross-linked with blocked isocyanates (Section 12.4.4) and the latter with epoxy resins (Chapter 11). Other cross-linkers include 2-hydroxyalklylamides (Chapter 13) and tetramethoxymethylglycouril (Chapter 6). (See Chapter 31 on Powder Coatings.)

Processes involving direct esterification of TPA are used, but since TPA has a very high melting point, it is common to make the hydroxy-terminated polyester by transesterification with dimethyl terephthalate. An appropriate transesterification catalyst is tetraisopropyl titanate. If desired the hydroxy-terminated TPA polyester can then be reacted with the other polyacids to form the carboxylic acid-terminated product in the second stage.

REFERENCES

1. P. J. Flory, *J. Am. Chem. Soc.*, **61**, 3334 (1939).
2. G. Odian, *Principles of Polymerization*, 3rd ed., Wiley, New York, 1991, pp. 40–125.
3. F. W. Billmeyer, Jr., *Textbook of Polymer Science*, 3rd ed., Wiley-Interscience, New York, 1984, pp. 25–48.
4. T. A. Misev, *J. Coat. Technol.*, **61** (772), 49 (1989).
5. E. T. Turpin, *J. Paint Technol.*, **47** (602), 40 (1975).
6. D. J. Golob, T. A. Odom, Jr., and R. W. Whitson, *Polym. Mater. Sci. Eng.*, **63**, 826 (1990).
7. M. L. Bender, F. Chloupek, and M. C. Neveu, *J. Am. Chem. Soc.*, **80**, 5384 (1958).
8. E. Kivinen and E. Tommila, *Suomen Kemistilehti*, **14B**, 7 (1941); G. V. Rao and N. Venkatasubramanian, *Aust. J. Chem.*, **24**, 201 (1971); A. Cambon and R. Jullien, *Bull. Soc. Chim. Fr.*, 2003 (1973).
9. S. N. Belote and W. N. Blount, *J. Coat. Technol.*, **53** (681), 33 (1981).
10. J. D. Hood, W. W. Blount, and J. T. Sade, *Proceedings of Water-Borne and Higher-Solids Coatings Symposium*, New Orleans, LA, 1986, p. 14.
11. L. J. Calbo, *Proc. Water-Borne and Higher-Solids Symp.*, New Orleans, LA (1986), p. 356.
12. Union Carbide, Specialty Polymers and Composites Division, *Tech Bull.*, TONE Polyols, 1986.
13. Amoco Chemicals Corp., *Tech. Bull. TMA-109e*, 1984.
14. F. N. Jones and G. Teng, unpublished results.

CHAPTER IX

Drying Oils

Among the oldest binders for paints are "drying oils," which are liquid vegetable or fish oils that can react with oxygen to form solid films. They have been used since prehistoric times, and, in the nineteenth and early twentieth centuries, binders of most paints were drying oils. Their use has decreased substantially; however, they still have some applications. Most importantly, they are raw materials for other binders such as alkyd resins (see Section 10.1), epoxy esters (sec 11.4), and uralkyds (see Section 12.3.1). These resins can, in many cases, be considered to be synthetic drying oils; an understanding of drying oil chemistry is a necessary foundation for understanding them. In this chapter drying oils themselves and modified drying oils are discussed. This chapter only summarizes the information available; see the general references for further details.

9.1. COMPOSITION OF NATURAL OILS

Naturally occurring oils and fats are *triglycerides*, triesters of glycerol and fatty acids. Some triglycerides are drying oils but many are not. The reactivity of drying oils with oxygen results from the presence of esters of fatty acids with two or more double bonds that are either separated by single methylene groups or are conjugated with each other. Esters of many different fatty acids occur in nature. Fatty acids with 18 carbon atoms are common; those found in oils and that are most important in coatings follow. The symbols Z and E stand for *zusammen* and *entgegen* and are equivalent to cis and trans, respectively.

Common Name	Structure
Stearic acid	$CH_3(CH_2)_{16}COOH$
Palmitic acid	$CH_3(CH_2)_{14}COOH$
Oleic acid	$CH_3(CH_2)_7CH{=}CH(CH_2)_7COOH$ (Z)
Linoleic acid	$CH_3(CH_2)_4CH{=}CHCH_2CH{=}CH(CH_2)_7COOH$ (both Z)
Linolenic acid	$CH_3CH_2CH{=}CHCH_2CH{=}CHCH_2CH{=}CH(CH_2)_7COOH$ (all Z)
Ricinioleic acid	$CH_3(CH_2)_5\overset{\underset{\mid}{OH}}{C}HCH_2CH{=}CH(CH_2)_7COOH$ (Z)
α-Eleostearic acid	$CH_3(CH_2)_3CH{=}CHCH{=}CHCH{=}CH(CH_2)_7COOH$ (E, Z, E)

The oils are mixtures of mixed triglycerides with different fatty acids distributed among the triglyceride molecules. An example of one of the many triglycerides in linseed oil follows:

A Triglyceride

Separation of the tens or hundreds of different triglyceride molecules present in an oil borders on the impossible. Therefore, oils are characterized by high-performance liquid chromatography (HPLC) or gas chromatographic (GC) analysis of the methyl esters obtained by transesterification [1]. Typical fatty acid contents of some widely used oils are given in Table 9.1. Compositions of the oils vary, sometimes quite widely, with variations in plant strain, climate, soil, and other

Table 9.1. Typical Fatty Acid Compositions of Selected Oils

Oil	Fatty Acids				
	Saturated[a]	Oleic	Linoleic	Linolenic	Other
Linseed	10	22	16	52	
Safflower	11	13	75	1	
Soybean	15	25	51	9	
Sunflower, MN	13	26	61	trace	
Sunflower, TX	11	51	38	trace	
Tung	5	8	4	3	80[b]
Tall oil fatty acids	8	46	41[c]	3	2[d]
Tall oil fatty acids	2.5	51	43[c]	trace	1.5[e]
Castor	3	7	3		87[f]
Coconut	91	7	2		

[a]Saturated fatty acids are mainly mixtures of stearic (C_{18}) and palmitic (C_{16}) acids; coconut oil also contains C_8, C_{10}, C_{12}, and C_{14} saturated fatty acids.
[b]α-Eleostearic acid.
[c]Linoleic plus geometric and conjugated isomers.
[d]Rosin.
[e]Unidentified.
[f]Ricinoleic acid.
Source: See Refs. 2–5.

growth conditions. Differences in sunflower oils are especially large as illustrated in Table 9.1 for sunflower oils from Minnesota and Texas. In general, oils derived from seeds grown in colder climates contain larger fractions of more highly unsaturated fatty acids as triglyceride esters.

Included in Table 9.1 is an important kind of fatty acid, called tall oil fatty acids (TOFAs). The word tall comes from the Swedish word for pine. Tall oil fatty acids are obtained as a byproduct of the sulfate pulping process for making paper. Oils present in the trees are saponified to soaps in the process. Tall oil fatty acids are obtained by acidification followed by fractional distillation to separate the fatty acids from rosin acids and other byproducts also present. These acids have a wide range of compositions; two examples are shown.

Animal oils and fats are also triglycerides, but the only animal oils used to any degree in coatings are fish oils. They contain triglycerides of a wide spectrum of fatty acids including C_{18}–C_{26} fatty acids with up to five nonconjugated double bonds separated by single methylene groups. They also contain large amounts of saturated fatty acid triesters. Raw fish oils are converted into useful drying oils by cooling and filtering off the solid, primarily saturated, triglycerides that freeze out. The composition of commercial fish oil varies widely depending on the variety of fish and the degree of removal of the saturated components.

9.2. AUTOXIDATION AND CROSS-LINKING

Oils are classified as drying oils, which form solid films on exposure to air; semidrying oils, which form tacky, that is, sticky, films; and nondrying oils, which do not show marked increase in viscosity on exposure to air. Drying oils are further differentiated as "nonconjugated" and "conjugated" oils depending on whether the double bonds in the predominant fatty acids are separated by a methylene group or are conjugated.

9.2.1. Nonconjugated Drying Oils

A useful empirical relationship is that nonconjugated oils having a "drying index" greater than 70 are drying oils [6]. The drying index is calculated as follows:

$$\text{Drying index} = (\% \text{ linoleic acid}) + 2(\% \text{ linolenic acid})$$

Based on this relationship a more general criterion for predicting drying capacity can be stated: Drying is related to the average number of methylene groups between double bonds ($CH=CHCH_2CH=CH$) per molecule. Empirically, if this number is greater than about 2.2, the oil is a drying oil; if it is moderately below 2.2, the oil is a semidrying oil; there is no sharp dividing line between semidrying and nondrying oils. These statements apply to synthetic drying oils as well as to natural oils. Note that linoleic and linolenic acids have one and two methylenes between double bonds, respectively.

Since the methylene groups between two double bonds are, in effect, reactive sites for cross-linking, it is convenient to relate the average number of such groups per molecule to the number average functionality $\bar{f}_n$ of the triglyceride or synthetic drying oil. We will do so to facilitate the ensuing discussion, although it is possible that some of the sites might be involved in more than one cross-linking reaction.

It is common to see definitions of drying, semidrying, and nondrying oils based on their "iodine value," that is, the number of grams of iodine required to saturate the double bonds of 100 g of an oil. Some authors classify oils as follows: drying oils, iodine value greater than 140; semidrying oils, iodine value, 125–140; and nondrying oils, iodine value less than 125 [7]. Although iodine values can serve as satisfactory quality control specifications, they are not useful, and can be misleading, as a means of defining a drying oil or for predicting reactivity.

As indicated above, the reactivity of nonconjugated drying oils is related to the average number of methylene groups between double bonds per molecule. Such methylene groups are activated by their allylic relationship to two double bonds and show much greater reactivity than methylene groups allylic to only one double bond. Evidence for these assertions is provided by the relative rates of autoxidation of synthetic triolein (glyceryl trioleate), trilinolein, and trilinolenein, which are 1:120:330, respectively [8]. The number of methylene groups between two double bonds ($\bar{f}_n$) of the three triglycerides are 0, 3, and 6, respectively; the theoretical iodine values are 86, 173, and 262. Clearly, the autoxidation rates are more closely related to the number of methylene groups between double bonds ($\bar{f}_n$) than they are to the iodine values, which are proportional to the average number of double bonds per molecule. Based on the data in Table 9.1, the $\bar{f}_n$ of the typical linseed oil is 3.6; it is a drying oil. The $\bar{f}_n$ of soybean oil is 2.07; it is a semidrying oil. The higher the $\bar{f}_n$ of a drying oil, the more rapidly a solvent-resistant, cross-linked film forms on exposure to air.

The reactions taking place during drying are complex. Most studies of the chemistry of drying were done without the benefit of modern analytical instrumentation, which has recently been brought to bear on this complex problem. For example, HPLC, NMR, and time-lapse Fourier transform infrared spectroscopy (FTIR) have been used to study the autoxidation of unsaturated fatty acids [5], as well as the drying of linseed oil and alkyd resins [9,10].

Cross-linked films form from linseed oil in the following stages: (1) an induction period during which naturally present antioxidants (mainly tocopherols) are consumed, (2) a period of rapid oxygen uptake with a weight gain of about 10% (FTIR shows an increase in hydroperoxides and appearance of conjugated dienes during this stage), and (3) a complex sequence of autocatalytic reactions in which hydroperoxides are consumed and the cross-linked film is formed. In one study [9], steps 1, 2, and 3 were far along in 4, 10, and 50 h, respectively, when catalyzed by a "drier" (see Section 9.2.2).

Cleavage reactions to form low molecular weight byproducts also occur during the latter stages of film formation. Slow continuing cleavage and cross-linking reactions through the lifetime of the film lead to embrittlement, discoloration, and slow formation of volatile byproducts. Oils with significant quantities of fatty acids with three double bonds, such as linolenic acid, discolor to a particularly marked degree.

The following scheme illustrates some of the many reactions that occur during cross-linking: Naturally present hydroperoxides decompose to form free radicals:

$$ROOH \longrightarrow RO\cdot + HO\cdot$$

At first, these highly reactive free radicals react mainly with the antioxidant, but as the antioxidant is consumed, the free radicals react with other compounds. Hydrogen atoms on methylene groups between double bonds are particularly susceptible to abstraction, yielding the resonance stabilized free radical **1**.

$$RO\cdot \ (or \ HO\cdot) \ + \ \ —CH{=}CH—CH_2—CH{=}CH— \ \longrightarrow$$

$$—CH{=}CH—\overset{\cdot}{\underset{\underset{H}{|}}{C}}—CH{=}CH— \ + \ ROH \ \ (or \ H_2O)$$

$$\mathbf{1}$$

Free radical **1** exists as three resonance contributors and reacts with oxygen to give predominantly a conjugated peroxy free radical such as **2**.

$$\overset{O\cdot}{\underset{\underset{\underset{\underset{H}{|}}{C}}{\overset{|}{O}}}{|}}$$
$$—\overset{}{C}—CH{=}CH—CH{=}CH—$$
$$\mathbf{2}$$

The peroxy free radicals can abstract hydrogen atoms from other methylene groups between double bonds to form additional hydroperoxides and regenerate free radicals like **1**. Thus a chain reaction is established resulting in autoxidation. At least part of the cross-linking occurs by radical–radical combination reactions forming C—C, ether, and peroxide bonds:

$$R\cdot \ + \ R\cdot \ \longrightarrow \ R{-}R$$
$$RO\cdot \ + \ R\cdot \ \longrightarrow \ R{-}O{-}R$$
$$RO\cdot \ + \ RO\cdot \ \longrightarrow \ RO{-}OR$$

These reactions correspond to termination by combination in free radical chain-growth polymerization (see Chapter 2).

Reactions analogous to the addition step in such polymerization could also produce cross-links. For example, the residual conjugated doubled bonds after intermediate **2** has cross-linked may undergo addition reactions:

$$R\cdot \ + \ —CH{=}CH—CH{=}CH— \ \longrightarrow \ —\overset{R}{\underset{\underset{H}{|}}{C}}—\overset{\cdot}{\underset{\underset{H}{|}}{C}}—CH{=}CH—$$

$$\mathbf{3}$$

Such reactions could yield C—C or C—O linkages depending on the structure of R·. Subsequently, free radical **3** can rearrange, add oxygen to form a peroxy free radical, abstract a hydrogen from a methylene between double bonds, combine with another free radical, or add to a conjugated double-bond system.

Recent studies of the reactions of ethyl linoleate with oxygen in the presence of driers (see Section 9.2.2) by ^{1}H and ^{13}C NMR indicated that the cross-linking reactions were only those that formed ether and peroxy cross-links; no new C—C bonds were detected [10]. These results with a model compound differ from the proposals that have been made for C—C cross-linking during the autoxidation of oils. Possibly the ratio of C—C/ether/peroxide bonds is related to the concentration of oxygen in the film as the film is forming. Further studies are needed to establish the structures of cross-links in drying oil films.

Rearrangement and cleavage of hydroperoxides to aldehydes and ketones, among other products, lead to low molecular weight byproducts. The characteristic odor of oil and alkyd paints during drying is attributable to such volatile byproducts, as well as to the odor of the organic solvents. This undesirable odor has been a factor motivating the replacement of oil and alkyds with latex vehicles for paints, particularly for interior applications. The reactions leading to odors have been extensively studied in connection with flavor changes of vegetable cooking oils [11]. Aldehydes have been shown to be major byproducts from the catalyzed autoxidation of methyl oleate, linoleate, and linolenate, as well as from curing of drying oil-modified alkyd resins [10,12].

9.2.2. Catalysis of Autoxidation and Cross-Linking

The rates at which nonconjugated drying oils dry are slow. Many years ago it was found that metal salts (driers) catalyze the drying rate. The most widely used driers are the oil-soluble cobalt, manganese, lead, zirconium, and calcium salts of octanoic or naphthenic acids. Salts of other metals, including rare earths, are also used. Cobalt and manganese salts, so-called "top driers" or "surface driers," primarily catalyze drying at the film surface. Lead and zirconium salts catalyze drying throughout the film and are called "through driers." Calcium salts show little, if any, activity alone but may reduce the amount of other driers needed. The surface-drying catalysis by cobalt and manganese salts presumably results from the catalysis of hydroperoxide decomposition:

$$Co^{2+} + ROOH \longrightarrow RO\cdot + OH^- + Co^{3+}$$

$$Co^{3+} + ROOH \longrightarrow ROO\cdot + H^+ + Co^{2+}$$

The net result is formation of water and a high concentration of free radicals that undergo reactions such as given above. Note that the cobalt cycles between the two oxidation states. The activity of through driers has not been adequately explained.

Combinations of metal salts are almost always used. Mixtures of lead with cobalt and/or manganese are particularly effective, but, as a result of toxicity control regulations, lead driers can no longer be used in consumer paints sold in interstate commerce in the United States. Combinations of cobalt and/or manganese with zirconium, and frequently with calcium, are commonly used. The amounts of driers needed are very system specific. Their use should be kept to the minimum possible level since they not only catalyze drying but also catalyze the postdrying embrittlement and discoloration reactions.

9.2.3. Conjugated Drying Oils

Oils containing conjugated double bonds, such as tung oil, dry more rapidly than any nonconjugated drying oil. Obviously, free radical polymerization of the conjugated diene systems can lead to chain-growth polymerization rather than just combination of free radicals to form cross-links. High degrees of polymerization are unlikely because of the high concentration of abstractable hydrogen atoms acting as chain-transfer agents. However, the free radicals formed by chain transfer also yield cross-links. In general, the water and alkali resistance of films derived from conjugated oils are superior, presumably because more of the cross-links are stable carbon–carbon bonds. However, since the α-eleostearic acid in tung oil has three double bonds, discoloration on baking or aging is severe.

9.3. SYNTHETIC AND MODIFIED DRYING OILS

Several types of chemical modification of drying oils are practiced. In addition to those discussed here, drying oil-modified alkyds, epoxy esters, and uralkyds are discussed in Sections 10.1, 11.4, and 12.4.1, respectively.

9.3.1. Heat Bodied Oils, Blown Oils, and Dimer Acids

Both nonconjugated and conjugated drying oils can be thermally polymerized by heating under an inert atmosphere to form so-called "bodied oils." Bodied oils have higher viscosities and are often used in oil paints to improve application and performance characteristics. Process temperatures may be as high as 300–320°C for nonconjugated oils and 225–240°C for conjugated oils [3], although the reactions occur at an appreciable rate at somewhat lower temperatures. At least in part, bodying may result from thermal decomposition of hydroperoxides, always present in natural oils, to give free radicals resulting in a limited degree of cross-linking. It has also been shown that thermal rearrangement to conjugated systems occurs followed by Diels–Alder reactions that lead to formation of dimers [13]. Since tung oil has a high concentration of conjugated double bonds, it undergoes thermal polymerization much more rapidly than nonconjugated oils (e.g., linseed oil). In fact one must be careful to control the heating of tung oil or the polymerization will lead to gelation.

Viscosity of drying oils can also be increased by passing air through oil at relatively moderate temperatures, 140–150°C to produce *blown oils* [3]. Presumably reactions similar to those involved in cross-linking cause autoxidative oligomerization of the oil.

Polyunsaturated acids will also dimerize or oligomerize by heat treatment; the reactions are acid catalyzed. For example, the doubly unsaturated fatty acids of TOFA can be dimerized or oligomerized to a major degree followed by removal of the residual monobasic acids by distillation. Under the high reaction temperatures, some decarboxylation occurs. Decarboxylation can be minimized by heating under pressure in the presence of a small percentage of water and activated clay. The products obtained are called "dimer acids." They are predominantly C_{36} dicarboxylic acids, with small fractions of monocarboxylic acid and some C_{54} tricarboxylic acids. For some uses, it is desirable to eliminate residual double bonds by hydrogenation. Dimer acids are used in making polyesters as discussed in Section 8.1.2 and polyamides as discussed in Section 11.2.2.

9.3.2. Varnishes

The drying rate of drying oils can be increased by dissolving a solid resin in the oil and diluting with a hydrocarbon solvent. Such a solution is called a "varnish." The solid resin serves to increase the T_g of the solvent-free film so that film hardness is achieved more rapidly. There is not an increase in the rate of cross-linking so the time required for the film to become solvent resistant is not shortened. Essentially, any high melting thermoplastic resin that is soluble in drying oil will serve the purpose. The higher the *melt point* of the resin, the greater is the effect on drying time. Naturally occurring resins, such as congo, copal, damar, and kauri resins, synthetic resins, such as ester gum (glyceryl esters of rosin), phenolic (see Section 11.8), and coumarone–indene resins, have been used.

In varnish manufacture, the drying oil (usually linseed oil, or tung oil, or mixtures of the two) and the resin are cooked together to a high temperature to obtain a homogeneous solution of the proper viscosity. The varnish is then thinned with hydrocarbon solvents to application viscosity. During cooking, some dimerization or oligomerization of the drying oil occurs; in some cases, reaction between the oil and resin has been demonstrated. Varnishes were widely used in the nineteenth and early twentieth centuries, but have been almost completely replaced by a wide variety of other products especially alkyds, epoxy esters, and uralkyds. The term *varnish* has come to be used more generally for transparent coatings even though few of them today are varnishes in the original meaning of the word.

9.3.3. Synthetic Conjugated Oils

Tung oil dries rapidly, but it is expensive and the films discolor rapidly due to the presence of three double bonds. These shortcomings of tung oil led to efforts to synthesize conjugated oils, especially those containing esters of fatty acids with two conjugated double bonds.

One approach is to dehydrate castor oil using acid catalysis. A major component (87%) of the fatty acid content of castor oil triglycerides is ricinoleic acid, 12-hydroxy-(Z)-9-octadecenoic acid, which dehydrates to mixed geometric isomers of the 9,11-conjugated and the 9,12-nonconjugated fatty acid esters. "Dehydrated castor oil" dries relatively rapidly at room temperature but on further exposure to air the surface becomes tacky. This "aftertack" has been attributed to the presence of various geometric isomers formed during the dehydration. Dehydrated castor oil and its fatty acids are mainly used in preparing alkyds and epoxy esters for baking coatings where aftertack does not occur.

Nonconjugated oils can be partially isomerized to conjugated oils by heating with a variety of catalysts, mostly alkaline hydroxides. These modified oils contain a mixture of geometric isomers and, like dehydrated castor oil, exhibit aftertack. A similar process can be used to partially conjugate the double bonds of TOFAs. Synthesis of conjugated fatty acids by treatment of oils at high temperature with aqueous alkali hydroxides accomplishes isomerization and saponification simultaneously [14]. The principal use of such conjugated oils and fatty acids has been in making alkyds and epoxy esters for baking coatings.

9.3.4. Esters of Higher Functionality Polyols

As mentioned in Section 9.2.1, the time required for nonconjugated oils to form a solvent-resistant, cross-linked film decreases as the average number of methylene groups between double bonds per molecule, $\bar{f}_n$, increases. When oil-derived fatty acids are reacted with polyols having more than three hydroxyl groups per molecule, the number of cross-linking sites per molecule increases relative to the corresponding, natural triglyceride oil. As shown earlier, soybean oil is a semidrying oil since $\bar{f}_n$ is 2.07. However, the pentaerythritol (PE) tetraester of soybean fatty acids has an $\bar{f}_n$ of 2.76 and is a synthetic drying oil. The PE ester of linseed fatty acids has an $\bar{f}_n$ of about 5 and gives dry, solvent-resistant films more rapidly than linseed oil. Still faster drying rates can be achieved with still higher functionality polyols, such as di- and tripentaerythritol. Alkyds, epoxy esters, and uralkyds (see Sections 10.1, 11.4, and 12.4.3) made with fatty acids from such oils as soybean and linseed oils can be considered as higher functionality synthetic drying oils.

Another useful type of polyol for synthetic drying oils are low molecular weight, alternating copolymers of styrene and allyl alcohol. Grades with different molecular weights and hydroxyl contents are commercially available. The relatively high functionality of drying oil fatty acid esters of styrene–allyl alcohol copolymers results in rapid cross-linking during film formation. Furthermore, the presence of a rigid aromatic ring for each long aliphatic chain gives a higher T_g and hence more rapid formation of tack-free films. This characteristic is analogous to that of alkyds discussed in Section 10.1. However, the ester backbone of alkyds results in lower hydrolytic stability relative to the styrene–allyl alcohol backbone that cannot be hydrolyzed.

9.3.5. Maleated Oils

Both oils with conjugated and with nonconjugated double bonds react with maleic anhydride to form adducts. Conjugated oils, such as dehydrated castor oil, react at moderate temperatures by a Diels–Alder reaction:

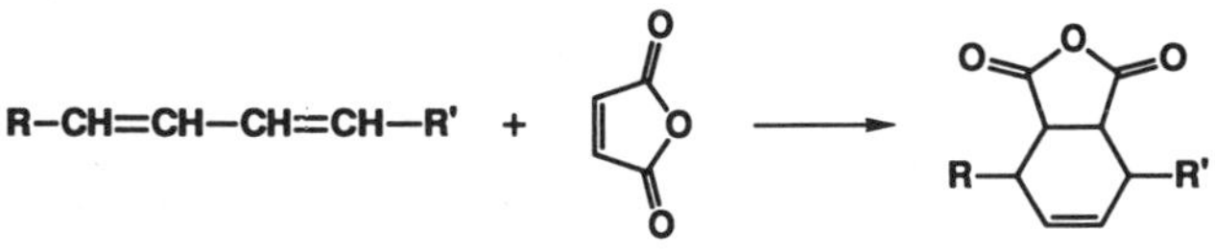

Nonconjugated oils such as soybean and linseed require higher temperatures (>100°C) and form a wide variety of adduct structures. Model compound studies with methyl linoleate demonstrate that maleic anhydride undergoes an "ene reaction" to give succinyl anhydride adducts. When the ene reaction results in conjugated bonds as shown in the general structure **4**, subsequent Diels–Alder reaction with a second maleic anhydride gives a dianhydride of general structure **5** [15].

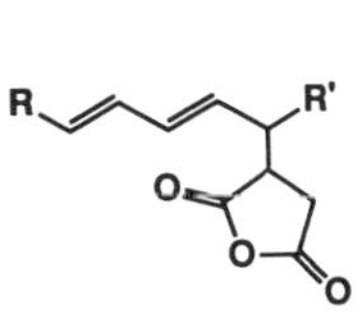

4

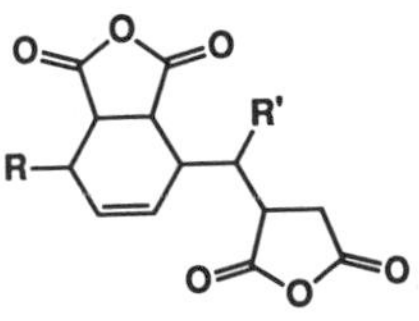

5

The products of these reactions, termed "maleated oils" or sometimes "maleinized oils," react with polyols to give moderate molecular weight derivatives that dry faster than the unmodified drying oils. For example, maleated, esterified soybean oil dries with a rate comparable to that of a bodied linseed oil having a similar viscosity.

Maleated linseed oil can be converted to a water-reducible form by hydrolysis with aqueous ammonium hydroxide to convert the anhydride groups to the ammonium salts of the diacid. Such products have not found significant commercial use but, as we will see in Sections 10.3 and 11.4.1, a similar process is used to make water-reducible alkyds and epoxy esters.

9.3.6. Vinyl-Modified Oils

Both conjugated and nonconjugated drying oils react in the presence of a free radical initiator with such unsaturated monomers as styrene, vinyltoluene, acrylic esters, and cyclopentadiene. High degrees of chain transfer cause the formation of wide varieties of products, for example, low molecular weight homopolymer of the monomer, short-chain graft copolymers, and dimerized drying oil molecules.

The reaction of drying oils with such monomers, except cyclopentadiene, is not commercially important; but the same principle is widely used in making modified alkyds (see Section 10.4). Linseed oil modified with cyclopentadiene has found fairly sizable commercial use. This product is made by heating a mixture of linseed oil and dicyclopentadiene above 170°C, the temperature above which the reverse Diels–Alder reaction liberates monomeric cyclopentadiene at an appreciable rate. The product is inexpensive and dries faster than linseed oil. However, its odor and dark color limit applications.

GENERAL REFERENCES

M. W. Formo, "Drying Oils," in D. Swern, Ed., *Bailey's Industrial Oil and Fat Products*, Wiley, New York, Vol. I, 1979, pp. 177–232 and 687–816; Vol. II, 1982, pp. 343–406.

F. L. Fox, *Oils for Organic Coatings*, Federation of Societies for Coatings Technology, Philadelphia, 1965.

A. E. Rheineck and R. O. Austin, "Drying Oils," in R. R. Myers and J. S. Long, Eds., *Treatise on Coatings*, Vol. I, No. 2, Marcel Dekker, New York, 1968, pp. 181–248.

Z. W. Wicks, Jr., "Drying Oils," *Encyclopedia of Polymer Science and Engineering*, 2nd ed., **5**, 203–214 (1986).

REFERENCES

1. R. G. Ackman, *Prog. Chem. Fats Other Lipids*, **12**, 165 (1972); G. R. Khan and F. Scheinmann, *Prog. Chem. Fats Other Lipids*, **15**, 343 (1977); J. W. King, E. C. Adams, and B. A. Bidlingmeyer, *J. Liq. Chromatography*, **5**, 275 (1982).

2. Z. W. Wicks, Jr., *Encyc. Polym. Sci. Eng.*, 2nd ed. **5** (1986) p. 205.

3. M. W. Formo, "Drying oils," in D. Swern, Ed. *Bailey's Industrial Oil and Fat Products*, Wiley, New York, Vol. I, 1979, pp. 177–232 and 687–816; Vol. II, 1982, pp. 343–406.

4. F. L. Fox, *Oils for Organic Coatings*, Federation of Societies for Coatings Technology, Philadelphia, 1965.

5. N. A. Porter, L. S. Lehman, B. A. Weber, and K. J. Smith, *J. Am. Chem. Soc.*, **103**, 6447 (1981).

6. J. H. Greaves, *Oil Colour Trades J.*, **113**, 949 (1948).

7. A. E. Rheineck and R. O. Austin, "Drying Oils," in R. R. Myers and J. S. Long, Eds., *Treatise on Coatings*, Vol. 1, No. 2, Marcel Dekker, New York, 1968, pp. 181–248.

8. J. R. Chipault, E. E. Nickell, and W. O. Lundberg, *Off. Dig. Fed. Paint Varn. Prod. Clubs*, **23**, 740 (1951).

9. J. H. Hartshorn, *J. Coat. Technol.*, **54** (687), 53 (1982).

10. W. J. Muizebelt, J. W. van Velde, and F. G. H. van Wijk, *Proc. XVth Intl. Conf. Org. Coatings Sci. Techn.*, 299 (1989).

11. E. N. Frankel, *Prog. Lipid Res.*, **19**, 1 (1980).

12. R. A. Hancock, N. J. Leeves, and P. F. Nicks, *Prog. Org. Coat.*, **17**, 321, 337 (1989).

13. D. H. Wheeler and J. White, *J. Am. Chem. Soc.*, **44**, 298 (1967).

14. T. F. Bradley and G. H. Richardson, *Ind. Eng. Chem.*, **34**, 237 (1942).

15. A. E. Rheineck and T. H. Khoe, *Fette Seifen Anstrichm.*, **71**, 644 (1969).

CHAPTER **X** _____

Alkyd Resins

While no longer the largest volume class of vehicles used in coatings, alkyds still are of major importance. Alkyds are prepared from polyols, dibasic acids, and fatty acids. As such, they are polyesters, but in the coatings field the term polyester is reserved for the "oil-free polyesters" discussed in Chapter 8. The term alkyds is derived from their preparation from *alc*ohols and ac*ids*. Chronologically alkyds preceded polyesters by over 20 years.

Alkyds tend to be lower in cost than most other vehicles and tend to give coatings that exhibit fewer film defects during application (see Chapter 23). However, durability of alkyd films, especially outdoors, tends to be poorer than that imparted by other major classes of vehicles such as acrylics and polyurethanes.

There are many ways of classifying alkyds. One system classifies alkyds into *oxidizing* and *nonoxidizing* types. Oxidizing alkyds cross-link by the same mechanism as drying oils, as discussed in Chapter 9. Nonoxidizing alkyds are used as polymeric plasticizers or as hydroxy-functional resins, which are cross-linked by MF or UF resins, or by isocyanate cross-linkers.

A second method of classification is based on the ratio of monobasic fatty acids/ dibasic acids utilized in their preparation. The terminology used in this method was adapted from the terminology used in classifying varnishes. Varnishes with high ratios of oil/resin were called long oil varnishes, those with a lower ratio were called medium oil varnishes, and those with a very low ratio were called short oil varnishes. The oil length of an alkyd is calculated by dividing the amount of *oil* in the final alkyd by the total weight of the alkyd solids, expressed as a percentage as shown in Eq. 10.1. The amount of oil is defined as the triglyceride equivalent to the amount of fatty acids in the alkyd. The 1.04 factor in Eq. 10.2 approximately converts the weight of fatty acids to the corresponding weight of triglyceride oil. Alkyds with an oil length greater than 60 are called long oil alkyds. Those with oil lengths from 40 to 60 are called medium oil alkyds and those with less than 40 are called short oil alkyds. There is some variation in the dividing lines between these classes set by different workers in the field.

$$\text{Oil length} = \frac{\text{Weight of } oil}{\text{Weight of alkyd } - \text{ water evolved}} \times 100 \qquad (10.1)$$

$$\text{Oil length} = \frac{1.04 \times \text{Weight of fatty acids}}{\text{Weight of alkyd } - \text{ water evolved}} \times 100 \qquad (10.2)$$

Another classification is *unmodified* or *modified alkyds*. Modified alkyds contain other monomers in addition to polyols, polybasic acids, and fatty acids. Examples are *styrenated alkyds* (see Section 10.4) and *silicone alkyds* (see Section 13.3.1).

10.1. OXIDIZING ALKYDS

Oxidizing alkyds can, in a very real sense, be considered as synthetic drying oils. They are polyesters of one or more polyols, one or more dibasic acids, and fatty acids from one or more drying or semidrying oils. The most commonly used polyol is glycerol, the most commonly used dibasic acid is phthalic anhydride (PA), and a very widely used oil is soybean oil.

Let us consider a very simple, idealized example of the alkyd prepared from 1 mol of PA, 2 mol of glycerol, and 4 mol of soybean fatty acids. Using the fatty acid composition data for soybean oil from Table 9.1, we can calculate that this alkyd would have an average of 2.76 methylene groups between double bonds $-(CH{=}CHCH_2CH{=}CH)-$ and, therefore, would dry to a solid film. It would form a solvent resistant film in about the same time as a pentaerythritol (PE) ester of soybean fatty acids since it has the same average functionality $\bar{f}_n$. However, it would form a tack-free film faster than the PE ester since the rigid aromatic rings from the PA serve to increase the T_g of the film.

If the mole ratio of PA/glycerol were 2:3, theoretically 5 mol of soybean fatty acid could be esterified, to yield an alkyd with $\bar{f}_n$ of 3.45. This alkyd would cross-link more rapidly than the 1:2:4 mole ratio alkyd because of the higher functionality and would also form tack-free films still faster because the ratio of aromatic rings to long aliphatic chains would be 2:5 instead of 1:4. As the ratio of PA/glycerol is decreased further, the average functionality for autoxidation increases and the T_g after solvent evaporation also increases because of the increasing ratio of aromatic to long aliphatic chains. For both reasons the films will dry faster.

A theoretical alkyd prepared from 1 mol each of glycerol, PA, and fatty acid would have an oil length of about 60. A highly idealized structure of such an alkyd is shown below. However, if one tried to prepare such an alkyd the resin would gel prior to complete reaction. Gelation would result from reaction of a sufficient number of trifunctional glycerol molecules with three difunctional PA molecules to form a cross-linked continuous polymer molecule, swollen with partially reacted components. Gelation can be avoided by using a sufficient excess of glycerol to reduce the extent of cross-linking. When the reaction is carried to near completion with excess glycerol, there are few unreacted carboxylic acid groups but many unreacted hydroxyl groups.

Idealized structure of an alkyd made from 1 mol of PA, 1 mol of glycerol, and 1 mol of linoleic acid

There have been many attempts, none fully successful, to calculate the ratios of functional groups and the extent of reaction that can be reached without encountering gelation. The problem is very complex. The reactivity of the hydroxyl groups can be different; for example, glycerol contains both primary and secondary alcohol groups. Under the conditions of esterification, polyol molecules can self-condense to form ethers and, in some cases, can dehydrate to form volatile aldehydes (see Section 10.6.2). Reactivity of the carboxylic acids also varies. The rate of formation of the first ester group from a cyclic anhydride is much more rapid than formation of the second ester group. Aliphatic acids esterify significantly more rapidly than aromatic acids. Polyunsaturated fatty acids and their esters can dimerize or oligomerize to form cross-links. Of the many papers in the field, that by Blackinton [1] perhaps recognizes the complexities best. In addition to the above complexities, he puts particular emphasis on the extent of formation of cyclic compounds by intramolecular esterification reactions.

In practice, alkyd resin formulators have found that the *mole* ratio of dibasic acid/polyol should be less than 1 to avoid gelation. How much less than 1 depends on many variables. Composition variables are discussed in Sections 10.1.1 through 10.1.3; the effect of variables in reaction conditions are discussed in Section 10.6 on synthesis.

For medium oil alkyds, the ratio of dibasic acid/polyol is not generally changed much relative to alkyds with an oil length of about 60, but the fatty acid content is reduced to the extent desired. Of course, this means that there is a larger excess of hydroxyl groups in the final alkyd. It is commonly said that as the oil length of an oxidizing alkyd is reduced below 60, the drying time decreases to a minimum at an oil length of about 50. However, this conventional wisdom must be viewed cautiously. It is true that the ratio of aromatic rings to aliphatic chains continues to increase, increasing the T_g after the solvent evaporates from the film tending to shorten the time to form a tack-free film. However, at the same molecular weight, the number of fatty acid esters groups per molecule decreases as the oil length decreases below 60 since more hydroxyl groups are being left unesterified. Therefore, the time required to achieve sufficient cross-linking for solvent resistance will increase.

Long oil alkyds are soluble in aliphatic hydrocarbon solvents. As the oil length decreases, mixtures of aliphatic and aromatic solvents are required and oil lengths below about 50 require aromatic solvents, which are more expensive than aliphatics. The viscosity of solutions of long oil alkyds, especially for those with oil lengths below 65, is higher in aliphatic than in aromatic solvents; in medium oil alkyds, which require mixtures of aliphatic and aromatic solvents, viscosity decreases as the proportion of aromatic solvents increases. In former days, and to some degree still today, it was considered desirable to use a solvent mixture that gave the highest possible viscosity; then, at application viscosity, the solids were lower and the raw material cost per unit volume was less. Accordingly, alkyds were designed to have high dilutability with aliphatic solvents. Obviously, this is false economy, but it was a common practice and is still being practiced to some extent now. Increasingly, the emphasis is more on reducing VOC, so the question becomes how to design alkyds with low solvent requirements rather than high dilutability potential. High solids alkyds are discussed in Section 10.2.

10.1.1. Monobasic Acid Selection for Oxidizing Alkyds

As explained above, drying alkyds can be made from the fatty acids in semidrying oils ($\bar{f}_n$ can be well above 2.2). For alkyds made by the monoglyceride process (see Section 10.6.1) soybean oil is used in the largest volume; soybean oil is economical and supplies are dependable because it is a large scale agricultural commodity; alkyd production takes only a few percent of the world supply. For alkyds made by the fatty acid process (again, see Section 10.6.1) tall oil fatty acids (TOFA) are more economical than soybean fatty acids. Both soybean oil and TOFA contain roughly 40–60% linoleic acid and significant amounts of linloenic acid (see Table 9.1). White coatings containing linolenic acid esters gradually turn yellow. Premium cost "nonyellowing" alkyds are made from safflower or sunflower oils, which are high in linoleic acid but contain very little linolenic acid.

Applications in which fast drying and high cross-link density are of paramount importance require alkyds made from drying oils. The rate of oxidative cross-linking will, obviously, be affected by the functionality of the drying oils used. At the same oil length and molecular weight, the time required to achieve a specific degree of cross-linking will decrease as the number of methylene groups between double bonds increases. Linseed long oil alkyds, therefore, cross-link more rapidly than soybean long oil alkyds. Tung oil based alkyds, as would be expected from the high proportion of esters with three conjugated double bonds, dry still faster. The effect is especially large in very long oil alkyds and becomes less noticeable in alkyds with oil lengths around 60, where $\bar{f}_n$ is very high even with soybean oil and the effect of further increase in functionality by using linseed oil is small.

Because of the large fraction of fatty acid esters with three double bonds in linseed alkyds, their color and color retention is poorer than that of soybean alkyds. Tung oil alkyds also exhibit a high degree of yellowing. Dehydrated castor alkyds have fairly good color retention since they contain only a small proportion of fatty acid esters with three double bonds; but they have the disadvantage of exhibiting aftertack (see Section 9.3.3) in air dry coatings and are, therefore, used primarily in baking coatings.

As discussed in Section 9.3.1, drying oils and drying oil fatty acids undergo

dimerization at elevated temperatures. Dimerization occurs concurrently with esterification during alkyd synthesis. This dimerization in effect generates difunctional
acids, increasing the mole ratio of dibasic acids to polyol. The rate of dimerization
is faster with drying oils having a higher average number of methylene groups
between double bonds per molecule and those having conjugated double bonds.
Thus the molecular weight and, therefore, the viscosity of an alkyd made with the
same ratio of reactants will depend on the fatty acid composition. The higher the
degree of unsaturation, the higher the viscosity because of the greater extent of
dimerization. Linseed alkyds have higher viscosities than soybean alkyds made with
the same monomer ratios under the same conditions. The effect is particularly
marked with tung oil. In fact, it is difficult to prepare straight tung alkyds because
of the risk of gelation; commonly, mixed linseed–tung alkyds are used when high
oxidative cross-linking functionality is desired.

A critical factor involved in the choice of fatty acid is cost. Drying oils are
agricultural products and hence tend to be volatile in price. By far the major use
of vegetable oils is for foods. Depending on relative prices, one drying oil is often
substituted for another in certain alkyds. By adjusting for functionality differences,
substitutions can frequently be made without significant changes in properties.

Fatty acids are not the only monobasic acids that can be used in making alkyds.
Benzoic acid is also widely used, especially to esterify some of the excess hydroxyl
groups remaining in the preparation of medium oil alkyds. The benzoic acid increases the ratio of aromatic/aliphatic chains in the alkyd, thus contributing to a
higher T_g of the solvent-free alkyd and more rapid formation of a tack-free film.
At the same time, the reduction in the free hydroxyl content in the unreacted film
may somewhat reduce the water sensitivity of the dried films. Rosin can also be
used in the same fashion. Although rosin is not an aromatic acid, its polynuclear
ring structures are rigid enough to increase T_g. In any of these cases, it must always
be borne in mind that, if the critical requirement in drying is rapid development
of solvent resistance, such benzoic acid and rosin modifications do not serve the
purpose. They only increase the rate of reaching a tack-free state. Frequently,
benzoic acid modified alkyds are called "chain-stopped" alkyds. The implication
of the terminology is that the benzoic acid stops chain growth. This is not the case;
the benzoic acid simply esterifies hydroxyl groups that would not have been esterified if the benzoic acid were absent. The effect on degree of polymerization is
negligible.

10.1.2. Polyol Selection

Glycerol is the most widely used polyol because it is present in the naturally
occurring oils from which alkyds are commonly synthesized (see Section 10.6.1).
The second most widely used is pentaerythritol (PE). In order to avoid gelation,
the tetrafunctionality of PE must be taken into account when replacing glycerol
with PE. If the substitution is made on a mole basis rather than an equivalent basis,
chances for gelation will be minimized. As mentioned earlier, the ratio of *moles*
of dibasic acid to polyol should be less than 1, and generally a slightly lower mole
ratio is required with PE than with glycerol. At the same mole ratio of dibasic
acid/polyol, more moles of fatty acid can be esterified. Hence, in long oil alkyds,
the average functionality for cross-linking will be higher, and the time to reach a

given degree of solvent resistance will be shorter for a PE alkyd as compared to a glyceryl alkyd. Due to this difference, one must be careful in comparing oil lengths of glycerol and PE alkyds.

$$CH_2OH$$
$$HOCH_2-\overset{\displaystyle CH_2OH}{\underset{\displaystyle CH_2OH}{C}}-CH_2OH$$

Pentaerythritol
PE

$$HOCH_2-\overset{\displaystyle CH_2OH}{\underset{\displaystyle CH_2OH}{C}}-CH_2-O-CH_2-\overset{\displaystyle CH_2OH}{\underset{\displaystyle CH_2OH}{C}}-CH_2OH$$

Dipentaerythritol
DiPE

$$HOCH_2-\overset{\displaystyle CH_2OH}{\underset{\displaystyle CH_2OH}{C}}-CH_2-O-CH_2-\overset{\displaystyle CH_2OH}{\underset{\displaystyle CH_2OH}{C}}-CH_2-O-CH_2-\overset{\displaystyle CH_2OH}{\underset{\displaystyle CH_2OH}{C}}-CH_2OH$$

Tripentaerythritol
TriPE

When PE is synthesized, di- and tripentaerythritol are obtained as byproducts and commercial PE contains some of these higher polyols. Consequently, care must be exercised in changing sources of PE, since the amount of the higher polyols in one manufacturer's grade of PE may be different than that from another manufacturer. Because of their very high functionality ($F = 6$ and 8, respectively) diPE and triPE are useful in making fast drying, low molecular weight alkyds. This will be discussed in Section 10.2 on high solids alkyds.

To reduce cost, it can sometimes be desirable to use mixtures of PE with ethylene or propylene glycol. Obviously, a 1:1 mole ratio of tetra- and difunctional polyols will give an average functionality of 3, corresponding to glycerol. The corresponding alkyds can be expected to be similar, but not identical.

Sorbitol has been recommended as a polyol, but its utility is limited since four of the six hydroxy groups are secondary and, therefore, less reactive. Perhaps even more limiting, cyclic ether formation occurs under the conditions required for esterification, which reduces the functionality of a significant but unpredictable fraction of the sorbitol to four.

Trimethylolpropane (TMP) can also be used in making alkyds. The rate of esterification is slower than that experienced with glycerol. Although all of TMPs alcohol groups are primary, they are somewhat sterically hindered by the neopentyl structure [2]. Trimethylolpropane, however, gives a narrower molecular weight distribution, which provides alkyds with a somewhat lower viscosity than comparable glycerol-based alkyds. A recent kinetic study demonstrated that esterification of one or two of the hydroxyl groups of TMP has little effect on the rate constant for esterification of the third hydroxyl group [3]. It can be speculated that PE behaves similarly.

10.1.3. Dibasic Acid Selection

Dibasic acids used to prepare alkyds are usually aromatic. Their rigid aromatic rings increase the T_g of the resin. Aromatic acids esterify more slowly than aliphatic acids so that the small amount of acid left unreacted at the end of the esterification is essentially all aromatic carboxylic acid groups. In Europe cycloaliphatic anhydrides, such as hexahydrophthalic anhydride, are also used. While they are not as rigid as aromatic rings, the cycloaliphatic rings also increase T_g.

Phthalic Anhydride   Isophthalic acid   Terephthalic acid
PA   IPA   TPA

By far the most widely used dibasic acid is PA. It has the advantage that the first esterification reaction proceeds rapidly by opening the anhydride ring. The amount of water evolved is, of course, lower, which also reduces reaction time. The relatively low melting point (the pure compound melts at 131°C) is desirable since the crystals melt and dissolve readily in the reaction mixture. In large scale manufacturing, hot, molten PA is used, which substantially reduces packaging, shipping, and handling costs.

The next most widely used dibasic acid is isophthalic acid (IPA). As discussed in Section 8.1.2, esters of IPA are more resistant to hydrolysis than are those of PA in the pH range of 4 to 8, which is probably the most important range for exterior durability. On the other hand, under alkaline conditions esters of phthalic acid are more resistant to hydrolysis than isophthalic esters. The raw material cost is not particularly different from PA (even after adjusting for the extra mole of water that will be lost), but the manufacturing cost is higher. Both of the carboxylic acid groups have to esterify rather than one ester forming by the anhydride reaction. The high melting point of IPAs (330°C) leads to problems getting it to dissolve in the reaction mixture so that it can react. High temperatures are required for longer periods of time than with PA; hence more dimerization of fatty acids occurs resulting in higher viscosity. The longer time at higher temperature also leads to greater extents of side reactions of polyol components [4], which are discussed in Section 10.6.2. Thus, in substituting IPA for PA, one must use a lower mole ratio of IPA/polyol in order to make an alkyd of similar viscosity.

Terephthalic acid (TPA) can be used but its melting point is even higher than that of IPA. It is generally preferable to incorporate terephthalate by transesterification of the low melting dimethyl terephthalate. Alkyds derived from TPA have similar hydrolytic stability to those from IPA and their cost is lower. However, there is a tendency for partial crystallization of terephthalate diesters from the final solution, giving hazy resins and possibly leading to film defects.

10.2. HIGH SOLIDS OXIDIZING ALKYDS

The need to minimize VOC emissions has led to substantial efforts to reduce solvent, requiring an increase in the solids content of alkyd resin coatings. Some increase in solids can be realized by a change of solvents. Aliphatic (and to a somewhat lesser degree, aromatic) hydrocarbon solvents promote intermolecular hydrogen bonding especially between carboxylic acids but also between hydroxyl groups, thereby increasing viscosity. Use of at least some hydrogen-bond acceptor solvent such as a ketone or hydrogen-bond acceptor–donor solvent such as an alcohol will give a significant reduction in viscosity at equal solids.

An obvious approach to increasing solids is to decrease molecular weight, easily accomplished by decreasing the dibasic acid/polyol ratio and going to longer oil length alkyds. However, making a really significant reduction in VOC by this route would require having an alkyd with lower functionality for cross-linking and a lower ratio of aromatic–aliphatic chains. Both changes would increase the time for drying. The effect of longer oil length on functionality can be minimized by using drying oils with higher average functionality. Unfortunately, use of oils containing linolenic or α-eleostearic acid is limited by their tendency to discolor. One can use safflower oil, which has a higher linoleic acid content and less linolenic acid than soybean oil. Proprietary fatty acids with 78% linoleic acid are commercially available.

Increasing the concentration of driers (see Chapter 9) can accelerate drying but also accelerates yellowing and probably other film deterioration processes. Current high solids alkyd formulations often contain much higher levels of zirconium than conventional formulations.

Solids can be increased by making resins with narrower molecular weight distributions. For example, one can add a transesterification catalyst near the end of the alkyd cook; this will give more uniform molecular weight and a lower viscosity product. In order to study the effect of molecular weight distribution, model alkyds with very narrow molecular weight distribution were synthesized by using dicyclohexylcarbodiimide, which allows low temperature esterification [2]. With the same ratio of reactants, the $\overline{M}_n$ and polydispersity were lower than the conventional alkyd cook control. These differences presumably result from less dimerization through reactions of the double-bond systems of the fatty acids and avoidance of self-etherification of polyol in the low temperature preparation. It was found that the solids could be 2–10% higher than with the conventionally prepared alkyd of the same raw material composition. The model alkyds dried more rapidly, but the film properties, especially impact resistance, were greatly inferior to those obtained with control resins having the usual broad molecular weight distribution [5]. It is of interest to note that conventionally prepared TMP alkyds had lower molecular weights and viscosities than the glycerol alkyds. This may result from less self-etherification of TMP as compared to glycerol.

High solids alkyds for baking applications have been made on a large scale using tripentaerythritol. The high functionality obtained using this polyol ($F = 8$) gives alkyds that cross-link as rapidly as shorter oil length, higher viscosity glycerol alkyds [6]. However, for air dry applications the lower aromatic/aliphatic ratio lengthens the tack-free time. Presumably, progress could be made using a high functionality polyol with some combination of phthalic and benzoic acid together with fatty acids having as high a functionality fatty acids as possible. The cost of such a system, would be high.

Another approach to high solids alkyds is to use so-called "reactive diluents" in place of part of the solvent. The idea is to have a component of lower molecular weight and much lower viscosity than the alkyd resin, which will react with the alkyd during drying, so that it is not part of the VOC emissions. At least four types of reactive diluents are used commercially in formulating high solids alkyd coatings.

Polyfunctional methacrylate monomers (e.g., trimethylolpropane trimethacrylate) are the first type of reactive diluents and they have been used in "force dry" coatings (coatings designed to be cured in the range of 60 to 80°C). Although polymerization of these monomers is inhibited by oxygen and there is a substantial degree of termination by chain transfer in the coreaction of such polyfunctional monomers, cross-linking still occurs. Polyfunctional acrylates can be used at ambient temperatures [7], but some people exhibit skin irritation on repeated exposure to polyfunctional acrylates.

A second example is the use of dicyclopentadienyloxyethyl methacrylate as a reactive diluent [8]. It is bifunctional due to the easily abstractable hydrogen on a tertiary carbon allylic to a double bond in the dicyclopentadiene ring structure and the acrylate double bond. The compound coreacts with the drying oil groups in the alkyd during drying of films.

Dicyclopentadienyloxyethyl methacrylate

A third example is the use of mixed acrylic and drying oil fatty acid amides of hexa(aminomethoxymethyl)melamine resins [9,10]. Such reactive diluents contain high functionalities of $>NCH_2NHCOCH{=}CH_2$ and $>NCH_2NHCOC_{17}H_x$ moieties and promote fast drying.

A fourth type is an oligomeric ester of trimethylolpropane diallyl ether with trimellitic anhydride [11]. The allyl ethers having methyl groups between a double bond and ether oxygen atoms coreact with the alkyds during drying.

Trimethylolpropane diallyl ether

Using optimized resins, and, in some cases, reactive diluents, good quality air drying and baking alkyd coatings can be formulated with VOC levels of 280 to 350 g/L of coating. As of 1992, the 250-g/L level required in California is attainable only with some sacrifice in application and film properties. Still lower limits of permissible VOC are projected for the 1990s, which has prompted current work on further solvent reduction.

10.3. WATER-REDUCIBLE ALKYDS

As with almost all other resin classes, work has been done to make alkyd resins for coatings that can be reduced with water. One approach is to emulsify the alkyd by adding a surfactant. It is common to add a few percent of an alkyd–surfactant blend to latex paints in order to improve adhesion to chalky surfaces and, in certain circumstances, to metals. It is important to use alkyds that are as resistant as possible to hydrolysis. Vehicles that are not really alkyds but are drying oil esters of styrene–allyl alcohol copolymers are widely used (see Section 9.3.4). Use of emulsified alkyds as the primary vehicle is apparently feasible but is discouraged by the excellent properties and low costs (at 1991 petroleum prices) of synthetic latexes.

Another approach has been to make alkyds with an acid number in the range of 50 using secondary alcohols or secondary ether alcohols as solvents. The acid groups are neutralized with ammonia or an amine. The resultant solution can be diluted with water to form a dispersion of solvent swollen aggregates in water. Such resins are analogous to the water-reducible acrylics discussed in Section 7.3, and their behavior during water dilution is similar. Using this approach it is possible to make high gloss water-reducible air dry coatings; high gloss is a problem area for latexes. Note that the use of primary alcohol solvents must be avoided because they can more readily transesterify with the alkyd during resin production and storage leading to reduction in molecular weight and $\bar{f}_n$ [12].

Hydrolytic stability is also a critical problem with the second approach. It is particularly important that the carboxylic acid groups on the alkyd are not lost by hydrolysis. For example, if the carboxylic acid groups are half esters of phthalic or trimellitic acid the hydrolytic stability would be poor—probably inadequate for paints that require a shelf life of more than a few months. Due to the anchimeric effect of the neighboring carboxylic acid group such an ester is relatively easily hydrolyzed. As hydrolysis occurs, the solubilizing acid salt is detached from the resin molecules, and the aqueous dispersion loses stability.

A more satisfactory way to introduce free carboxylic acid groups is by reacting a completed alkyd with maleic anhydride. Part of the maleic anhydride will add to the unsaturated fatty acid esters as discussed in Section 9.3.5. These anhydride groups are then hydrolyzed with amine and water to give the desired carboxylate salt groups that are attached resin molecules with C—C bonds and cannot be hydrolyzed off. There is sill a hydrolytic stability problem with the alkyd backbone but at least the hydrolysis does not result in destabilization of the dispersion. Better yet, fatty acid esters of styrene–allyl alcohol (see Section 9.3.4) can be maleated to make water-reducible *alkyds*.

After the film is applied, the water, solvent, and amine evaporate, and the film cross-links by autoxidation. Since there are a fairly large number of residual carboxylic acid groups left in the cross-linked binder, the water resistance and particularly the alkali resistance of the films are not as good as those from solvent-borne alkyds [13]. However, for many applications they are still satisfactory.

There can be a problem of "early water resistance" if, for example, a freshly painted surface is rained on before all the amine has evaporated from the film. Commonly, ammonia is used as the neutralizing amine because it is assumed that ammonia will volatilize faster than any other amine. This is not necessarily so. If the T_g of the alkyd film is sufficiently high before all of the amine has volatilized, the loss of amine can become diffusion rate controlled. In this instance the rate of diffusion of amine through the carboxylic acid-functional film will be affected by the base strength of the amine. A less basic amine such as morpholine may leave the film before ammonia even though its volatility is considerably lower.

A related consideration in designing water-borne alkyds is that it may be necessary to compromise between time required to develop a tack-free film and time required to achieve water resistance. Some formulators have jumped to the conclusion that shorter tack-free time will mean faster development of water resistance, but the contrary can be true.

10.4. STYRENATED ALKYDS

Oxidizing alkyds can be modified by reaction with vinyl monomers. The most widely used monomers are styrene, vinyl toluene, and methyl methacrylate but essentially any vinyl monomer can be reacted in the presence of an alkyd to give a modified alkyd. Methyl methacrylate imparts better heat resistance than styrene but at higher cost.

In making styrenated alkyds, an oxidizable alkyd is prepared in the usual way and cooled to about 130°C in the reactor; then styrene and a free radical initiator such as benzoyl peroxide are added. The resulting free radical chain process leads to a variety of reactions, including formation of low molecular weight homopolymer of styrene, grafting of polystyrene onto the alkyd, and dimerization of alkyd molecules. The reaction is generally carried out at about 130°C, which favors decomposition of the benzoyl peroxide to form phenyl free radicals; phenyl free radicals have a greater tendency to abstract hydrogen atoms, which favors grafting more than the less reactive benzoyl free radicals. After the reaction is complete, the resin is diluted with solvent—usually toluene or xylene.

The ratio of alkyd/styrene can be varied over a wide range; commonly 50% styrene—50% alkyd is used. The ratio of aromatic rings/aliphatic chains is, obviously, greatly increased and, as a result, the T_g of styrenated alkyds will be substantially higher than that of the unmodified alkyd. Therefore, the tack-free time will be substantially shorter. Commonly styrenated alkyds will give a *dry* film in 1 h or less instead of 4 to 6 h for the counterpart nonstyrenated alkyd. However, the average functionality for oxidative cross-linking is substantially reduced, not just by dilution with styrene but also because the free radical reactions involved in the styrenation consume some of the activated methylene groups. As a result, the time required for the extent of cross-linking to be sufficient to develop solvent resistance is longer than for the counterpart alkyd. The fast drying and low cost make styrenated alkyds very attractive for some applications but, in other cases, the longer time required for cross-linking is more critical and styrenated alkyds are not appropriate.

In the case of air dry primers based on a styrenated alkyd as the vehicle, one must be very careful to apply the top coat almost immediately or not until after the film has had ample time to cross-link. During the intermediate time interval, application of a top coat is likely to give nonuniform swelling of the primer leading to what is called "lifting" of the primer. The result of lifting is the development of wrinkled areas in the surface of the dried film. End-users who are accustomed to using alkyd primers, which do not give a hard film until a significant degree of cross-linking has occurred, are particularly likely to encounter problems of lifting if they switch to styrenated alkyd primers.

10.5. NONOXIDIZING ALKYDS

Certain low molecular weight short-medium and short oil alkyds are compatible with such polymers as nitrocellulose and thermoplastic polyacrylates. Therefore, such alkyds can be used as plasticizers for these polymers. They have the distinct advantage over plasticizers (e.g., dibutyl or dioctyl phthalate) in that they do not volatilize appreciably when the coatings are baked. It is generally not desirable to use oxidizing alkyds, which would cross-link and lead to embrittlement of the films, especially on outdoor exposure. Therefore, nondrying oil fatty acids (or oils) are used in the preparation of alkyds for such applications. For exterior acrylic lacquers, pelargonic acid, $(n\text{-}C_8H_{17}COOH)$ derived alkyds combine excellent resistance to photodegradation with good compatibility with the thermoplastic acrylic resins. An interesting sidelight on terminology is that these pelargonic alkyds have been called polyesters rather than alkyds because the word polyester connotes higher quality than the word alkyd. Castor oil derived alkyds are particularly appropriate for nitrocellulose lacquers for interior applications since the hydroxyl group on ricinoleic acid promotes compatibility.

All alkyds, but particularly short-medium oil and short oil alkyds, are made with a large excess of hydroxyl groups to avoid gelation. These hydroxyl groups can be cross-linked with melamine–formaldehyde (MF) resins or with polyisocyanates. In some cases, relatively small amounts of MF resin are used to supplement the cross-linking during baking of medium oil oxidizing alkyds. In order to achieve compatibility, butylated MF resins are used. Such coatings provide somewhat better durability and faster curing than alkyd resins alone with little increase in cost. The important advantages of relative freedom from film defects common to alkyd coatings can be retained (see Chapter 23). However, the high levels of unsaturation remaining in the cured films reduce resistance to discoloration on overbake and exterior exposure and cause loss of gloss and embrittlement on prolonged exterior exposure, limiting applications.

These difficulties can be substantially reduced by use of nondrying oils with minimal levels of unsaturated fatty acids as raw materials in making the alkyds. Coconut oil has been widely used; its performance can be further enhanced by hydrogenation of the small amount of unsaturated acids present in coconut oil. Since IPA esters are more stable to hydrolysis in the pH range of 4 to 8 than phthalate esters, the highest performance exterior alkyd–MF enamels use nonoxidizing IPA alkyds. Exterior durability of such coatings is satisfactory for automobile top coats with opaque pigmentation. The films have greater appearance of "depth" than acrylic–MF coatings. The films are perceived to be thicker than films of acrylic–MF coatings of comparable thickness and pigmentation. However, for many applications, alkyd–MF coatings have been replaced with acrylic–MF or polyester–MF coatings to improve the overall balance of film properties.

10.6. SYNTHETIC PROCEDURES FOR ALKYD RESINS

Various synthetic procedures, each involving many variables, are used to produce alkyd resins. Alkyds can be made directly from oils or by using free fatty acids as raw materials.

10.6.1. Synthesis from Oils or Fatty Acids

Monoglyceride process. In the case of glycerol alkyds, it would obviously be absurd to first saponify an oil to obtain fatty acids and glycerol and then reesterify the same groups in a different combination. Rather, what is done commercially is to react the oil first with sufficient glycerol to achieve the total desired glycerol content, including the glycerol in the oil. Since PA is not soluble in oils but is soluble in glycerol the transesterification of oil with glycerol must be carried out as a separate step before the PA is added; otherwise glyceryl phthalate gel particles would form early in the process. This two-stage procedure is often called the "monoglyceride process." The transesterification reaction is run at 230 to 250°C in the presence of a catalyst. Many catalysts have been used. Before the strict regulation of lead in coatings, litharge (PbO) was widely used; the residual transesterification catalyst also acted as a drier. Examples of catalysts now used in the United States are tetraisopropyl titanate, lithium hydroxide, and lithium ricinoleate. The reaction must be carried out under an inert atmosphere such as CO_2 or N_2 to minimize discoloration and dimerization of drying and semidrying oils.

While the process is called the monoglyceride process, the transesterification reaction actually results in a mixture of unreacted glycerol, monoglycerides, diglycerides, and unconverted oil. The composition depends on the ratio of glycerol/ oil and on the reaction conditions: catalyst, time, and temperature. In general, the reaction is not continued until equilibrium is reached. At some relatively arbitrary point the PA is added, beginning the second stage. The viscosity and properties of the alkyd can be affected by the extent of reaction before the PA addition. While many tests have been devised for evaluating the extent of transesterification, none is very general because the starting ratio of glycerol/oil varies over a considerable range depending on the oil length of the alkyd being made. A useful empirical test is to follow the solubility of molten PA in the reaction mixture. This test has the advantage that it is directly related to a major requirement that must be met.

In the first stage, it is common to transesterify the oil with PE or occasionally with other polyols to obtain mixed partial esters. Pentaerythritol is usually less expensive than glycerol.

The second stage, esterification of the *monoglyceride* with PA, is carried out at a temperature of 220 to as high as 255°C. (In calculating the mole ratio of dibasic acid/polyol, the glycerol already esterified in the oil must also be counted.)

Fatty acid process. It is often desirable to base the alkyd on PE or on any polyol alone other than glycerol. In this case fatty acids must be used instead of oils, and the process can be performed in a single step with reduced time in the reactor. Any drying, semidrying, or nondrying oil can be saponified to yield the fatty acids, but the cost of separating the fatty acids from the reaction mixture substantially increases the cost of the alkyd.

A more economical alternative is the use of TOFA, which have the distinct advantage that they are produced as fatty acids. Tall oil fatty acid composition is fairly similar to that of soybean fatty acids (see Table 9.1). Specially refined tall oils with higher linoleic acid content are available as are other grades that have been treated with alkaline catalysts to partially isomerize the double bonds to conjugated structures. When fatty acids are used, generally the polyol, fatty acids, and dibasic acid are all added at the start of the reaction and the esterification of both aliphatic and aromatic acids is carried out simultaneously in the range of 220 to as high as 255°C.

10.6.2. Process Variations

Esterification is a reversible reaction, therefore an important factor affecting the rate of esterification is the rate of removal of the water from the reactor. As a result, most alkyds are produced using a *reflux solvent*, such as xylene, to promote the removal of water by azeotroping. Since the reaction is run at a temperature far above the boiling point of xylene, less than 5% of xylene is used. The amount is very dependent on the reactor and is set empirically such that there is enough to reflux vigorously, but not so much as to cause flooding of the condensor. Some of the xylene is distilled off along with the water, the water is separated and the xylene is returned to the reactor. A schematic diagram of equipment for alkyd preparation when using refluxing solvents is given in Figure 10.1. The presence of the solvent is desirable for other reasons: The vapor serves as an inert atmosphere, reducing the amount of inert gas needed, and the solvent serves to avoid accumulation of sublimed solid monomers, mainly PA, in the reflux condensor.

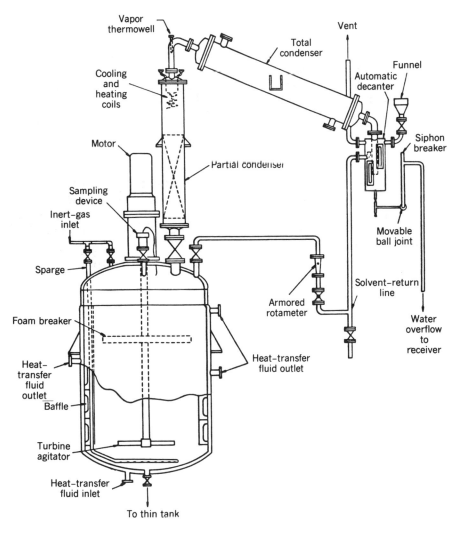

Figure 10.1. Schematic diagram of an alkyd reactor. (With permission from Shell Development Co. US 2643 272, 1947, US 3763 699, 1956 (C. P. Van Dejk) F. J. F. Van der Plas) GB 765 765, 1955; GB 790 166, 1956.)

Reaction time is also strongly affected by reaction temperature. Figure 10.2 shows the relationship among temperature and time and viscosity in preparing an alkyd. Figure 10.3 shows the relationship with acid number (i.e., the milligrams of KOH required to titrate 1 g of resin solids). There are substantial economic advantages in short reaction times. Operating costs are reduced and the shorter times permit more batches of alkyd to be produced in a year, in effect increasing capacity without capital investment in more reactors. Therefore, it is desirable to operate at as high a temperature as possible without risking gelation of the resin in the reactor.

A critical aspect of alkyd synthesis is deciding when the reaction is completed. The disappearance of carboxylic acid is followed by titration and the increase in molecular weight is followed by viscosity. Determination of acid number and viscosity both take some time. Meanwhile, in the reactor, the reaction is continuing. After it is decided that the extent of reaction is sufficient, the reaction mixture must be "dropped" into a larger tank containing solvent. When a 40,000-L batch of alkyd is being made, a significant time is required to get the resin out of the reactor into the reducing tank and, meanwhile, the reaction is continuing. The decision to start dropping the batch must be made so that the acid number and viscosity of the batch will be right after the continuing reaction, which will occur between the time of sampling, determination of acid number and viscosity, and discharging of the reactor.

The time for these determinations becomes the rate controlling step in production. If they can be done rapidly enough, the reaction can be carried out at 240°C or even higher without overshooting the target acid number and viscosity. On the other hand, if the control tests are done slowly, it may be necessary to run the reaction at only 220°C, which may require 2 h or more of additional reaction time (see Figs. 10.2 and 10.3). Automatic titration instruments permit rapid determination of acid number, so the usual limit on time required is viscosity determination. While attempts have been made to use the viscosity of the solvent-free resin at

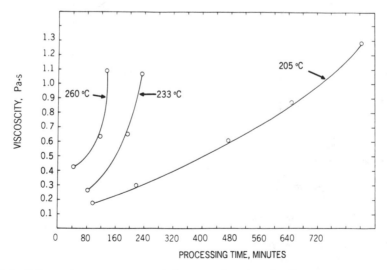

Figure 10.2. Effect of temperature and time on viscosity. (Adapted from Ref. [15], with permission.)

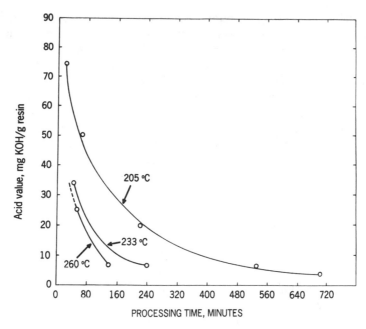

Figure 10.3. Effect of temperature and processing time on acid value of a typical medium oil linseed alkyd (Adapted from Ref. [15], with permission.)

reaction temperature to monitor change in molecular weight, the dependence of viscosity on molecular weight at that high temperature is not sensitive enough to be very useful. The viscosity must be determined on a solution at some lower standard temperature. Since viscosity is very dependent on both solution concentration and temperature, care must be used so that the concentration of the solution is relatively close to the standard and, particularly, that the temperature of the solution has been brought down to the standard temperature.

In alkyd production, viscosity is commonly determined using Gardner bubble tubes. A series of standard tubes with liquids of known viscosity is used to compare the time required for a bubble to rise with that for a sample of the batch in a tube of the same dimensions. The longer the time, the higher the viscosity. One finds the standard tube in which the bubble takes the same time to rise as the batch. The standards are designated by letters of the alphabet: A, B, C, and so on; after Z, there are Z_1, Z_2, and so on. For example, the specification might be Z at 60% solids in mineral spirits (a mixture of aliphatic hydrocarbons). The cook is continued until the viscosity is high enough so that by the time the resin batch in the factory is dropped into the solvent and the batch cooled, its viscosity will be Z. This means starting to discharge the reactor when the test sample is at some lower viscosity. It is not possible to generalize how large this difference should be; it depends on the specific alkyd composition, the temperature at which the reaction is being run, the time required to do the determination, the time required to empty the reactor, and so on. Viscosities can be determined more rapidly using a cone and plate viscometer (see Chapter 19) than with bubble tubes; the very small sample required for the cone and plate viscometer can be cooled and equilibrated at the measurement temperature more quickly.

Many variables affect the acid numbers and viscosities of alkyds. One, obviously, is the ratio of reactants. The closer the moles of dibasic acid approach the moles of polyol, the higher the molecular weight of the backbone of the resin but also the greater the likelihood of gelation. A useful rule of thumb for a starting point is to use a mole ratio of 0.95. The final ratio is determined by adjustments such that the combination of acid number and solution viscosity come out at the desired levels. The greater the ratio of hydroxyl groups to carboxylic acid groups the faster the acid groups are reduced to a low level. The degree of completion of the reaction is an important factor controlling the viscosity as well as the acid number. It is usually desirable to have a low acid number, typically in the range of 5 to 10.

The compositions of the fatty acids can also be a major factor affecting the viscosity, and compositions of a given oil or grade of TOFA can be expected to vary somewhat from lot to lot. Dimerization and oligomerization of unsaturated fatty acids occur in the same temperature range at which the esterification is being carried out (see Section 9.3.1). Fatty acids with conjugated double bonds dimerize more rapidly than those with nonconjugated bonds and dimerization rates increase with the level of unsaturation. At the same ratio of phthalic/polyol/fatty acids, alkyds of the same acid number and solution concentration will increase in viscosity in the order soybean < linseed < tung.

There will be some volatilization of polyol, PA, and fatty acids out of the reactor. Depending on the design of the reactor, the rate of reflux of the azeotroping solvent, the rate of inert gas flow, and the reaction temperature, among other variables, the amount and ratio of these losses will vary. Obviously, variations in such losses will affect the viscosity at the standard acid number. The exact ratio of reactants must be set in the reactor that is actually used for the synthesis. Since gelation can occur if the ratio of dibasic acid/polyol is too high, it is well not to put all of the PA into the reactor in the beginning. If, when the acid number is getting down near the standard, the viscosity is too low, more PA can easily be added. The amount held back can be reduced as experience is gained cooking a particular alkyd in a particular reactor.

Side reactions can affect the viscosity–acid number relationship. Glycerol and other polyols will form ethers to some degree during the reaction. Glycerol can form acrolein by successive dehydrations.

$$HOCH_2-\underset{\underset{OH}{|}}{CH}-CH_2OH \longrightarrow HOCH_2-\underset{\underset{OH}{|}}{CH}-CH_2-O-CH_2-\underset{\underset{OH}{|}}{CH}-CH_2OH$$

$$HOCH_2-\underset{\underset{OH}{|}}{CH}-CH_2OH \longrightarrow H_2C{=}CH-\underset{\overset{O}{\|}}{C}-H$$

As these reactions occur, the mole ratio of dibasic acid/polyol increases. The number of hydroxyl groups decreases; therefore, at the same acid number, the molecular weight will be higher. Excessively high viscosity and even gelation can result. Ether formation is catalyzed by strong protonic acids; therefore, it is desirable to avoid strong acid catalysts for the esterification. Monobutyltin oxide has been used as an esterification catalyst, which presumably does not significantly catalyze ether formation. As noted earlier, pentaerythritol and trimethylopropane seem less vulnerable than glycerol to undesirable side reactions such as ether formation, and glycerol is the only polyol that can decompose to form acrolein.

A hydroxyl group on one end of a growing polyester chain can react with a carboxylic acid group on another end of the same molecule, leading to ring formation. Transesterification of chain linkages can have the same result. Since cyclization reactions reduce chain length, their net effect is to reduce viscosity.

Many alkyd resins have broad and uneven molecular weight distributions (see Fig. 2.3). Kumanotani has shown that even modest changes in reaction conditions can cause large differences in molecular weight distribution, which can have significant effects on final film properties [16]. In many alkyds, very small gel particles, microgels, are formed, probably in the later stages of the process. Kumanotani [16] has shown that these microgels play an important role in giving greater strength properties to final films. It is speculated that process changes that may make alkyd polymers more uniform may be undesirable. For example, allowing the glycerolysis stage to approach equilibrium before addition of PA and use of transesterification catalysts in the final stages of esterification probably favor narrower molecular weight distributions and lower viscosities, but films made from the more uniform alkyds may actually exhibit inferior mechanical properties.

GENERAL REFERENCES

Anonymous, *The Chemistry and Processing of Alkyd Resins*, Monsanto Chemical Co., St. Louis, MO, 1962.

F. N. Jones, "Alkyd Resins," in *Ullmann's Encyclopedia of Industrial Chemistry*, 6th ed. (1st Intl ed.) Vol. A-1, VCH Verlagsgesellschaft mbH, Weinhein, Germany, (1985), pp. 409–423.

K. F. Lin, "Alkyd Resins," in *Kirk-Othmer Encyclopedia of Chemical Technology*, 4th ed., Vol. 2, Wiley, New York (1992), pp. 53–85.

REFERENCES

1. R. J. Blackinton, *J Paint Technol.*, **39** (513), 606 (1967).

2. S. L. Kangas and F. N. Jones, *J. Coat. Technol.*, **59** (744), 89 (1987).

3. R. Bacaloglu, M. Maties, C. Csunderlik, L. Cotaraca, A. Moraru, J. Gros, and N. Marcu, *Angew. Makromol. Chem.*, **164**, 1 (1988).

4. R. Brown, H. Ashjian, and W. Levine, *Off. Digest*, **33**, 539 (1961).

5. S. L. Kangas and F. N. Jones, *J. Coat. Technol.*, **59** (744), 99 (1987).

6. P. Kass and Z. W. Wicks, Jr., U.S. Patent 2, 577, 770 (1951).

7. E. Levine, *Proc. Water-Borne Higher-Solids Coat. Symp.*, New Orleans, LA, 1977, p. 155.

8. D. B. Larson and W. D. Emmons, *J. Coat. Technol.*, **55** (702), 49 (1983).

9. Monsanto Chemical Co., *Technical Bulletin, Resimene AM-300 and AM-325*, January 1986.

10. W. F. Strazik, J. O. Santer, and J. R. LeBlanc, U.S. Patent 4, 293, 461 (1981).

11. C. J. Coady, Amoco Chemical Co., *Technical Service Report*, GTSR 75A (1990).

12. C. J. Bouboulis, *Proc. Water-Borne Higher-Solids Coat. Symp.*, New Orleans, LA, 1982, p. 18.

13. R. Hurley and F. Buona, *J. Coat. Technol.*, **54** (694), 55 (1982).

14. F. N. Jones, "Alkyd Resins," in *Ullmann's Encyclopedia of Industrial Chemistry*, 6th ed. (1st Intl. ed.) Vol. A-1, VCH Verlagsgesellschaft mbH, Weinhein, Germany (1985) p. 416.

15. Anonymous, *The Chemistry and Processing of Alkyd Resins*, Monsanto Chemical Co., St. Louis, MO, 1962.

16. J. Kumanotani, H. Hironori, and H. Masuda, *Adv. Org. Coatings Sci. Tech. Ser.*, **6**, 35 (1984).

CHAPTER XI _____

Epoxy and Phenolic Resins

Epoxy and phenolic resins are two other classes of step-growth polymers that are important in coatings.

11.1. EPOXY RESINS

The use of epoxy resins in coatings dates back to 1939. Before discussing these resins, we must spend a moment on terminology. Epoxy groups (also called epoxides) are three-membered cyclic ethers; in the International Union of Pure and Applied Chemistry (IUPAC) nomenclature and Chemical Abstracts (CA) they are called oxiranes. Most commercially important epoxy resins are derived from (chloromethyl)oxirane, more commonly known as epichlorohydrin (ECH). The resins generally contain oxiranylmethyl ethers or esters, usually called glycidyl ethers or esters. A variety of epoxy resins is used in both baked and ambient cure coatings, with about 70% of the usage in baked coatings.

$$H_2C\overset{O}{\diagdown}CH-CH_2-Cl \qquad H_2C\overset{O}{\diagdown}CH-CH_2-OR \qquad H_2C\overset{O}{\diagdown}CH-CH_2-O-\overset{O}{\overset{\|}{C}}-R$$

Epichlorohydrin   Glycidyl Ether   Glycidyl Ester

11.1.1. Bisphenol A Epoxy Resins

The first epoxy resins used in coatings, and still by far the largest in volume, are bisphenol A (BPA) epoxies made by reacting BPA with ECH. Under basic conditions, the initial reaction is formation of a BPA anion (BPA$^-$), which attacks ECH and results in the formation of a new oxirane ring with elimination of chloride anion (Cl$^-$), as shown in Scheme 11.1.

The initial product shown in Scheme 11.1 is the monoglycidyl ether of BPA (MGEBPA). Analogous reaction of the phenolic group of MGEBPA with NaOH and ECH gives the diglycidyl ether of BPA (DGEBPA), as shown. The epoxy groups of MGEBPA and DGEBPA react, analogously to ECH, with BPA$^-$ to extend the chain, as shown for DGEBPA. These reactions introduce hydroxyl groups on the backbone, as shown. Continuation of these reactions results in linear polymers, since both the BPA and ECH are difunctional. Bisphenol A epoxy resins

162

SCHEME 11.1

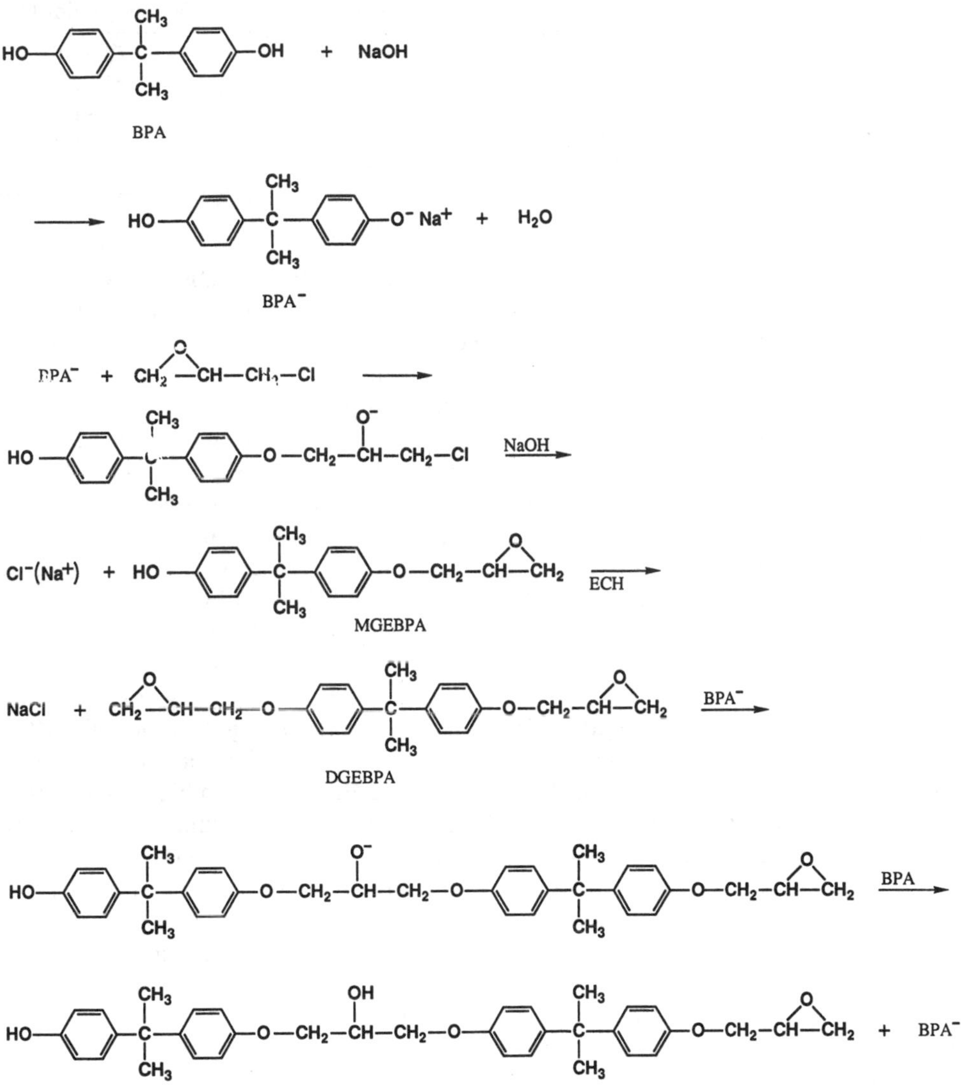

BPA

BPA⁻

MGEBPA

DGEBPA

are made with excess ECH so that the end groups are glycidyl ethers. The polymers may be represented by the following general formula, where the molar ratio of ECH/BPA determines the average n value: ECH/BPA $= (n + 2)/(n + 1)$.

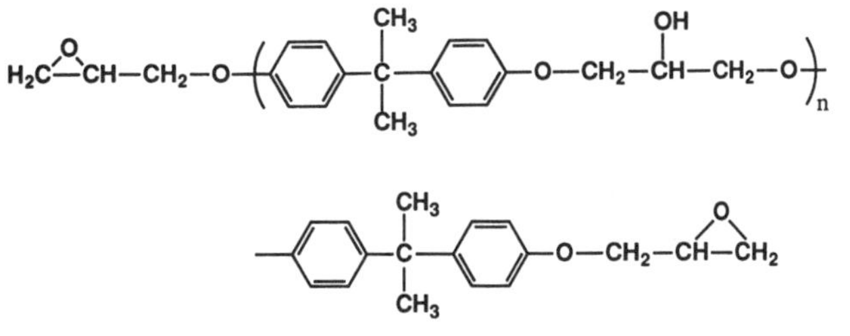

Bisphenol A Epoxy Resins

Table 11.1. Characterization of Commerical BPA Epoxy Resins

Resins	n Value	EEW	Melting Point (°C)
Standard liquid	0.13	190	Liquid
1001 Type	2	500	65–75
1004 Type	5.5	950	95–105
1007 Type	14.4	2250	125–135
1009 Type	16	3250	145–155

The molecular weight of the polymer is controlled by the ratio of ECH/BPA. With a large excess of ECH, it is possible to make a resin that is dominantly DGEBPA, that is, where $n = 0$ in the general formula. The pure $n = 0$ compound is a crystalline solid but commercial grades of low molecular weight resin are liquids with n values of 0.11 to 0.15 (so-called standard liquid resin).

As the ratio of ECH/BPA is reduced, that is, as ECH/BPA approaches 1:1, the molecular weight and the n value of the epoxy resin increase. Viscosity also increases with molecular weight. Above an average n value of 1, the resins are largely amorphous solids with increasing T_g. Although the resins are said to have melting points, they do not melt in the sense that a crystalline solid does. Rather, under specified test conditions, the resins flow to some standard extent at the so-called melting point. For this reason the term softening point rather than melting point is sometimes used. The commercially available higher molecular weight resins are often designated as types 1001, 1004, 1007, and 1009. Table 11.1 gives average n values, epoxy equivalent weights (EEW), and *melting points* for commercial BPA epoxy resins [1]. Of course, as molecular weight increases, epoxy equivalent weight and the average hydroxy functionality also increase. In some even higher molecular weight epoxy resins, the amount of epoxy groups present is so small the resins are really just polyfunctional alcohols, commonly called *phenoxy resins*.

It is important to recognize that, to a minor but sometimes significant degree, side reactions occur. The most important of these is the ring-opening reaction of a phenoxide anion polymer end group with ECH at the less favored, more sterically hindered position (Eq. 11.1). The resulting 1,3-chlorohydrin derivative (after proton transfer) cannot ring close to give an oxirane and is relatively stable under the reaction conditions. This group is the main site of "nonhydrolyzable chlorine" that is commonly reported in specifications of commercial epoxy resins.

$$\tag{11.1}$$

Another side reaction is hydrolysis, probably HO^- catalyzed, of the epoxy groups to form the corresponding glycol, as shown in Eq. 11.2. Two percent of the chain ends in a representative liquid resin have glycol groups.

$$\text{Ar}-O-CH_2-CH\overset{O}{\overbrace{}}CH_2 \; + \; H_2O \longrightarrow$$

$$\text{Ar}-O-CH_2-\underset{\underset{OH}{|}}{CH}-CH_2-OH \tag{11.2}$$

Both of the above side reactions tend to reduce the epoxy functionality $\bar{f}_n$ in the resins. On the other hand, reaction of alcohol groups on the polymer backbone with ECH could result in side chain glycidyl ether groups, which would tend to increase $\bar{f}_n$. If there were no side reactions, the functionality of BPA epoxy resins would be 2, however, the net effect of the side reactions is that commercial BPA epoxy resins generally have $\bar{f}_n$ less than 2, commonly about 1.9. In some cases, this lower $\bar{f}_n$ can have an important effect on film properties.

The presence of terminal glycol groups can give lower viscosity resins since they result from chain termination, reducing molecular weight. It has been demonstrated that the presence of small controlled amounts of terminal glycol groups can have beneficial effects on adhesion [2].

The described procedure (Scheme 11.1) for making BPA epoxy resins is called the *taffy process*. Stoichiometric amounts of NaOH are required, resulting in formation of relatively large quantities of NaCl, which must be removed from the resin by washing with water. The washing step is relatively easy for the standard liquid resin because of its low viscosity. However, as the ratio of ECH/BPA is decreased, higher average molecular weight products are produced and the reaction mixture becomes highly viscous, and water washing becomes difficult. Also, as the molecular weight and viscosity increase, the probability of branching increases. For these reasons the taffy process is now used only for resins with average n values less than 4.

Higher molecular weight resins are prepared by reacting the standard liquid epoxy resin ($n = 0.13$) with BPA in the presence of proprietary catalysts, which are generally nucleophiles. The catalyzed reaction of BPA with both epoxide groups of the standard resin results in a higher molecular weight resin with BPA end groups, which further react with standard resin to give epoxide end groups, and so on. The average molecular weight depends, of course, on the ratio of liquid resin to BPA. There can be variations in the resins obtained depending on the specific catalyst used. This procedure for making epoxy resins is called the *advancement process*. The reaction is carried out at higher temperature, hence the viscosity is lower, mixing is better, and there is less branching. No NaCl is produced; therefore, the product does not have to be washed free of salt, considerably simplifying the manufacturing process as compared to the taffy process for synthesis of higher n value resins.

The development of advanced HPLC analytical techniques has permitted substantial improvements in process development and control. Not only can the individual oligomers ($n = 0,1,2,3$, etc.) be separated, but also the oligomers where one or both ends have 1,3-chlorohydrin or 1,2-dihydroxy groups [3] (see Eqs. 11.1 and 11.2). Analysis of the products resulting when minor process changes are made in reaction conditions permits establishing process parameters to meet relatively narrow product specifications. Tight process control is particularly critical for epoxy resins that will be used in electronic applications and powder coatings (see Chapter 31).

It is of interest that resins made by the taffy process consist of oligomers with $n = 0,1,2,3,4,5$, and so on, whereas resins made by the advancement process have largely even number n value oligomers, that is, $n = 0,2,4,6$, and so on. The predominance of oligomers with even number n values in resins from the advancement process follows from starting with the diglycidyl ether of BPA. Roughly 10 wt% of odd n molecules result from the presence of about 10% of $n = 1$ resin in the liquid resin. It follows that epoxy resins from the taffy and advancement processes with the same average n value have different proportions of oligomers; this can be expected to affect their properties.

Bisphenol A epoxy resins perform especially well in coatings applications where excellent adhesion and corrosion resistance are required. A major limitation in their use is poor exterior durability primarily resulting from direct absorption of UV radiation by the aromatic ether groups, which ultimately leads to photoxidative degradation (see Chapter 25).

11.1.2. Other Epoxy Resins

A wide variety of other epoxy resins is manufactured.

Bisphenol F (BPF) epoxy resins (from reaction of ECH with BPF in place of BPA) have the advantage of lower viscosities at the same n value. Standard liquid BPA epoxies have viscosities of the order of 12 Pa·s at 25°C, whereas, the viscosities of comparable BPF epoxies are 6 Pa·s or less.

BPF

Still lower viscosities can be obtained by reacting BPA epoxies with an alcohol such as n-butyl alcohol followed by reacting the resultant hydroxyl groups with ECH, using $(CH_3)_4NCl$ as catalyst [4]. A new epoxy resin is obtained, a major component of which is 2,2-bis[4-(3-n-butoxy-2-glycidyloxy-1-propxy)phenyl] propane. Viscosities of the order of 1 Pa·s are reported [5]. The effect of the butyl ether group is to lower the T_g and hence the viscosity. Cross-linked films are, as would be expected from the lower T_g, softer and more impact resistant than comparable films using BPA resins.

2,2-Bis[4-(3-n-butoxy-2-glycidyloxy-1-propoxy)phenyl] propane

Epoxy resins are also prepared from reaction of ECH with novolac phenolic resins (see Section 11.8.2). The resulting *novolac epoxy resins* are useful in applications where more than two epoxy groups per molecule are desirable. Epoxy resins derived from the reaction of *o*- or *p*-cresol–formaldehyde novolacs and ECH are available, with an epoxy $\bar{f}_n$ of 2.2 to 5.5. A general structure of novolac epoxies is shown below.

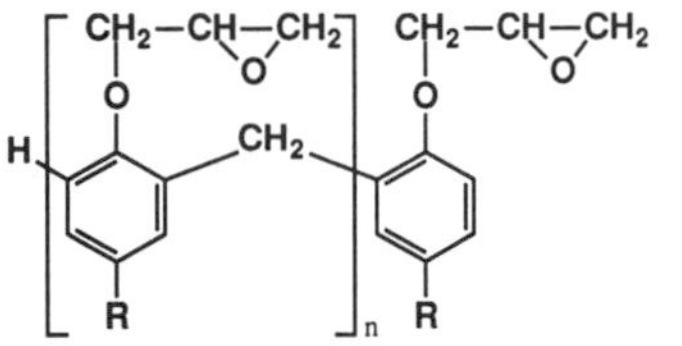

Epoxy Novolac Resin

Epoxy resins derived from hydrogenated-BPA (HBPA) and ECH also exhibit lower T_g and hence lower viscosity at the same *n* value. Hydrogenated-BPA epoxy resins also exhibit substantially better exterior durability, relative to both BPA and BPF resins, resulting from the absence of the UV absorbing aromatic ether groups.

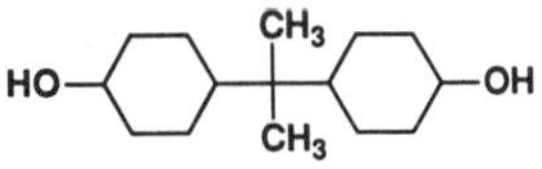

Hydrogenated-Bisphenol A (HBPA)

Triglycidylisocyanurate (TGIC) is a solid trifunctional epoxy cross-linker that is finding wide acceptance in powder coatings. The presence of three functional groups gives higher cross-link density than can be obtained with BPA epoxy resins, and the photochemical stability of the cured coatings is far superior. However, there are concerns that use of TGIC may present toxic hazards.

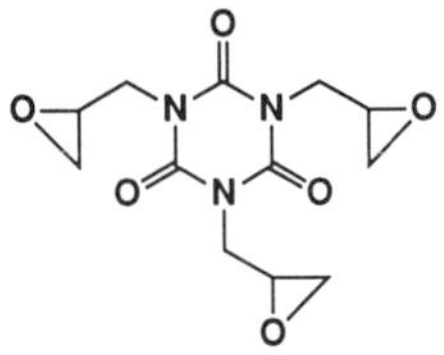

Triglycidylisocyanurate

A variety of other aliphatic epoxy products is available. Some are made by the reaction of polyols, such as glycerol, sorbitol, and polyethylene or polypropylene glycols, with ECH in the presence of a Lewis acid catalyst. These aliphatic epoxy resins give films with better exterior durability than can be obtained with BPA epoxy resins.

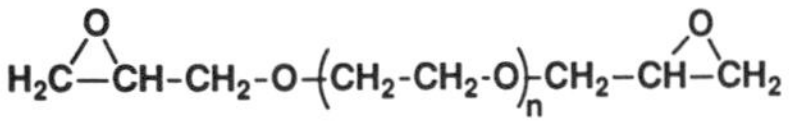

Polyethylene Glycol Diglycidyl Ether

Epoxy resins with lower viscosity than BPA epoxy resins, and yet, with many of the excellent properties characteristic of BPA epoxies can be made by substituting a flexible diol for BPA in the advancement process to make what might be called copolymer epoxy resins. Examples are advancement resins made from combinations of propylene or dipropylene glycol with BPA [6]. When cross-linked with phenolic resins such resins form films that are more flexible than films from homopolymer BPA epoxy resins and have excellent adhesion.

Also available are low molecular weight cycloaliphatic diepoxy compounds such as **1**, prepared by epoxidation of the corresponding alkenes, generally with peracetic acid. While the amounts used in coatings are small, such low molecular weight epoxy derivatives are particularly useful as reactive diluents in cationic UV curable coatings (see Chapter 32).

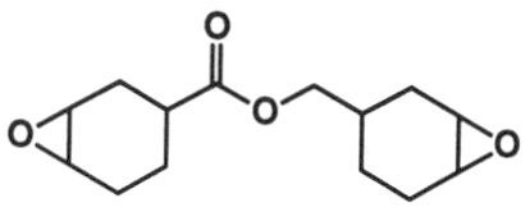

1

An increasingly important approach to making epoxy-functional resins is by radical polymerization of acrylic esters with glycidyl methacrylate (GMA) as a comonomer. By appropriate selection of comonomers, GMA content, and molecular weight, a wide range of products can be made. Exterior durability and acid resistance can be excellent. Glycidyl methacrylate modified acrylic resins are finding substantial use in top coats for automobiles, among other applications.

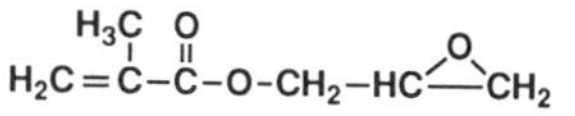

Glycidyl Methacrylate (GMA)

11.2. EPOXY–AMINE SYSTEMS

Epoxy groups react at ambient temperatures with primary aliphatic amines to form secondary amines and with secondary amines to form tertiary amines. Tertiary amines can react at higher temperatures to form quaternary ammonium compounds.

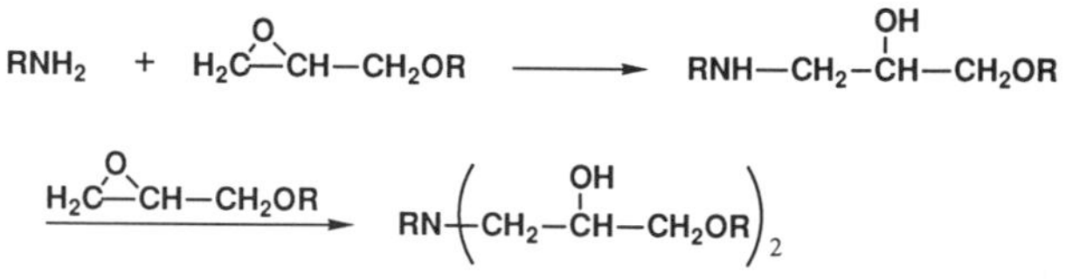

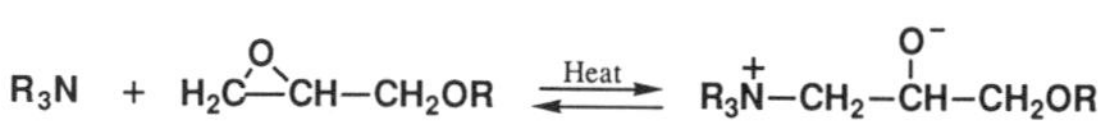

Reaction rates depend on epoxy and amine structure and concentration, catalysis, and media effects. Terminal epoxy groups, such as glycidyl ethers and esters, are generally more reactive than internal epoxy groups (e.g., the cycloaliphatic diepoxide **1**), which are more sterically hindered.

Reactivity of amines tends to increase with base strength and decrease with steric crowding. The general order of reactivity, primary > secondary >> tertiary amines can be attributed to steric effects, as well as to the absence of a transferable proton in the case of tertiary amines. Aliphatic amines are much more reactive than aromatic amines, which are much less basic.

The reaction is catalyzed by alcohols and by weak acids, most notably by phenols, which promote ring opening by proton complexation with the epoxide oxygen. Strong acids are not effective catalysts for the reaction. Hydrogen-bond acceptor solvents tend to reduce reaction rates, probably by complexing with hydrogen donors in competition with the epoxy group.

SCHEME 11.2

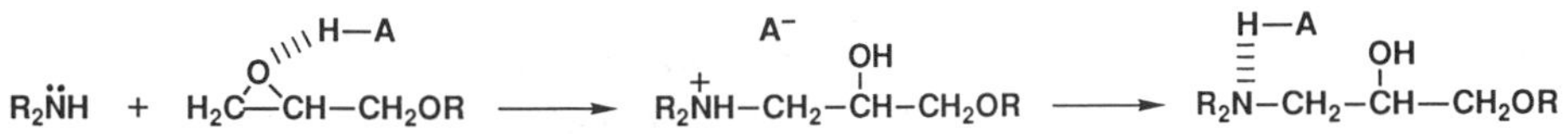

$$R_2\ddot{N}H \; + \; H_2C\overset{O}{\overset{|}{-}}CH\!-\!CH_2OR \;\xrightarrow{\;\;H\!-\!A\;\;}\; R_2\overset{+}{N}H\!-\!CH_2\!-\!\underset{OH}{CH}\!-\!CH_2OR \;\xrightarrow{\;\;A^-\;\;}\; R_2\overset{H\!-\!A}{N}\!-\!CH_2\!-\!\underset{OH}{CH}\!-\!CH_2OR$$

Scheme 11.2 shows the attack of an epoxy group by the nonbonded electron pair of a secondary amine. Attack is primarily at the less sterically hindered terminal end of the epoxy group. The scheme also shows catalysis with a weak acid (HA) by hydrogen complexation and hydrogen donation to the epoxide oxygen, which facilitates ring opening. The catalyst is regenerated by removing the proton from the amine nitrogen, which takes on a positive charge during the reaction. The conjugate base A^- could also participate in the catalysis by assisting removal of the amine proton *during* the ring-opening reaction. Based on this reasonable possibility, an explanation for the more effective catalysis by weak acids relative to strong acids noted earlier can be hypothesized. A strong acid (e.g., HCl) will exist predominately as the amine salt ($R_2NH_2^+Cl^-$). Because of the large excess of amine reactant as compared to the catalyst, even weaker acids such as phenols are also expected to exist largely as the corresponding salts ($R_2NH_2^+ArO^-$). These considerations suggest that the proton donor in the reaction is the same amine ion in both cases. On the other hand as a proton acceptor ArO^- is a substantially stronger base than Cl^-, because conjugate base strength increases as acid strength decreases. The proton acceptor ArO^- more effectively promotes the second reaction shown in Scheme 11.2. It is interesting that 2,4,6-[tris(dimethylaminomethyl)]phenol, which has both phenolic and tertiary amine groups, is an important commercial catalyst for epoxy–amine reactions.

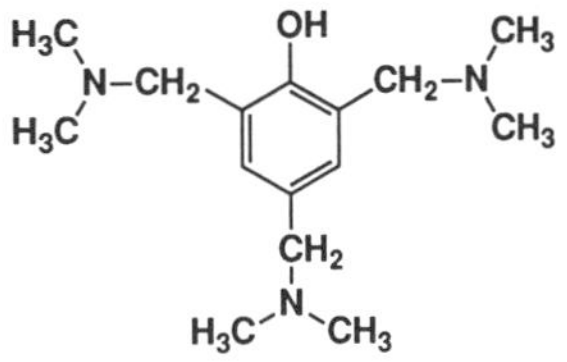

2,4,6-[tris(dimethylamino)]phenol

Although we are unaware of experimental evidence directly supporting this hypothesis in epoxy–amine reactions, the general importance of concerted weak acid–weak base catalysis is well established [7]. Furthermore, other curing reactions used in coatings are more effectively catalyzed by weak acids or bases than by their stronger counterparts. For example, the weak acid catalysis of the reaction of Class II melamine resins with alcohols, discussed in Section 6.1.2, may fall into this category.

Epoxy–amine systems are important in several coatings applications. Epoxy–amine reactivity is too high at ambient temperatures to allow sufficient storage stability of a coating containing polyamine and polyepoxide in the same package. Two-package (2K) coatings are required. With many aliphatic amines the pot life is limited to a few hours and the coating will generally take about a week to cure at ambient temperature.

11.2.1. Pot Life–Cure Time Considerations

It is usually a major goal in epoxy coatings to maximize pot life and minimize cure time (see Section 3.2 for a discussion of kinetic limitations that prevent both long-term, one-package stability and moderate cure temperatures and of ways to obviate these limitations). Many factors must be considered, including reactive group concentrations, the structural effects of amine, epoxy, and solvents on reaction rates, as well as the equivalent and molecular weights, and the $\bar{f}_n$ of the reactants. Using concentration as an example, as the molecular weight of a BPA epoxy resin is increased, the number of equivalents per liter of epoxy groups will decrease; therefore, the rate of reaction will be lower. Furthermore, as the molecular weight increases, the viscosity increases so that, in order to formulate to the same viscosity, the amount of solvent must be increased. This leads to a decrease in concentration of both the amine and epoxy groups and to a lengthening of pot life. Unfortunately, from this point of view, the increasing cost of solvents and the need to reduce VOC emissions, because of air quality standards, are forcing formulation of higher and higher solids coatings, which can be expected to have shorter and shorter pot lives.

An approach to the problem of lengthening pot life without importantly reducing cure time is the use of blocked amine cross-linkers. Ketones will react with primary amines to give ketimines that will not readily react with epoxy groups. However, ketimines will hydrolyze with water to release the free amine plus the ketone, which is the reverse reaction of ketimine formation, as shown. This reversible reaction constitutes the basis for ketimine–epoxy moisture cure coatings, which are stable in a moisture-free environment but cure after application and exposure to ambient moisture. More commonly, methyl ethyl ketone (MEK) is used. The high volatility of MEK from thin-coating films minimizes reverse reaction with amine.

$$RN=\underset{\underset{CH_3}{|}}{C}-CH_2CH_3 \; + \; H_2O \; \rightleftharpoons \; RNH_2 \; + \; H_3C-\overset{\overset{O}{\|}}{C}-CH_2CH_3$$

Ketimine   MEK

Ketimine–epoxy systems are stable indefinitely in the absence of water and should, therefore, permit formulation of one-package coatings. Nevertheless, they are most commonly used in long pot life two-package coatings. This occurs because of the difficulty of drying all of the components in a coating. Most epoxy–amine coatings are pigmented, and pigment surfaces have water on them before they are dispersed. This water remains in the coating and will hydrolyze the ketimine. While removal of water from the surface of the pigment is possible, it adds to the cost. The amount of water is not high, usually less than 1% of the pigment weight, but the molecular weight of water is so low that a little water will hydrolyze a large amount of ketimine. While pigment is the major source of water, solvents also contain water, and use of anhydrous solvents would also increase costs.

An intriguing approach to increasing the storage stability of epoxy coatings sufficiently to permit formulation of 1K coatings that can be cured at moderately elevated temperatures is the use of a cross-linker that must undergo a phase change in order to react. A commercially available solid polyamide cross-linker, prepared from PA and diethylene triamine, is an example [8]. A fine particle size dispersion of the cross-linker in an epoxy resin is also available commercially and has been shown to be stable for at least 6 months at 25°C. When a film is applied and heated above 100°C, the solid polyamide liquefies, which greatly facilitates intramolecular cyclization to yield epoxy-soluble amine-functional products. The amine groups that are generated react with epoxy groups to give a cross-linked film. This process is shown below. Note that the imide, amine (DETA), and bis-imide products have functionalities of 3, 5, and 1, respectively. Thus, the bis-imide cannot participate in cross-linking.

Phthalic Anhydride/DETA Polyamide →

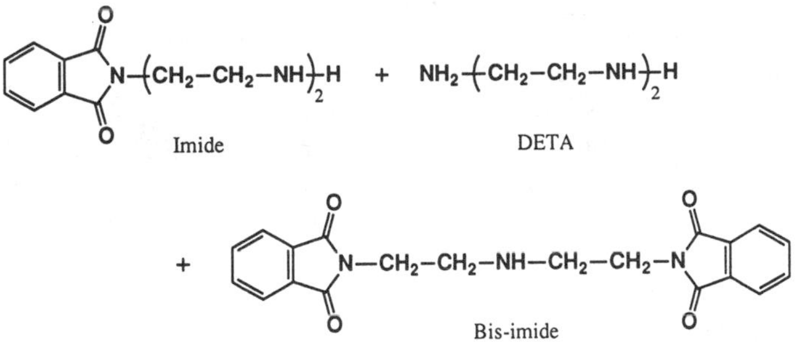

Imide   DETA

Bis-imide

Since the reaction cannot occur when the temperature is substantially below the T_g of the solid cross-linker, package stability is indefinitely long. But when the temperature is increased above the T_g, free volume increases sufficiently for the cyclization reaction to occur, yielding the soluble amine-functional products. Thus, the reaction rate during storage is not controlled by the kinetics of the reaction, so the Arrhenius restrictions on package stability versus cure schedule (see Section 3.2.2) do not apply.

A commonly used cross-linker for epoxy powder coatings is dicyandiamide (dicy), a crystalline compound (mp 205°C), which also provides latency by insolubility.

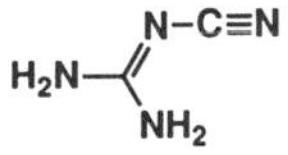

Dicyandiamide

Dicyandiamide apparently decomposes into soluble reactive products below its melting point, and cure temperatures can be reduced substantially by using as "accelerators" tertiary amines and imidazoles, such as 2-methylimidazole. The reactions with dicyandiamide are complex; various proposals have been made. It has been suggested that dicy initially decomposes into two cyanamide (H_2N—$C\equiv N$) molecules that can then react with an epoxy resin. A more recent study indicates that the initial reaction is between an amino group of the dicy and an epoxy group to form an *N*-alkyl dicyandiamide followed by a complex series of reactions [9].

11.2.2. Toxicity and Stoichiometry Considerations

Other factors are involved in selecting amines besides reaction rates. Many amines are toxic. While they are easily handled safely in a chemical plant, toxic hazards can arise if inexperienced, careless, or uninformed personnel mix and apply two-package epoxy coatings incorporating certain amines.

For example, diethylene triamine (DETA) is a very efficient cross-linker for epoxies, but the handling hazard is high. In general, toxic hazards are reduced by increasing molecular weight and reducing water solubility. As molecular weight increases, volatility decreases, which reduces the chances of inhaling dangerous amounts of amine. Also, as water solubility decreases and molecular weight increases, permeability through body membranes such as skin decreases, generally reducing the toxic hazard. There are, of course, exceptions to these broad generalizations. Safety data sheets supplied by the manufacturer should always be read for safe handling recommendations.

Other disadvantages of utilizing low molecular weight, highly functional amines are their low equivalent weights and viscosities. The equivalent weight of pure DETA is 21. If DETA is used with an epoxy resin having an equivalent weight of about 500, the stoichiometric weight ratio of the two components would be about 25:1. This disparate amount of the two components would increase the difficulty of obtaining proper mixing and would also result in a high risk of significant error in mixing stoichiometric amounts in two-package coatings. The difference in viscosity between the DETA and epoxy resin also makes uniform mixing difficult.

One approach to designing amine cross-linkers with higher equivalent weight and lower toxic hazard is to make so-called *amine adducts*. Standard liquid BPA epoxy ($n = 0.13$) is reacted with an excess of a polyamine such as DETA. The excess polyamine is then removed by vacuum distillation leaving the amine-terminated adduct, as shown by the following idealized structure. The low molecular weight DETA is handled only in the chemical factory where proper precautions are easily taken.

Epoxy Amine Adduct

Similar adducts are prepared using a variety of amines to provide adducts that will permit a range of cure rates and pot lives.

A second important approach is to react the polyamine (e.g., DETA) with dimer acids to form amine-terminated polyamides. Dimer fatty acids are complex mixtures, predominately C_{36} dicarboxylic acids, which are made by acid catalyzed dimerization of unsaturated C_{18} fatty acids (see Section 9.3.1). The reaction yields amine-functional polyamides, often called polyamides, another example of confusing terminology, since the amide nitrogen groups are unreactive. Furthermore, there are similar polyamides, which are not amine-functional and not useful as cross-linkers for epoxy resins. To minimize confusion the products are sometimes called "amido-amines."

An example of a widely used amine-functional polyamide is made by reacting dimer acids with excess DETA. A mixture of products is obtained including the simplest amido-amine shown below. The amine equivalent weight is the corresponding molecular weight divided by the number of reactive NH groups (six in the amido-amine shown below).

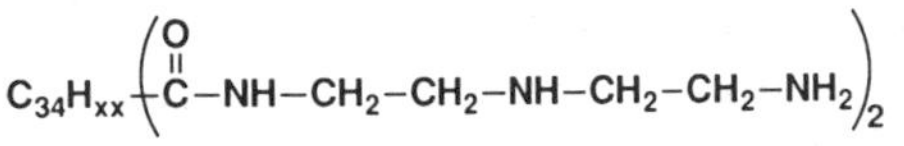

Amido-amine

In the preparation of polyamides using a polyamine such as DETA, the amido-amine groups can react further by eliminating water and forming a terminal imidazoline group. The extent of this reaction varies depending on reaction conditions. Imidazoline formation reduces viscosity but also reduces average functionality [10].

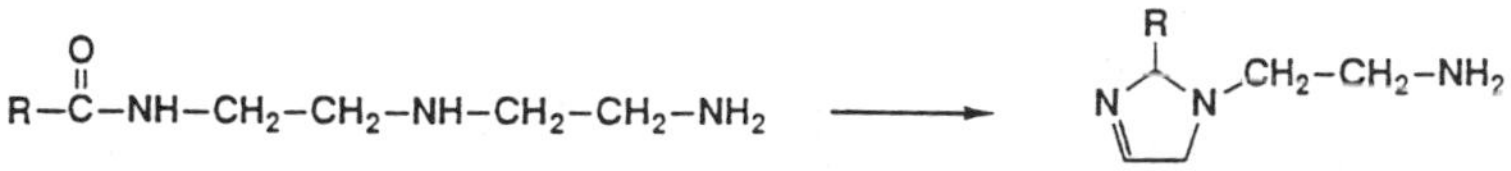

Imidazoline

A wide range of polyfunctional amines is used in making amido-amines, including triethylene tetramine (TETA) and aromatic diamines, such as *m*-phenylenediamine. Some suppliers have built a specialty business selling proprietary amine cross-linkers, fairly commonly called *epoxy hardeners*.

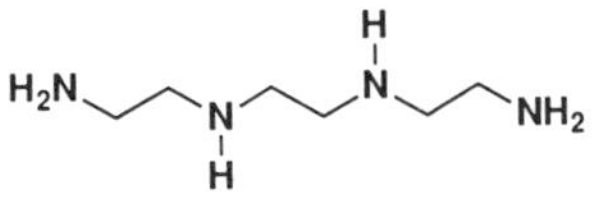

Triethylene tetramine (TETA) 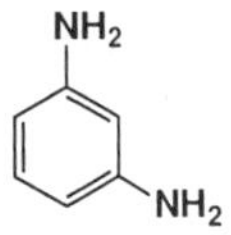   m-Phenylenediamine

11.2.3. Graininess and Blushing

Bisphenol A epoxy resins and polyamides are mutually soluble in the solvents (often xylene) used in epoxy–amine coatings, but most are not compatible in the absence of solvent. Thus, as solvent evaporates phase separation can occur, resulting in a rough surface, called "graininess" and possibly in altered film properties. Graininess can be avoided by allowing the coating to stand for 30 min to 1 h after the two packages are mixed. Partial reaction of the two components takes place during the waiting period. The elimination of graininess can be attributed to increased viscosity and/or the formation of reaction products that improve compatibility of the mixture. When standard liquid resin is used, longer times are required after mixing to avoid phase separation. Proprietary amine cross-linkers are available that exhibit better compatibility, minimizing the problem [10].

Another phenomenon, called "blushing," or "scumming," is the appearance of a grayish, greasy deposit on the surface of films, usually accompanied by incomplete surface cure. Blushing is thought to result from the formation of relatively stable bicarbonate salts of some amine groups on exposure to carbon dioxide and water vapor in the atmosphere.

$$RNH_2 + CO_2 + H_2O \longrightarrow R\overset{+}{N}H_3 \, HCO_3^-$$

As with graininess, it is often possible to minimize this type of blushing by mixing the epoxy and amine components 1 h or so before application. It can be speculated that during the waiting period some of the most reactive amine groups (those most likely to form carbonates) react with epoxy groups, so that when the film is applied after this waiting period, blushing is less likely to occur. If the waiting period is too long, the viscosity increase will be excessive.

In coatings the term blushing is also used to describe other phenomena, for example, the whitish appearance sometimes encountered when coatings with fast evaporating solvents are sprayed on humid days; like many terms in the field the same word may be used to mean quite different phenomena. One must be careful to be sure how a term is being used.

11.2.4. T_g Considerations

It is critical to select a combination of epoxy resin and amine cross-linker that will lead to a final T_g that will permit relatively complete reaction of the amine and epoxy groups at the temperature that will be encountered in actual application. T_g increases as polymerization and cross-linking proceed. If the T_g of the final reacted coating is higher than the cure temperature, the reaction may become very slow while a substantial fraction of the functional groups are unreacted. As the T_g of a homogeneous network approaches the cure temperature, $(T - T_g)$ and free volume decrease, and the reaction rate becomes limited by the mobility of the reactants rather than by their reactivity. Limited mobility would result in a substantial decrease in reaction rate. If the T_g reaches a value of about 40 to 50°C above the reaction temperature, the reaction will probably essentially cease (see Section 3.2.3).

It follows that the combination of epoxy resin and amine cross-linker should be chosen so that the T_g of the cured coating is below or only a little higher than the cure temperature. Otherwise, unreacted functional groups will be present, which may cause adverse effects on mechanical properties and solvent resistance. For example, if a coating, which cures well when applied to an oil rig in the Caribbean, is applied to the support structure of an offshore oil rig in the North Sea, where the water temperature even in summer does not exceed 4°C, adequate curing may not occur.

An example indicating the importance of such considerations is a report that, after 7 days curing at 25°C, films from the butyl ether modified BPA epoxy resin

(Section 11.1.2) cross-linked with an amine adduct have better methyl alcohol resistance than films made from BPA epoxy resins and the same amine adduct [5]. There is no evident explanation for this result based on the chemical compositions. One can speculate that the superior methyl alcohol resistance may result from a greater extent of reaction of the lower T_g butylated derivative before mobility limitation slows down the reaction.

11.2.5. Other Formulating Considerations

If the coating is to be applied under water, it is essential that the polyamine not be soluble in water and that solubility of water in the polyamine is minimal.

Epoxy–amine coatings tend to exhibit limited solvent resistance and are particularly susceptible to attack by acidic solvents, such as acetic acid. At least in part, this sensitivity probably results from diffusion of acetic acid into the film followed by formation of acetate salts with the amine groups in the system. The hydrophilic salt groups then increase the solubility of water in the film, thereby increasing water permeability and softening the film, making it more susceptible to damage.

This situation is exacerbated when the cross-link density is low. Bisphenol A epoxy resins have an $\bar{f}_n$ of about 1.9 so that, even with highly functional amine components, cross-link density is limited, especially if there is deviation from stoichiometry. Since the amines are usually polyfunctional, it is generally best to formulate with a small excess of amine cross-linker (~10%) to assure that the epoxy groups are fully reacted. The problem can also be alleviated by use of higher functionality epoxy resins (e.g., novolac epoxies), which have $\bar{f}_n$ up to 5. With the higher functionality and more viscous novolac epoxies, care must be exercized in the selection of the amine cross-linker to ensure that adequate reaction can occur in spite of changing the epoxy resin. Sometimes blends of BPA and novolac epoxy resins are used.

Another factor to be considered in formulating epoxy–amine coatings is the possible effect of solvent composition on the coating. As noted earlier, hydrogen-bond acceptor solvents will extend pot life. However, esters should be avoided since they undergo aminolysis, especially with primary amines, even at room temperature. Alcohol solvents and water can affect the stability of the epoxy component package. Alcohols react very slowly with epoxy groups at room temperature, but, over a period of months, appreciable reaction is possible. When the reactant is a monoalcohol, there is little change in viscosity. Coating formulators generally judge package stability by changes in viscosity; but, in this case, there could be a significant change in epoxy functionality without a significant change in viscosity. To the extent that epoxy groups react with an alcohol, the potential for cross-linking is decreased. As repeatedly noted, the $\bar{f}_n$ of a BPA epoxy resin is somewhat less than 2, and further loss of functionality could cause the final film to have inferior properties. Apparently, in many systems, there is no difficulty. For example, some epoxy resins are sold as solutions in glycol monoethers. However, in other cases, a decrease in epoxy content has been observed when stored over a period of several months.

A similar situation can arise with water. Again, reaction of an epoxy group with water does not lead to cross-linking; however, epoxy groups are used up. In some cases, where the epoxy package is pigmented with TiO_2, it has been shown that epoxy content decreases with storage time, presumably as a result of the reaction of water from the surface of the TiO_2 with the epoxy groups. This reaction, as well as that with alcohols, may be catalyzed by basic and/or acidic impurities, as well as alumina (basic) and silica (acidic), generally present as surface treatments on TiO_2. Again, in many cases, epoxy-package pigmented coatings have been used without problems. Apparently, the effects of both alcohols and water on package stability are system dependent. It is recommended, in view of such reports, that package stability be checked periodically by epoxy group analysis rather than by relying solely on viscosity changes.

11.2.6. Water-Borne Epoxy–Amine Systems

Because of the need to reduce the solvent content of coatings, water-borne epoxy–amine coatings have been widely investigated. One approach has been to make emulsion systems. Incorporation of emulsifying agents in either or both the amine and the epoxy package permits the addition of water during the mixing. Although the application solids of the resulting water-borne coating will be lower than that of an organic solvent-soluble system, the VOC content will also be lower since a substantial fraction of the volatile material will be water. Some loss of epoxy groups can be expected during the pot life of the mixed coating due to the high water content. While such systems have been extensively studied, their commercial uses are limited by the adverse effect of surfactant on the coatings performance.

A novel approach to the surfactant problem is the utilization of a weakly acidic solvent, such as nitroalkanes [11,12]. Nitroalkanes form salts of amines, as shown below, and the salt groups stabilize epoxy–amine emulsions in water. Following application, the nitroalkane solvent evaporates, shifting the acid–base equilibrium to the free amine, that is, to the left in the equation. Thus, the amine–nitroalkane combination functions as a transient surfactant, which stabilizes the emulsion during storage and yet is not present to adversely effect the final film properties. The conversion of amine groups to salts is also expected to prolong the pot life of the mixed composition, since the polar salt groups are oriented outwards into the water phase while the epoxy groups are in the interior of the emulsion particles.

$$R-NH_2 \ + \ R_2CH-NO_2 \ \rightleftharpoons \ R-\overset{+}{N}H_3 \ \ R_2C=NO_2^-$$

Another approach to water-borne systems is through the use of salts of amine-functional resins. The resins in commercial use are proprietary. The amine groups of a concentrated solution of the resin in organic solvents are neutralized with hydrochloric acid. When the solution is diluted with water, polymer aggregates, swollen with solvent and water, are formed with the amine salt groups on the outer periphery of the aggregates suspended in a continuous water phase. The behavior is analogous to that of water-reducible resins discussed in Section 7.3. A solution of epoxy resin in solvent can now be mixed into the system. The epoxy resin will enter the inside of the resin aggregates. The epoxy groups are, in this way, kept separate from the amine hydrochloride groups permitting a pot life of several days. After the coating is applied, the water and solvent evaporate leaving the amine hydrochloride and epoxy groups in the same phase. They react to yield a chlorohydrin and a free primary amine. Then the amine can react twice with two additional epoxy groups. Since BPA epoxy resins have less than two epoxy groups per molecule and about one in three of these groups will be converted to chlorohydrins, it is desirable to use at least some novolac epoxy with an $\bar{f}_n$ up to 5. Then, when one-third of the groups are converted to chlorohydrins, there would still be an average of more than two epoxy groups per molecule to cross-link with amine groups on the resin.

Resins used in automotive primers applied by cathodic electrodeposition are prepared by the reaction of epoxy and amine resins. The use of aqueous dispersions of carboxylate salts of such resins are discussed in Chapter 30.

11.3. OTHER CROSS-LINKING AGENTS FOR EPOXY RESINS

While amines are the most widely used cross-linking agents for epoxy resins in coatings applications, a wide variety of other reactants is used.

11.3.1. Phenols

As is evident from the way in which BPA epoxy resins are made, they can be cross-linked with phenolic resins. Both resole and novolac phenolic resins, which are discussed in Sections 11.8.1 and 11.8.2, respectively, can be used. Structures of resole and novolac phenolics are shown in Section 11.8.

The reaction with phenols occurs predominantly at the less hindered CH_2 site of unsymmetrical epoxides (e.g., glycidyl derivatives) as shown.

The phenolic hydroxyl groups of both resole and novolac phenolics react accordingly with epoxy groups on the epoxy resins. In addition, the methylol groups of the resole phenolics undergo self-condensation and probably also react with hydroxyl groups on the epoxy resin. Thus, cross-link density tends to be higher with resole phenolics. Bisphenol A epoxies as well as polyfunctional epoxy resins (e.g., novolac epoxies) can be used. The systems require baking, yet package stability is relatively limited.

Package stability is enhanced with etherified resole resins. Increased solids combined with high functionality are reported by using butoxymethylolated BPA as the phenolic resin [5]. For other examples see Section 11.8.3.

Unpigmented epoxy–phenolic coatings have been used in linings for beverage cans and for some types of food cans. Pigmented epoxy–phenolic coatings are used as high-performance primers. In both cases, the major advantages are adhesion to metals, even in the presence of water, and complete resistance to hydrolysis. In both of these applications neither the discoloration, which occurs on baking, nor the poor exterior durability is an important factor.

11.3.2. Carboxylic Acids

Carboxylic acids are effective cross-linkers for baked epoxy coatings. The reaction of a carboxylic acid and epoxy group yields a hydroxy ester. Ring opening occurs predominantly at the less-hindered CH_2 carbon, although reaction at the more hindered CH—R site is significant.

$$R'-\overset{\overset{O}{\parallel}}{C}-OH \quad + \quad H_2C\overset{O}{\diagup}CH-R \quad \longrightarrow \quad R'-\overset{\overset{O}{\parallel}}{C}-O-CH_2-\overset{\overset{OH}{\mid}}{C}H-R$$

Without catalyst, relatively high curing temperatures (~150°C for 30 min) are required. In spite of this sluggish reactivity, there can be storage stability problems. This signifies that the temperature dependence of the reaction rate is low; in other words, the reaction has a low activation energy, E_a (see Section 3.2.2). The reaction has been shown to be second order in carboxylic acid. Most likely, one of the acid groups functions as a nucleophile by attack at the CH_2 group, whereas the other functions as an electrophile and assists ring opening by complexation with the epoxy oxygen (see Scheme 11.2). The second-order dependence of reaction rate on acid concentration also results in a rapidly decreasing rate with conversion, which exacerbates the difficulty of achieving high conversions.

$$\text{Rate} = k[\text{epoxy}]\,[\text{RCOOH}]^2$$

The third-order dependence of the reaction (first order in epoxy and second order in carboxylic acid) also results in a small Arrhenius A value owing to the high molecular ordering of the epoxide and two carboxylic acid groups required in the transition state, further reducing reactivity at all temperatures.

Tertiary amines catalyze the reaction of carboxylic acids with epoxies. Triphenylphosphine is reported to be a particularly effective catalyst. With triphenylphosphine catalyst and an excess of epoxy groups to carboxylic acid groups, coatings can be formulated that will cross-link at 25°C [13].

Apparently, the reaction of hydroxyl groups with epoxies competes with that of carboxylic acid groups, at least at high temperatures. Therefore, when carboxylic acids are used as cross-linking agents, at least part of the cross-links result from the reaction of epoxies with hydroxyl groups, which are originally present on the epoxy resin or are generated in the epoxy-carboxylic acid reaction. Esterification of carboxylic acid groups with hydroxyl groups may also occur.

Carboxylic acid-functional acrylics can be cross-linked with BPA epoxy resins. Furthermore, acrylic copolymers with pendant epoxy groups, made using glycidyl methacrylate (GMA) as a comonomer, will cross-link with carboxylic acid-functional acrylate copolymers. Self-cross-linking acrylics can be made by incorporating both (meth)acrylic acid and GMA in the same polymer. Although at first blush this idea seems attractive, it is seldom as desirable as using two acrylic resins, one with the epoxy groups and the other with the carboxylic acid groups. Blending two resins provides more latitude to the formulator and better resin stability. While the reaction rate of epoxy groups with carboxylic acid groups at storage temperatures is slow, it is not zero. When self-cross-linking resins are made, the storage stability clock starts when the resin is made. On the other hand, when the functional groups are on two different resins, the clock does not start until the resins are mixed to make the liquid coating.

Increasing use is being made of the epoxy–carboxylic acid reaction to cross-link aliphatic and cycloaliphatic epoxy resins. In some cases coatings with excellent exterior durability have been claimed.

11.3.3. Anhydrides

Closely related to carboxylic acids is the use of cyclic anhydrides, such as 1,2,4,5-benzenetetracarboxylic acid dianhydride (pyromellitic anhydride), as cross-linking agents for epoxy resins. While epoxy–anhydride compositions are widely used in some plastics applications, their use in the coatings field is largely limited to powder coatings.

Reaction of anhydrides with epoxy resins can occur initially with the epoxy resin hydroxyl groups, yielding esters and carboxylic acids. The resulting carboxylic acid groups then react with epoxy groups, as shown earlier. This reaction generates a new hydroxyl group, and so on. Epoxy groups can also react directly with anhydrides in catalyzed reactions. Tertiary amines are generally used as catalysts; they probably function primarily by reacting with the epoxy forming a transient zwitterion that then reacts with the anhydride.

11.3.4. Mercaptans

Mercaptans (thiols, RSH) react with epoxies to yield sulfides. The reaction is strongly catalyzed by tertiary amines, which convert the mercaptan into the more highly reactive mercaptide anion RS^-.

$$RS-H + R_3N \rightleftharpoons RS^- R_3\overset{+}{N}H$$

$$RS^- + H_2C\overset{O}{\overset{\diagup\!\!\!\diagdown}{-}}CH-R \longrightarrow RS-CH_2-\underset{\underset{OH}{|}}{CH}-R$$

The reactivity of the mercaptide anion is sufficient that ambient temperature cure two-package coatings can be prepared. Low molecular weight polysulfide rubbers are mercaptan-terminated polymers; in coatings, they have been used to cross-link with BPA epoxy resins in primers for aircraft. The unpleasant odor of mercaptans is a drawback for some applications.

11.3.5. Homopolymerization

Epoxy groups undergo homopolymerization in the presence of very strong acids (super acids). The acid can be generated thermally or photochemically from acid precursors. Generally, acid precursors that function thermally are called blocked or latent acids; those that function photochemically are called cationic photoinitiators. Ultraviolet curable epoxy coatings are discussed in Chapter 32. Acid-catalyzed homopolymerization of epoxies yield polyethers.

$$R-CH-CH_2 + H^+ \rightleftharpoons R-CH-CH_2 \xrightleftharpoons{H_2C-CH-R}$$

$$R-CH-CH_2-O{\overset{+}{\underset{CH-R}{\diagdown}}}{\overset{CH_2}{|}} \xrightleftharpoons{H_2C-CH-R} R-CH-CH_2-O-CH-CH_2-O{\overset{+}{\underset{CH-R}{\diagdown}}}{\overset{CH_2}{|}}$$

Suitable "super acids" are trifluoromethylsulfonic acid (triflic acid) (F_3CSO_3H), hexafluoroantimonic acid ($HSbF_6$), hexafluoroarsenic acid ($HAsF_6$), and hexafluorophosphoric acid (HPF_6). Since the stronger an acid, the weaker is its conjugate base, the corresponding counterions, in these cases, are highly nonbasic and nonnucleophilic. Only super acids are effective for homopolymerization of epoxies. Even relatively strong acids, such as HCl and p-toluene sulfonic acid (pTSA), are ineffective because the conjugate base of such acids are nucleophilic enough to add to the protonated epoxy group, preventing addition of a second epoxy group as in the homopolymerization reaction. The result is overall addition to form a chlorhydrin rather than polymerization.

$$H_2C-CH-R + HCl \longrightarrow R-CH-CH_2-Cl \\ \underset{OH}{|}$$

11.3.6. Cross-Linking by Reactions with Hydroxyl Groups

The cross-linkers discussed thus far react primarily with epoxy groups on the epoxy resins. Many epoxy resins have multiple hydroxyl groups on each molecule; in fact, higher molecular weight BPA epoxy resins have many more hydroxyl groups per molecule than epoxide groups. It has already been pointed out that cross-linkers for epoxy groups such as resole phenolic resins, carboxylic acids, and anhydrides may well react not only with epoxy groups but also with hydroxyl groups. Some cross-linkers react predominantly with the hydroxyl groups rather than with epoxy groups.

Amino resins, both MF and UF resins, are used; cross-linking occurs mainly between the activated ether groups of the MF or UF resin and the hydroxyl groups of the epoxy by transetherification (see Chapter 6). Generally, blocked acids, amine salts, or esters of pTSA or some other sulfonic acid, are used as latent catalysts.

Polyisocyanates also cross-link epoxies. Blocked isocyanates are preferred since they permit one-package stability (see Section 12.4.4 for a discussion of blocked isocyanates).

11.4. EPOXY ESTERS

Another use of epoxy resins in coatings is their conversion into what are commonly called *epoxy esters*. Epoxy esters are made by reacting BPA epoxy resins with fatty acids. Drying oil or semidrying oil fatty acids are generally used so that the products can be cross-linked by autoxidation. As shown in Section 11.3.2, the epoxy groups will undergo a ring-opening reaction with carboxylic acids to generate an ester and a hydroxyl group. These hydroxyl groups, as well as the hydroxyl groups originally present on the epoxy resin, can undergo esterification with fatty acids.

Epoxy esters are generally made by starting with a low molecular weight epoxy resin (i.e., the standard liquid resin, $n = 0.13$) and chain extending with BPA (advancement process) to the desired average molecular weight. The fatty acids are added to the molten, hot resin, and the esterification reaction is continued until the acid number is low, usually less than 7 mg of KOH per gram of resin.

In the esterification reaction with fatty acids, the average number of sites for reaction is the n value, corresponding to the number of hydroxyl groups, plus two times the number of epoxy groups. The esterification is carried out at high temperatures ($\sim$220–240°C). The rate of esterification slows as the concentration of hydroxyl groups diminishes, and side reactions, especially dimerization of the drying oil fatty acids (or their esters), can occur. For this reason, it is not practical to make epoxy esters with more than about 90% of the potential hydroxyl groups (including those from ring opening the epoxy groups) esterified. The lower useful limit of the extent of esterification is about 50%; this is required to ensure sufficient fatty acid groups for oxidative cross-linking.

Tall oil fatty acids are commonly used because of their low cost. Linseed oil fatty acids give faster cross-linking coatings due to the higher average functionality of activated methylene groups. However, the viscosity is higher because of the greater degree of dimerization during esterification; the cost is also higher. For still faster cross-linking, part of the linseed fatty acids can be substituted with tung oil fatty acids, but the viscosity and cost will be still higher. The color of epoxy esters from linseed and linseed–tung oil fatty acids is darker than the tall oil esters. Depending on cost and rate of cross-linking considerations, any fatty acids can be used.

As with alkyds (see Section 10.1) the rate of formation of a dry film from epoxy esters depends on two factors: the average number of methylene groups between double bonds $-(CH{=}CHCH_2CH{=}CH)-$ per molecule (from the fatty acids) and the ratio of aromatic rings/long aliphatic chains. Increasing the number of methylene groups activated by two double bonds increases $\bar{f}_n$, resulting in faster formation of solvent resistant films by oxidative cross-linking. The $\bar{f}_n$ can be maximized by using higher molecular weight BPA epoxy resin as a starting material and using enough fatty acid to react with a large fraction of the epoxy and hydroxyl groups. The ratio of aromatic rings/fatty acids can also be increased by using higher molecular weight epoxy resins and will be maximized by esterifying a smaller fraction of the available hydroxyl groups.

Epoxy esters are used in coatings where adhesion to metal is of major importance. While the reasons are not completely understood, it is common for epoxy coatings, including epoxy esters, to have good adhesion to metals and to retain that adhesion reasonably well after exposure of the coated metal to high humidity, a critical factor in corrosion protection (see Chapter 26 for a discussion on adhesion).

A distinct advantage of epoxy esters over alkyd resins is their greater resistance to hydrolysis and to saponification of coatings made with epoxy esters. The backbone of the alkyds is held together with esters from PA and the polyol whereas, in the case of epoxy esters, the backbone is held together entirely with C—C and ether bonds. Of course, the fatty acids are bonded to the backbone with ester groups in both cases, but the fraction of polymer bonds subject to hydrolysis or saponification is substantially lower in the case of epoxy esters. This is a critical advantage in corrosion resistance.

On the other hand, exterior durability of epoxy ester coatings is poor, as is the case with all films made from BPA epoxy resins. As a result of these advantages and disadvantages, the major uses of epoxy esters are in primers for metal and in metal decorating coatings, such as for crowns (bottle caps), where the important requirements are adhesion to metals and hydrolytic stability. In baking primers, it is sometimes desirable to supplement the cross-linking through oxidation of the drying oil fatty acids by including a relatively small amount of etherified MF resin in the formulation to cross-link with part of the free hydroxyl groups on the resin.

Epoxy ester resins with quite good exterior durability (better than that of alkyds) can be prepared by reacting epoxy-functional acrylic copolymers (made with glycidyl methacrylate as a comonomer) with fatty acids. The product is an acrylic resin with multiple fatty acid ester side chains. By appropriate selection of acrylate ester comonomers and molecular weight, the T_g of the resin can be designed so that a tack-free film is obtained by solvent evaporation. Subsequently, the coating will cross-link by autoxidation. For an application like repainting an automobile at ambient temperatures, if the T_g is high enough, the cross-linking can proceed relatively slowly and need not be catalyzed by metal salt driers. Of course the rate of cross-linking is slower without driers, but so is the rate of film degradation on exterior exposure.

Reaction of epoxy resins with acrylic acid to form acrylic esters gives another type of epoxy ester derivative used in coatings. These acrylate-functional resins are discussed in Section 13.2.

11.4.1. Water-Reducible Epoxy Esters

Epoxy esters can also be made water-reducible for similar end use applications. The most widely used water-reducible epoxy esters have been made by reacting maleic anhydride with epoxy esters prepared from dehydrated castor oil fatty acids (see Section 9.3.5). Subsequent addition of a tertiary amine, such as 2-(dimethylamino)ethanol (DMAE), in water results in ring opening of the anhydride to give amine salts. Analogously to other water-reducible resins discussed previously, these resins are not soluble in water but form a dispersion of resin aggregates swollen with solvent and water in an aqueous continuous phase. Storage stability of these coatings is superior to that of water-reducible alkyd coatings.

The hydrolytic stability of these epoxy esters is sufficient that they were used in electrodeposition primers on a very large scale until anionic primers were replaced by cationic electrodeposition primers (see Chapter 30). Water-reducible epoxy esters are still used in spray applied baking primers and primer surfacers. They are also used in dip coating primers where nonflammability is a compelling advantage. Their performance is, at best, equal to that of solvent-soluble epoxy ester primers.

11.5. WATER-REDUCIBLE EPOXY ACRYLIC GRAFT COPOLYMERS

Another very large scale use of epoxy resins in coatings is as a raw material in making acrylic graft copolymers for use as interior lining coatings for beverage cans

[14,15]. To prepare the graft copolymers a solution of a BPA epoxy resin in a glycol ether solvent is reacted with ethyl acrylate, styrene, and methacrylic acid using benzoyl peroxide (BPO) as the initiator. The reaction is carried out at about 130°C, where both benzoyloxy and phenyl radicals are generated. These radicals can initiate polymerization, but a substantial fraction of them abstract hydrogen atoms from the resin instead. The most readily abstractable hydrogen atoms on the BPA epoxy resin are on carbon atoms alpha to the ether and hydroxyl groups on the epoxy resin backbone, those underlined in the following partial structure.

Abstraction of one of these hydrogens results in a free radical on the epoxy resin backbone, as shown, which serves as an initiating site for polymerization of the vinyl comonomers. Thus, a graft copolymer is formed with acrylic–styrene side chains substituted with carboxylic acid groups (from the methacrylic acid). The product is a mixture of epoxy–acrylic graft copolymer, nongrafted acrylic copolymer, and unreacted epoxy resin. The complex resin mixture is neutralized with an amine such as DMAE. Class I MF resin is added as a cross-linker and the system is diluted with water. The result is a dispersion that is used as a spray applied coating for the interior of two piece beverage cans. Since the water-solubilizing groups are attached by C—C bonds rather than by ester groups, the resins are resistant to hydrolysis during storage. Sometimes latex resins are blended with the dispersion to reduce cost.

11.6. EPOXY RESIN PHOSPHATE ESTERS

Phosphoric acid reacts with BPA epoxy resins to generate phosphate esters. Low molecular weight *epoxy phosphates* have been used as adhesion promoters.

Higher molecular weight epoxy resins can also be modified by reacting with minor amounts of phosphoric acid and water. Complex reactions occur including formation of phosphate esters of a primary alcohol from the ring-opening reaction [16]. Some epoxy groups are hydrolyzed during the reaction to give the corresponding dihydroxyl derivative. Use of phosphoric acid modified epoxy resins in epoxy–phenolic formulations gives coatings with improved adhesion and flexibility as compared with corresponding unmodified epoxy–phenolic coatings. A phosphoric acid catalyst is not needed as it is with conventional epoxy–phenolic coatings.

11.7. EPOXY STABILIZERS FOR CHLORINATED POLYMERS

Epoxy resins are used as stabilizers for polychlorinated polymers (e.g., poly(vinyl chloride), PVC and chlorinated rubber). As explained in Section 4.2, these polymers undergo thermal and photoinduced degradation by an autocatalyzed elimination of HCl. Epoxy resins serve as scavengers for HCl, which reacts rapidly with the epoxy groups to yield chlorohydrins (see Section 11.3.5), thereby retarding autocatalysis. In some cases, propylene oxide is used as an additive to give package stability. Commonly, epoxidized soybean oil is used both as a plasticizer and stabilizer for PVC.

11.8. PHENOLIC RESINS

Although their importance in coatings has waned since the early twentieth century, phenolic resins still have significant uses. Phenolics are made by reactions of formaldehyde with phenol and substituted phenols. The products depend primarily on the phenol(s) utilized, the stoichiometric ratio of phenol/formaldehyde, and the pH during the reaction.

Phenolic resins are divided into two broad classes: *resole phenolics*, which are made utilizing alkaline catalysts and high ratios of formaldehyde to the phenol, and *novolac phenolics*, which are made utilizing acid catalysts and low ratios of formaldehyde to the phenol. Phenolic resins are used on a large scale in plastics and adhesives application (see Ref. [17] for details of the chemistry and the wide range of applications).

11.8.1. Resole Phenolic Resins

Under alkaline conditions, the initial reaction product of phenol and formaldehyde is a mixture of ortho and para methylolated phenols. The methylolated phenols are more reactive with formaldehyde than the unsubstituted phenol resulting in the rapid formation of 2,4-dimethylolphenol and, subsequently, 2,4,6-trimethylolphenol; the latter is the predominant product with a large excess of formaldehyde and a relatively short reaction time.

With lower ratios (but still molar excess) of formaldehyde/phenol and with longer reaction times, formation of higher molecular weight resole phenolic resins is favored. Polymerization occurs primarily by a methylol group on one phenol reacting at the ortho or para position of another phenol to form a methylene bridge connecting the two phenols. Dibenzyl ether bridges connecting two phenols form also by an etherification reaction of two methylol groups. With excess formaldehyde, methylol groups are present on the terminal phenol groups of such resins, which are called resole resins. Although not shown in the general structure, some of the aromatic rings have three substituent methylol groups.

Such phenol-based resole resins cross-link on heating and are used on a large scale in adhesive and plastics applications. However, they are not suitable for coatings applications primarily because the cross-link density that develops as the reaction proceeds is far in excess of that appropriate for any coating. Furthermore, the package stability of the resins is low.

Resole phenolics, which are useful in coatings applications, are made from monosubstituted phenols and mixtures of monosubstituted phenols with phenol. The use of substituted phenols reduces the potential cross-link density. There are two broad categories of such resins: (1) those that are soluble in alcohol and other low molecular weight oxygenated solvents (they are commonly called alcohol-soluble, heat-reactive phenolics); and (2) those that are soluble in vegetable oils and are called oil-soluble, heat-reactive phenolics.

Resole Phenolic Resin (from p-cresol)
(Idealized Structure)

Alcohol-soluble, heat-reactive resole resins are prepared by reacting phenol, o- or p-cresol, and formaldehyde in the presence of a base catalyst at less than 60°C while removing water under vacuum. The catalyst is neutralized, alcohol is added, and the salt, resulting from catalyst neutralization, is removed by filtration. Potential cross-link density is controlled by the ratio of phenol to cresol; molecular weight is controlled by the ratio of formaldehyde/phenols and by the reaction time.

Such resole phenolic resins are used in interior can coatings and tank linings. They require baking using an acid catalyst to cure in short times. To enhance flexibility and adhesion, they are commonly blended with low molecular weight poly(vinyl butyral) as a plasticizer. The films are very resistant to swelling by oils such as encountered in canned fish and are completely resistant to hydrolysis. These resins and other heat-reactive phenolics discolor during baking, which restricts their usage to applications where development of a yellow-brown color is permissible.

These resins are also blended with epoxy resins in thermosetting coatings in such applications as primers and can coatings (Section 11.3.1). The absence of hydrolyzable bonds and generally excellent adhesion are their chief advantages.

Oil-soluble, heat-reactive phenolics are prepared by reacting a para substituted phenol (e.g., p-phenylphenol, p-tert-butylphenol, or p-nonylphenol) with somewhat less than 2 mol of formaldehyde per mole of substituted phenol. The resulting resole phenolics are cast from the reactor after neutralization of the catalyst. They are solid, linear resins with terminal methylol groups, as illustrated above.

The most common use for such resins was in making varnishes with linseed oil and/or tung oil. However, since varnishes have largely been replaced by other vehicles, the consumption of these solid resins has declined markedly. Large volumes of somewhat similar types of resins are still used in linings for many vegetable and fruit cans. They are no longer processed as varnishes from solid phenolic resin and drying oil; rather the drying oils are incorporated into the phenolic resin in the latter stage of preparation.

11.8.2. Novolac Phenolic Resins

Novolac phenolics of interest for coatings are made with acid catalysts and o- or p-substituted phenols. Again the molecular weight is controlled by the molar ratio of the phenol/formaldehyde, which in this case is always greater than 1. Thus, in contrast to resole phenolics, the terminal phenol groups are not metholylated, as shown in the idealized structure.

Novolac phenolic resin

Three types of novolac resins are used in coatings.

1. Alcohol-soluble, nonheat-reactive, low molecular weight phenolics are derived from o- or p-cresol. An important use of these resins is in the preparation of novolac epoxy resins by reaction with epichlorohydrin (see Section 11.1.2).

2. Oil-soluble, nonheat-reactive phenolic resins are made using a low ratio of formaldehyde, an acid catalyst, and a substituted phenol (e.g., p-phenyl-, p-tert-butyl-, or p-nonylphenol). They are used together with drying oils particularly tung oil or tung/linseed oil mixtures in making varnishes. Such varnishes are still used in a few can coatings and as marine spar varnishes, where their reputation for durability maintains their position in the do-it-yourself marine yacht market. The durability of phenolic varnishes may well result, in part at least, from the antioxidant activity of the phenolic groups (see Chapter 25 for a discussion of phenolic antioxidants).

3. Rosin-modified phenolic resins are the principal type of modified phenolics still in use. The phenolic resin is prepared in the presence of rosin esters and/or zinc or calcium salts of rosin. The structures of the reaction products, which are high melting, hydrocarbon-soluble resins, are not completely known. In coatings they are used to a limited extent in low cost varnishes. Their largest volume use is in printing inks. An important example of their use is in publication gravure inks such as those used in mail-order catalogs and certain magazines. They are also used, to a degree, in heat-set letterpress printing inks for magazines and paper-back book covers.

11.8.3. Ether Derivatives of Phenolic Resins

The package stability of alcohol-soluble resole resins and their compatibility with epoxy resins can be improved by partial conversion of the methylol groups to ethers. The ether groups undergo exchange reactions with hydroxyl groups in the presence of acid catalysts. Allyl ethers have been blended with epoxy resins in interior can coatings for many years.

More recently, low molecular weight butyl ethers have been made available for use in cross-linking epoxy resins and other hydroxy-substituted resins, primarily by etherification and transetherification reactions [18]. A typical resin has an average of 2.2 aromatic rings per molecule. Due to its low molecular weight (320) the resin has a moderate viscosity as it is supplied in butyl alcohol solution. As might be expected with such a low molecular weight resin, some free phenol remains in the resin. The reactive groups in the resins are primarily butoxymethyl groups, but there are also benzyloxy groups and some free methylol groups. Furthermore, the phenol groups can react with epoxy groups. Acid catalysts, such as phosphoric or sulfonic acids, are required. Blocked acids may be utilized to extend shelf life.

REFERENCES

1. L. V. McAdams and J. A. Gannon, "Epoxy Resins," in *Encyclopedia of Polymer Science and Engineering*, 2nd ed., Vol. 6, Wiley, New York, 1986, pp. 322–382.

2. P. S. Sheih and J. L. Massingill, *J. Coat. Technol.*, **62** (781), 25 (1990).

3. D. R. Scheuing, *J. Coat. Technol.*, **57** (723), 47 (1985).

4. M. M. Bagga, *European Patent Application* 22,073 (1981).

5. E. G. Bozzi and D. Helfand, FSCT Symposium, Louisville, May 1990. See also K. L. Payne and J. S. Puglisi, *J. Coat. Technol.*, **59** (752), 117 (1987).

6. J. L. Massingill, P. S. Sheih, R. C. Whiteside, D. E. Benton, and D. K. Morisse-Arnold, *J. Coat. Technol.*, **62** (781), 31 (1990).

7. W. P. Jencks, *Catalysis in Chemistry and Enzymology*, McGraw-Hill, New York, 1969, pp. 199–211; W. P. Jencks, *Chem. Rev.*, **72**, 705 (1972).

8. M. Agostinho and V. Brytus, *J. Coat. Technol.*, **60** (764), 61 (1988).

9. M. D. Gilbert, N. S. Schneider, and W. J. MacKnight, *Macromolecules.*, **24**, 360 (1991).

10. V. Brytus, *J. Coat. Technol.*, **58** (740), 45 (1986).

11. R. Albers, *Proceedings of the 10th Water-Borne Higher-Solids Coatings Symposium*, New Orleans, LA, 1983, pp. 130–143; U.S. Patent 4, 352, 898 (1982).

12. J. A. Lopez, U.S. Patent 4, 816, 502 (1989).

13. M. D. Shalati, J. R. Babjak, R. M. Harris, and W. P. Yang, *Proc. 16th Intl. Conf. Coat. Sci. Technol.*, Athens, Greece, (1990) p. 525.

14. J. T. K. Woo, V. Ting, J. Evans, R. Marcinko, G. Carlson, and C. Ortiz, *J. Coat. Technol.*, **54** (689), 41 (1982).

15. J. T. K. Woo and A. Toman, *Polym. Mater. Sci. Eng.*, **65**, 323 (1991).

16. J. L. Massingill, *J. Coat. Technol.*, **63** (797), 47 (1991).

17. P. W. Kopf, "Phenolic resins," in *Encyclopedia of Polymer Science and Engineering*, 2nd ed., Vol. 11, Wiley, New York, 1988, pp. 45–95.

18. Monsanto Chemical Co., *Santolink EP 560*, Springfield, MA, 1990.

CHAPTER XII _____

Coatings Binders Based on Isocyanates—Polyurethanes

Polyurethanes are polymers that contain the urethane (carbamate) structure (NHCOO). Urethanes are usually formed by reaction of an alcohol with an isocyanate. In the coatings field, the term polyurethane is applied to a wide variety of binders derived from isocyanates.

Their usefulness is based on characteristics that are different from most other binders available for coatings. Urethane groups can form stable intermolecular hydrogen bonds between polymer molecules; they may be acyclic and/or cyclic, as shown.

$$
\begin{array}{cc}
\overset{O}{\underset{H}{\overset{\|}{-N-C-O-}}} & -N-C-O- \\
\vdots & \\
O & \\
\| & \\
-C-NH- & -C-N- \\
\text{Acyclic H bond} & \text{Cyclic H bonds}
\end{array}
$$

Under mechanical stresses, energy (~ 20–25 kJ mol^{-1} of hydrogen bonds) may be absorbed by separation of these hydrogen bonds, which can reform (probably in different positions) during stress and when the stress is removed. Energy absorption by this reversible bond-breaking/reforming process presumably reduces the likelihood of irreversible bond breaking of polymer covalent bonds that would lead to degradation. This characteristic makes it possible to design polyurethanes that are abrasion resistant while still resisting swelling with strong solvents. Polyurethanes tend to absorb a little water from the environment, and strong hydrogen bonding with water can reduce intermolecular hydrogen bonding; the result is plasticization of the coating.

The isocyanate group is highly reactive and, therefore, polyisocyanates can be used to make coatings that will cure at room temperature or at moderately elevated temperatures. Coatings based on aliphatic diisocyanates exhibit exceptional exterior durability especially when stabilized with hindered amine light stabilizers (see Section 25.3.3). There are indications that the resistance of urethane coatings to spotting by acid rain is superior to that of some MF cross-linked coatings.

The principal limitations of polyurethanes are their cost (they are generally more expensive than alkyds or MF cross-linked polyesters and acrylics) and toxicity (particularly of relatively low molecular weight compounds containing N=C=O groups). It must be added that any cross-linker that will react with hydroxyl, amine, and/or carboxylic acid groups near room temperature is likely to be toxic since the body contains proteins and other materials with just such substituents. The important question is not toxicity as much as toxic hazard. Since higher molecular weight reduces vapor pressure and permeability through body membranes, the toxic hazard decreases, in general, as molecular weight increases. With adequate ventilation and use of air masks and protective clothing, as are commonly used in chemical factories and resin manufacturing plants, even relatively low molecular weight isocyanates can be handled safely. Most low molecular weight diisocyanates are sensitizers, that is, after exposure some people become allergic to further exposure to isocyanate. The exposure leading to sensitization varies over a wide time period for individuals and may occur only after years of repeated exposure for some. The most common symptoms are hives and asthma. In extreme cases sensitized people cannot be in the same room or even in the same building where isocyanates are being handled.

Paint applicators and especially amateur users are less likely than professional chemists to have appropriate facilities or even to follow safety instructions. In recommending use of any highly reactive cross-linker, the competence of the user to handle toxic materials should be taken into consideration in establishing coatings formulas. Information on safety practices should be supplied.

12.1. CHEMICAL REACTIONS OF THE ISOCYANATE GROUP

Isocyanates react with essentially any "active hydrogen" compound. The reactions of greatest importance in coatings are discussed in this section.

Alcohols (and phenols) react with isocyanates to form urethanes (carbamates), where R and R' can be aromatic or aliphatic groups. The reaction is reversible at elevated temperatures.

$$R-N=C=O \ + \ H-O-R' \longrightarrow R-\overset{H}{\underset{}{N}}-\overset{O}{\underset{}{C}}-O-R'$$

In general terms, the rates of urethane formation decrease in the following order: primary alcohols > secondary alcohols > 2-alkoxyethanols > 1-alkoxy-2-propanols > tertiary alcohols. As a broad rule the ease of reversion is more or less the inverse of the reactivity, but there are probably exceptions to this rule. Urethanes made from tertiary alcohols are relatively unstable and, when heated, may decompose to give alkenes, carbon dioxide, and amines rather than alcohols and isocyanates.

Urethanes can react further with another molecule of isocyanate to form allophanates. This reaction is much slower than the original reaction of the isocyanate with the alcohol.

$$R-N=C=O \ + \ R'-O-\overset{O}{\underset{}{C}}-\overset{H}{\underset{}{N}}-R \longrightarrow R'-O-\overset{O}{\underset{}{C}}-\overset{\overset{O}{\parallel}\ \ C-NH-R}{\underset{}{N}}-R$$

Isocyanates react very rapidly with primary and secondary amines to form substituted ureas. This reaction is much faster than the reaction of isocyanates with alcohols. The reaction is normally so fast at ambient temperatures that the pot life of a polyamine–polyisocyanate system would be too short for adequate mixing prior to application. It can be used to prepare resins for use in coatings.

$$R-N=C=O \ + \ R'NH_2 \ \longrightarrow \ R-\overset{H}{\underset{\mid}{N}}-\overset{O}{\overset{\parallel}{C}}-\overset{H}{\underset{\mid}{N}}-R'$$

Isocyanates react with ureas to form biurets. Biuret formation is slower than urethane formation but is probably faster than allophanate formation.

$$R-NH-\overset{O}{\overset{\parallel}{C}}-\overset{H}{\underset{\mid}{N}}-R' \ + \ R-N=C=O \ \longrightarrow \ R'-\overset{H}{\underset{\mid}{N}}-\overset{O}{\overset{\parallel}{C}}-\overset{\overset{O}{\overset{\parallel}{C}-NH-R}}{\underset{\mid}{N}}-R$$

Isocyanates react with water to form unstable carbamic acids, which dissociate into carbon dioxide and an amine. The amine is so much more reactive than water that it reacts with a second isocyanate (in preference to water) to form a urea. The reactivity of water with isocyanates is similar to that of secondary alcohols, slower than that of amines, and more rapid than that of ureas. Relative reaction rates can change with different conditions or catalysts.

$$R-N=C=O \ + \ H_2O \ \longrightarrow \ \left[R-\overset{H}{\underset{\mid}{N}}-\overset{O}{\overset{\parallel}{C}}-OH \right] \ \longrightarrow \ R-NH_2$$

$$\underline{R-N=C=O} \ \ R-\overset{H}{\underset{\mid}{N}}-\overset{O}{\overset{\parallel}{C}}-\overset{H}{\underset{\mid}{N}}-R$$

Carboxylic acids react with isocyanates to form amides and CO_2. This reaction requires elevated temperatures in order to proceed rapidly.

$$R-N=C=O \ + \ R'COOH \ \xrightarrow{\text{Heat}} \ \left[R-\overset{H}{\underset{\mid}{N}}-\overset{O}{\overset{\parallel}{C}}-O-\overset{O}{\overset{\parallel}{C}}-R' \right] \ \longrightarrow$$

$$R-\overset{H}{\underset{\mid}{N}}-\overset{O}{\overset{\parallel}{C}}-R' \ + \ CO_2$$

In the absence of water, ketimines (see Chapter 11) react with isocyanates to yield a variety of products depending on the particular reactants and conditions [1]. For example, isobutyl isocyanate reacts with the ketimine derived from methylamine and acetone in 3 h at 60°C to yield isobutyl methylurea and a cyclic unsaturated urea. Note that multifunctional reactants would give a cross-linked network.

In the presence of water, of course, the principal product would be isobutyl methylurea because water rapidly hydrolyzes ketimines at 25°C. Thus, reactions of polyisocyanates with polyketimines are potentially useful for cross-linking at ambient temperature with water present (see Section 12.4.2) and at moderately elevated temperature when water is absent [1].

Isocyanate molecules can react with each other to form dimers (uretdiones) and trimers (isocyanurates). The formation of isocyanurates is catalyzed by some tertiary amines. In some cases, when alcohols are reacted with isocyanates, especially in the presence of amine catalysts, part of the isocyanate is converted to trimer rather than reacting with the alcohol to form a urethane.

Uretdione   Isocyanurate

Additional reactions of isocyanates are discussed in Section 12.4.4 on blocked isocyanates.

12.2. KINETICS OF ISOCYANATE REACTIONS WITH ALCOHOLS

While the reactions of isocyanates with alcohols have been extensively studied for many years, the mechanisms of reaction and catalysis are still not completely understood. One must use considerable caution in evaluating rate data from the literature; caution is especially needed in interpreting tables of rate constants. The original literature should be consulted to make sure that comparisons are based on studies done in the same solvents and at the same initial concentrations since reaction rates can depend strongly on these two variables. Furthermore, rates of urethane formation are commonly determined by following the disappearance of isocyanate. In the event that other products, such as allophanates and isocyanurates, are also formed, isocyanate disappearance will not correspond to urethane formation.

12.2.1. Noncatalyzed Reactions

One might expect the urethane forming reaction to follow second-order kinetics with the rate proportional to the concentration of each of the reactants, as shown in Eq. 12.1.

$$\text{Rate} = k[\text{R—N=C=O}] \, [\text{R'—OH}] \qquad (12.1)$$

In Eq. 12.1, and the equations that follow, k is the overall apparent rate constant for a specific reaction at a specific temperature.

Kinetic studies show that the real situation is considerably more complex. It has been reported that, in some cases, the reaction follows third-order kinetics; first order in isocyanate and second order in alcohol concentration. Rate equation 12.2

would apply to such cases. The implication of Eq. 12.2 is that two molecules of alcohol are involved in the reaction with one molecule of isocyanate.

$$\text{Rate} = k[\text{R—N=C=O}] \, [\text{R'—OH}]^2 \tag{12.2}$$

Such results can be reasonably interpreted in terms of the mechanism provided in Scheme 1. Reaction of isocyanate and alcohol (at a rate proportional to k_1) produces a zwitterionic reactive intermediate (**RI**), which can revert to starting materials (k_{-1}) or proceed to the product urethane (k_2). Product formation requires a proton transfer from oxygen to nitrogen. The role of the second molecule of alcohol is to facilitate this transfer by way of a six-membered cyclic activated complex **A**.

SCHEME 1

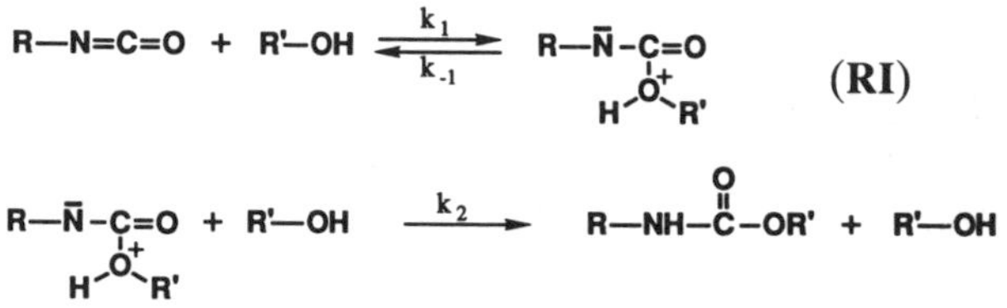

Equation 12.3 is a rate expression corresponding to the sequence of reactions in Scheme 1.

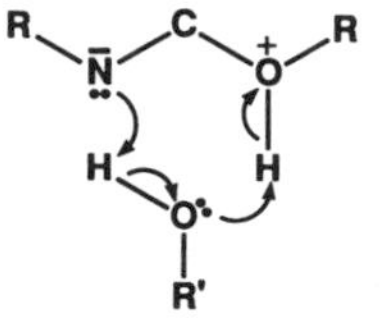

$$\text{Rate} = k_1[\text{RNCO}][\text{R'OH}] \, \frac{k_2[\text{R'OH}]}{k_2[\text{R'OH}] + k_{-1}} \tag{12.3}$$

If the term $k_2[\text{R'OH}]$ is substantially greater than k_{-1}, Eq. 12.3 simplifies to Eq. 12.1. This situation is favored when the alcohol concentration is high, such as at the start of the reaction. If, on the other hand, the term k_{-1} is substantially larger than $k_2[\text{R'OH}]$ Eq. 12.3 simplifies to Eq. 12.2, and second-order dependence on alcohol concentration is observed. The latter circumstance is more likely to occur as alcohol is consumed during the reaction, reducing [R'OH]. As a result, the reaction rate diminishes rapidly as the reaction proceeds (since the rate is proportional to the square of a small, diminishing alcohol concentration), thereby requiring long time periods and/or elevated temperatures for completion.

The mechanism in Scheme 1 provides an explanation for the strong solvent effects observed. Rates decrease as the hydrogen-bond acceptor abilities of solvents increase in the order aliphatic hydrocarbons, aromatic hydrocarbons, esters and ketones, ethers, and glycol diethers. As a result, rates in aliphatic hydrocarbons can be two orders of magnitude faster than in glycol diether solvents. Hydrogen-bonding of the zwitterionic **RI** and/or the alcohol with solvent may reduce the rate by reducing the concentration of the cyclic activated complex **A**.

The mechanism in Scheme 1 is also consistent with the observed higher reactivity of aromatic isocyanates. When R is an aromatic group (e.g., phenyl) the negative charge on nitrogen in **RI** is delocalized in the π-electron system of the aromatic ring, resulting in a lower energy (by resonance stabilization) for **RI** and a faster rate of formation relative to aliphatic isocyanates.

The rate constants reported in the literature for reactions of alcohols with isocyanates are most commonly apparent initial second-order rate constants. The utilization of initial rate constants provides a useful way of comparing reactivities without becoming involved in the complexities of the kinetics, provided comparisons are made in the same solvent and with the same initial concentrations of alcohols and isocyanates in the solvent. Unfortunately, it is fairly common for rate constants to be compared, which were determined in different solvents and/or at different initial concentrations. Such comparisons can be very misleading.

The kinetics of the isocyanate–alcohol reaction are even more complex than the third-order effect discussed thus far. While the reaction initially slows down more rapidly than would be predicted by second-order kinetics, it does not stay third order throughout the course of the reaction. It has been proposed that the reaction is autocatalyzed by the urethane being formed. Another cyclic transition state (**B**) could be involved in which a molecule of urethane facilitates proton transfer from oxygen to nitrogen in the zwitterionic **RI**.

The autocatalyzed reaction would follow third-order kinetics but would be first order in alcohol. The decline in alcohol concentration would be offset by the increase in urethane concentration as the reaction proceeds. The apparent rate constant would change through the course of the reaction depending on the rate constant for the reaction involving activated complex **A** in comparison with **B**.

Sato [2] studied the reaction of various isocyanates with methyl alcohol. In the absence of catalyst, his results can be expressed by Eq. 12.4, where a and b are the initial concentrations of isocyanate and methyl alcohol, x is the concentration of product (urethane), and k_2 represents the rate constant of the autocatalytic reaction.

$$\frac{dx}{dt} = k_1(a - x)(b - x)^2 + k_2 x(a - x)(b - x) \tag{12.4}$$

It is assumed that there are no side reactions; therefore, $(a - x)$ in Eq. 12.4 equals [R—N=C=O] in Eq. 12.2 and $(b - x)$ equals [R'OH]. In most cases studied by Sato, k_2 was larger than k_1, but in a few cases it was comparable. When k_2 is substantially larger, the second term in Eq. 12.4 dominates in the later stages of the reaction (as the urethane concentration builds up). Sato's studies were all done in di-*n*-butyl ether; it would be of interest to know what changes in the rate constants would result from changes in solvent.

Sato [2] also reported that autocatalysis is more important for aliphatic than aromatic isocyanates, although the rate constants for autocatalysis were similar. This result can be attributed to the higher reactivity of aromatic isocyanates with alcohols (k_1), which reduces the relative importance of autocatalysis. In general, one should anticipate that there may be differences in catalysis of aromatic and aliphatic isocyanate reactions.

12.2.2. Catalysts

Reactions of aromatic isocyanates with alcohols are catalyzed by a wide variety of compounds, including urethanes (as already mentioned), tertiary amines, metal salts and chelates, and organometallic compounds. The most widely used catalysts in coatings applications are tertiary amines, commonly diazabicyclo[2.2.2]octane (DABCO—a trademark of Air Products), and organotin compounds, most commonly dibutyltin dilaurate (DBTDL). Combinations of DABCO and DBTDL often act synergistically, that is, the effect of the combination is greater than would be predicted by the sum of the effect of the two catalysts alone.

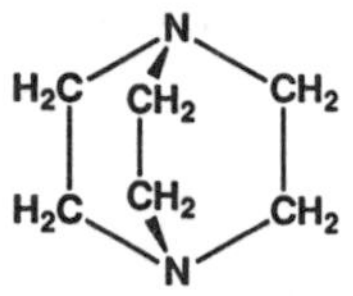

"DABCO" 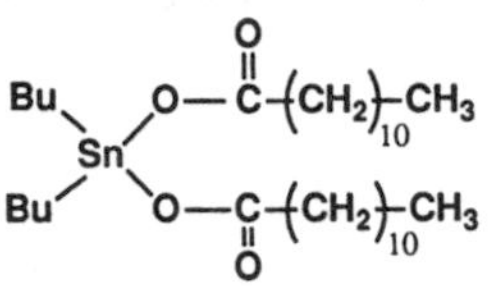   DBTDL

The mechanisms by which these and other catalysts operate are controversial. A reasonable explanation for catalysis by amines is that they facilitate proton transfer from the alcohol to the isocyanate (in a similar manner to alcohols and urethanes). Proton removal from the alcohol may occur during reaction with the isocyanate, thereby avoiding formation of the positive charge on oxygen and lowering the energy of the reactive intermediate, which may proceed to product by a proton transfer from the protonated amine, as shown in Scheme 2. Proton removal at an earlier stage than with alcohols and urethanes is reasonable because of the greater basicity of amines.

SCHEME 2

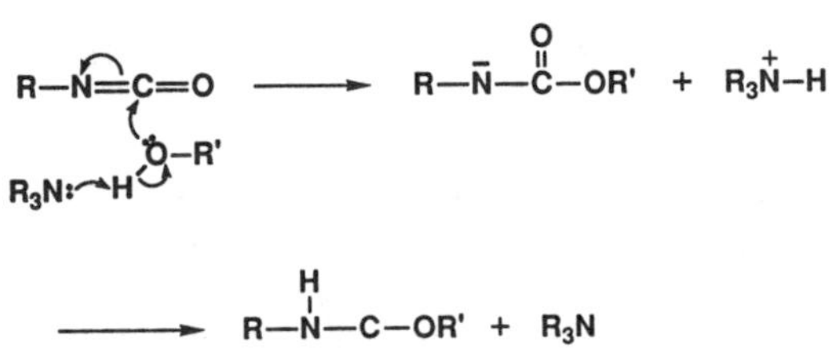

Sato [2] included catalysts in the study described in Section 12.2.1. For triethyl-amine catalyzed reactions of isocyanates with methyl alcohol, Sato's data fit Eq. 12.5, where k_3 and (cat) represent the rate constant of the catalyzed reaction and the concentration of catalyst, respectively.

$$\frac{dx}{dt} = k_1(a-x)(b-x)^2 + k_2x(a-x)(b-x) + k_3(\text{cat})(a-x)(b-x) \qquad (12.5)$$

If the catalyst is effective, k_3 is relatively large, and the rate is governed by the third term in Eq. 12.5; then the rate will be first order in alcohol. This is generally observed in the amine catalyzed reactions. The mechanism in Scheme 2 is also consistent with the third term of rate equation 12.5. However, amine basicity cannot be the only important factor since DABCO, a weaker base, is a much more active catalyst than triethylamine. On the other hand, the nitrogen electron pairs are more readily accessible with DABCO than with triethylamine, which could account for its higher catalytic activity. This possibility, that both nitrogen basicity and electron accessibility are important factors, is supported by the even higher catalytic activity (relative to DABCO) of 1-azabicyclo[2.2.2]octane (quinuclidine), which possesses both the nitrogen accessibility of DABCO and high basicity of triethylamine.

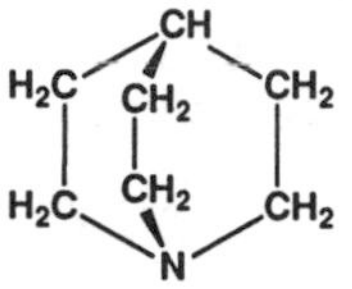

Quinuclidine 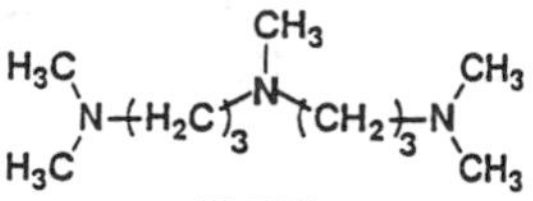   PMPTA

Comparisons are complicated because tertiary amines also catalyze the trimer-ization of isocyanates to form isocyanurates. For example, reaction of phenyl is-ocyanate with n-butyl alcohol (at 50°C in acetonitrile) in the presence of penta-methyldipropylenetriamine (PMPTA) yielded 30% urethane, while 70% of the isocyanate was converted into triphenylisocyanurate [3]. On the other hand, when DABCO was used as a catalyst, the expected urethane was the principal product, but a small amount of an allophanate also formed. A possible explanation for these results is that urethane formation may be favored by sterically accessible amines (e.g., DABCO); whereas isocyanurate formation is less sensitive to this factor. Clearly, the kinetics of isocyanate reactions can be misleading if they are based only on the disappearance of N=C=O without checking whether there is a cor-responding appearance of the expected product.

Of the many metal derivative catalysts for the reaction of isocyanates, DBTDL is the most widely used in coatings. It is soluble in a wide range of solvents, comparatively low in cost, colorless, and, in general, highly effective. Dibutyltin dilaurate is said to promote urethane formation without promoting allophanate formation [4] and/or trimerization [3]. For example, when DBTDL was used as the catalyst for the above reaction, the only product detected was urethane [3].

Dimethyltin diacetate (DMTDA) is usually a somewhat more effective catalyst than DBTDL and is particularly useful with sterically hindered isocyanates.

Dibutyltin dilaurate is an effective catalyst for the reaction of alcohols with aromatic isocyanates and is even more effective with aliphatic isocyanates. While aromatic isocyanates are substantially more reactive than aliphatic isocyanates in uncatalyzed reactions with alcohols, reactivity can be roughly equalized by adding sufficient DBTDL. On the other hand, amine catalysts are more effective with aromatic than aliphatic isocyanates, thereby widening the difference in reactivity.

Many mechanisms have been proposed for the catalytic activity of tin compounds, but none has been universally accepted. A plausible proposal by Van der Weij [5] was based on studies of the reaction of phenyl isocyanate with excess methyl alcohol using dibutyltin diacetate (DBTDA) as catalyst. He found that the reaction rate was first order in isocyanate and half-order in both alcohol and catalyst concentration. Restating this kinetic equation in the same form used thus far gives Eq. 12.6.

$$\frac{dx}{dt} = k_3(\text{cat})^{0.5}(a - x)(b - x)^{0.5} \tag{12.6}$$

Based on his kinetic results and on the observation that the rate was strongly suppressed by the addition of strong acid, Van der Weij proposed the mechanism provided in Scheme 3. The mechanism involves sequential complexation of alcohol (with loss of H^+) and isocyanate to the tin. The proposed proton loss is consistent with the observed reduction in activity of tin catalysts in the presence of carboxylic acids. Presumably, addition of H^+ would favor decomplexation of alcohol and reversion to starting materials.

SCHEME 3

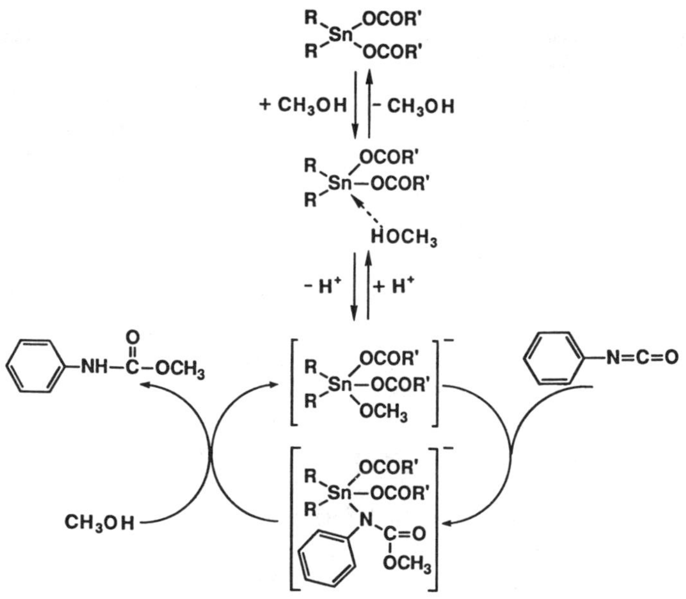

Coordination of the isocyanate with tin is consistent with the observed roughly equivalent reactivity of aliphatic and aromatic isocyanates with tin catalysts, since stabilization of the reactive intermediate by electron delocalization with the aromatic ring is eliminated as a major factor. Furthermore, tin activates both the alcohol and isocyanate, whereas amine catalysts probably activate only the alcohol by facilitating proton removal.

The rate dependence on alcohol concentration has important implications on the pot life versus curing schedule with isocyanates. If the dependence changes from first to second order, characteristic of the uncatalyzed reaction, the rate will diminish rapidly as the reaction nears completion. With less sensitive half-order dependence, as reported for tin catalysis, the rate at high concentrations (during storage) will be lower and will not slow down as much as the reaction proceeds to completion, thereby favoring both a longer pot life and shorter cure time (or lower cure temperature).

The first-order dependence with amine catalysis would, of course, result in an intermediate situation. Unfortunately, no reports have been found in the literature where the effect of solvent on rates of catalyzed reactions have been studied.

12.3. ISOCYANATES USED COMMERCIALLY IN COATINGS APPLICATIONS

Of the large number of commercially available isocyanates only a limited number is widely used in coatings. They are divided between aromatic and aliphatic isocyanates.

12.3.1. Aromatic Isocyanates

One of the lowest cost diisocyanates used in coatings is toluene diisocyanate (TDI), which consists of a mixture of about 80%, 2,4- and 20% 2,6-diisocyanato isomers. Nearly pure 2,4-TDI is also available at a premium price. Due to toxic hazards, TDI is used in manufacturing resins for coatings but is not used as such in formulated coatings. For coatings applications where unreacted isocyanate groups are needed, TDI is converted into derivatives of higher molecular weight and higher functionality. The higher molecular weight reduces the toxic hazard and the higher functionality permits formulation of coatings that will cross-link to solvent resistant films more rapidly.

2,4-TDI   2,6-TDI

Two types of higher molecular weight polyisocyanate derivatives of TDI are used in coatings. First, the isocyanurate derivative (see Section 12.1) made by trimerizing TDI is available. Second, so-called "prepolymers" made by reacting an excess of TDI with polyhydroxy compounds are widely used, as shown for an idealized reaction:

$$R{-}(OH)_n \;+\; n \text{ TDI} \longrightarrow R{-}\!\left(\!O{-}\overset{\overset{\text{O}}{\|}}{C}{-}NH{-}\!\!\bigcirc\!\!{-}CH_3 \atop \qquad\qquad N{=}C{=}O\right)_{\!n}$$

TDI Prepolymer

Toluene diisocyanate has the important advantage of a differential in reactivity between the ortho- and the para-isocyanate groups, which makes possible the synthesis of isocyanurates and prepolymers with narrower molecular weight distribution than with other aromatic diisocyanates in which the isocyanate groups are equally reactive.

At 40°C, the para-isocyanate group of TDI is about seven times more reactive than the ortho group. Furthermore, no matter which isocyanate group reacts first, the second group is less reactive than the first. Overall, it has been reported that, following reaction of the para-isocyanate, the remaining ortho group is 20 times less reactive than a para-isocyanate on a second TDI. It is important to recognize, however, that this difference in reactivity decreases as the temperature increases. At temperatures above 100°C, the ortho- and para-isocyanate groups have similar reactivities. Thus, for maximum selectivity the preparation of the prepolymer should be done at low temperature. Of course, this means that the reaction is slow. Catalysts can be used but the catalyst stays in the product and, therefore, sets a lower limit on the amount of catalyst the formulator can have in the final formula.

Any polyhydroxy compound can be reacted with TDI to make prepolymers. Low molecular weight hydroxy-terminated polyesters or mixtures of diols and triols are commonly used. For safety the levels of unreacted TDI in the prepolymer must be very low. Low levels of TDI can be assured by using an excess of polyester or diol–triol and pushing the reaction to completion, but some chain extension (caused by reaction of both N=C=O groups of some of the TDI molecules) will inevitably increase the molecular weight of the product. An alternative process, used when low molecular weight prepolymers are needed, is to react the polyhydroxy compound, often trimethylolpropane, with a large excess of 2,4-TDI (preferably the high 2,4-isomer grade, although the mixed isomer grade can be used) and then removing the excess TDI using a vacuum wiped-film evaporator. Very low levels of free TDI and minimal chain extension are attainable; this process affords relatively low molecular weight prepolymers suitable for high solids coatings.

Several other aromatic diisocyanates are commercially available. The next most widely used in coatings is bis(4-isocyanatophenyl)methane (MDI):

$$O{=}C{=}N{-}\!\!\bigcirc\!\!{-}CH_2{-}\!\!\bigcirc\!\!{-}N{=}C{=}O$$

12.3.2. Aliphatic Isocyanates

Aliphatic isocyanates are more expensive than aromatic isocyanates and are used where the greater color stability and exterior durability, which they impart, are required. The principal aliphatic isocyanates used as coatings intermediates are 1,6-hexamethylene diisocyanate (HDI), isophorone diisocyanate (IPDI), bis(4-isocyanatocyclohexyl)methane ($H_{12}MDI$), tetramethyl-*m*-xylidene diisocyanate (TMXDI), and isopropenyldimethylbenzylisocyanate (TMI).

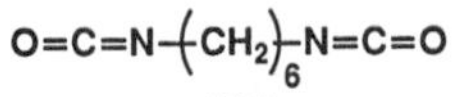

HDI

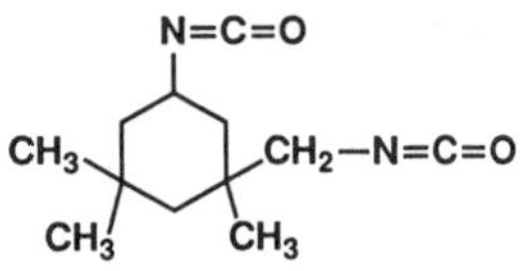

IPDI

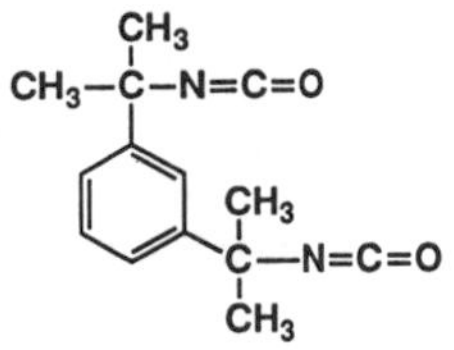

TMXDI

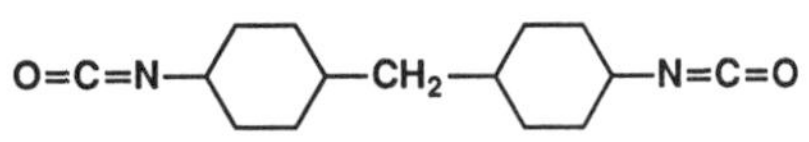

$H_{12}MDI$

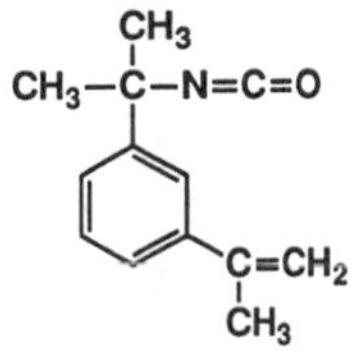

TMI

1,6-Hexamethylene diisocyanate is especially hazardous and is handled on a large scale only in specialized chemical plants. A less hazardous derivative, a biuret made by reacting HDI with a small amount of water and removing the excess HDI, is widely used. The structure of HDI biuret shown below is idealized; commercial products contain varying fractions of oligomeric biurets. (Presumably the presence of oligomeric biurets makes the average functionality higher than 3.) These polyfunctional isocyanates give coatings with good color retention and weather resistance. The viscosity of an early commercial product was about 11.5 Pa·s at 20°C. Recently, grades with lower number average molecular weights (and presumably average functionalities nearer to 3) have become available with viscosities as low as 1.4 Pa·s, which are used in high solids coatings.

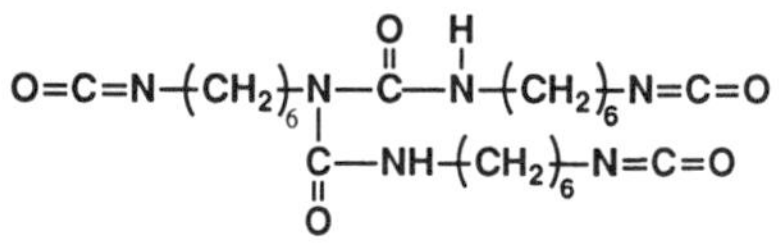

HDI - Biuret

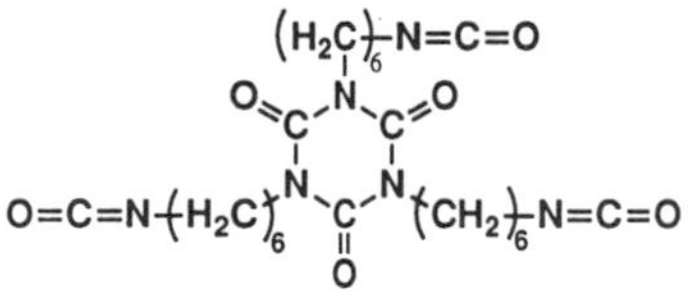

HDI - Isocyanurate

1,6-Hexamethylene diisocyanate isocyanurates are also available; these derivatives give coatings with greater heat resistance and even better long-term exterior durability than HDI biuret. Commercial products presumably contain oligomeric material, and average functionality is probably over 3.

Bis(4-isocyanatocyclohexyl)methane is much less volatile than HDI and is sometimes used as a free diisocyanate in coatings to be applied by roller coating but not for spray coating. The compounds $H_{12}MDI$, IPDI, and TMXDI are primarily used in making prepolymers, hydroxy-terminated polyurethanes, and blocked isocyanates (see Section 12.4.4). Isocyanurate derivatives of IPDI are commercially available. Tetramethyl-*m*-xylidene diisocyanate has been offered as a low molecular weight, essentially diisocyanate-free prepolymer with trimethylolpropane. Isopropenyldimethylbenzylisocyanate is used as a comonomer with acrylic esters to make low molecular weight copolymers with ratios such that few, if any, monofunctional isocyanate molecules are present, and where essentially no volatile diisocyanate molecules can be detected even at elevated temperatures.

Commercial IPDI is a mixture of Z (cis) and E (trans) isomers in a 70:30 ratio [6]. The isomers are difficult to separate. Isophorone diisocyanate is unique among the commercial aliphatic diisocyanates in that it has two different types of N=C=O groups. Two studies, performed under different conditions, showed that with DBTDL catalysis, the secondary N=C=O groups of the both the Z and the E isomers are more reactive than the primary N=C=O groups [7,8]. On the other hand, with an amine catalyst (DABCO), the primary N=C=O groups are the more reactive. With both catalysts the relative reactivities of the primary and secondary N=C=O groups of the two isomers are similar [8].

Tetramethyl-*m*-xylidene diisocyanate has an aromatic ring, but it behaves like an aliphatic isocyanate from the standpoints of color retention and weather resistance. The exterior durability of TMXDI-derived urethanes probably results from the absence of isocyanate groups directly substituted on the aromatic ring, as well as from the absence of abstractable hydrogen atoms on the carbon atoms adjacent to nitrogen. Since the isocyanate group is on a tertiary carbon, the reactivity is lower than that of less sterically hindered aliphatic isocyanates. However, this difference can be offset by using higher catalyst levels and sterically accessible tin catalysts such as DMTDA instead of DBTDL.

12.4. APPLICATIONS OF ISOCYANATES IN COATINGS

Isocyanates are used in many different ways in coatings. Essentially, any coating in which an isocyanate is used is called a *urethane* or a *polyurethane* coating even though in some cases the primary cross-links are urea or other linkages rather than urethane groups. In addition to the uses discussed in this section, see Section 13.2 for a discussion of the use of isocyanate-terminated resins for making acrylated urethanes and Section 13.4.2 for mention of the use of isocyanates as cross-linkers for acetoacetylated resins.

12.4.1. Uralkyds

Uralkyds are also called "urethane alkyds" or "urethane oils." In effect, they are alkyd resins in which a diisocyanate, usually TDI, has fully or partly replaced the PA usually used in the preparation of alkyds. One transesterifies a drying oil with a polyol such as glycerol or pentaerythritol to make a *monoglyceride* (see Section 10.6.1) and reacts with some PA (if desired) and then with somewhat less diisocyanate than the equivalent amount of N=C=O based on free OH content. To assure that no N=C=O groups remain unreacted, methyl alcohol is added at the end of the process.

Analogously to alkyds, uralkyds dry faster than the drying oil from which they were made since they have a higher average functionality (more activated methylene groups between two double bonds, —CH=CH—CH$_2$—CH=CH—, per average molecule). The rigidity of the TDI aromatic rings also speeds up the drying by increasing the T_g of the resin.

Two principal advantages of uralkyd over alkyd coatings are superior abrasion resistance and resistance to hydrolysis. Disadvantages are inferior color retention (when TDI is used) of the films, higher viscosity of resin solutions at the same percent solids, and somewhat higher cost than alkyds. Uralkyds made with aliphatic diisocyanates have better color retention but are more costly.

The largest use of uralkyds is in architectural coatings. Most so-called varnishes sold to the consumer today are based on uralkyds; they are not really varnishes in the original sense of the word (see Section 9.3). Uralkyd *varnishes* are used as transparent coatings for furniture, woodwork, and floors, applications where good abrasion resistance is important. They are generally made from TDI and tend to turn yellow and then light brown as they age; yellowing is acceptable in clear *varnishes* but would be a substantial drawback in light colored pigmented paints.

12.4.2. Moisture Curing Polyisocyanate Coatings

Another important class of polyurethanes is moisture curing coatings, which cross-link under ambient conditions by reaction of isocyanate groups with atmospheric water. The resins used as binders have multiple isocyanate groups per molecule. The coatings are stable when stored in the absence of water but cross-link after application because of the reaction of the isocyanate groups with water to form amines which then, in turn, react with another isocyanate to form substituted urea linkages (see Section 12.1). They are called urethane coatings even though the cross-links are urea groups rather than urethane groups. In most cases, the isocyanate-functional resins are made from hydroxy-terminated polyesters by reacting the terminal hydroxyl groups with excess diisocyanate like MDI or TDI or, if color retention is needed, with an aliphatic diisocyanate. So that the coating does not contain unreacted diisocyanate, the resin is prepared utilizing a ratio of N=C=O/OH, which is significantly less than 2:1. The isocyanate-terminated resins of course contain a significant fraction of urethane links.

Moisture cure urethane coatings are used in applications such as floor coatings where exceptional abrasion resistance and hydrolytic stability are important. Since the urea groups also can form intermolecular hydrogen bonds, presumably they can affect resistance to mechanical stress similarly to urethane groups. Hydrolytic resistance of coatings is affected by a wide range of variables, including cross-link density and free volume availability, as well as the functional groups present. Studies of two pairs of model compounds showed that analogous substituted ureas are somewhat more easily hydrolyzed under both neutral and acidic (pH 1.34) conditions than are urethanes [9]. Ureas and urethanes from aliphatic isocyanates are somewhat more resistant to hydrolysis than those from aromatic isocyanates. Both urea and urethane groups are generally thought to be more stable to hydrolysis than unhindered ester groups.

In general, these coatings are designed for ambient temperature cure with the water for the cross-linking coming from the water vapor in the air. Cure rates, of course, depend on the relative humidity as well as the temperature. As long as the relative humidity is greater than 30%, useful cure rates can be achieved. As in any other ambient cure system, care has to be exercised in resin design so that the T_g of the cross-linked resin will be such that adequate cross-linking will be possible before the free volume limitation reduces the reaction rate substantially or even prevents the reaction from proceeding further.

Moisture cure coatings have the advantage that they are less susceptible to errors in application than the two-package coatings discussed in Section 12.4.3, where there is a chance of wrong ratios of packages affecting the stoichiometric ratios. A disadvantage is that the solvents, pigments, and other coatings components must be essentially water-free. The use of moisture cure coatings is primarily in clear, gloss coatings since it is very costly to remove adsorbed water from commercial pigments. Pigmented moisture cure coatings can be made using drying agents such as an alkyl orthoformate or p-toluenesulfonylisocyanate as water scavengers. Such water scavengers, especially sulfonylisocyanates must be handled with care because of their toxicity. Carbon dioxide is generated by the cross-linking reaction and can lead to bubbling, especially in thick films; in pigmented coatings bubbling can sometimes be minimized by using some $Ca(OH)_2$ in the pigment combination. The $Ca(OH)_2$ reacts with CO_2 to form calcium carbonate.

Another example of a moisture cure urethane coating system is based on using polyfunctional oxazolidines as coreactants. In the presence of an acid catalyst the oxazolidine rings open with water to give amino alcohols. The amine groups, and more slowly the hydroxyl groups, react with the isocyanates to give a cross-linked film.

12.4.3. Two-Package (2K) Polyisocyanate–Polyol Coatings

The largest volume of urethane coatings is two-package (2K) coatings that are mixed just before application. One of the packages contains the polyol, pigments, solvents, catalyst(s), and additives; the other contains the polyisocyanate and moisture free solvents. Sometimes the catalyst is in a separate third package so that cure rate can be adjusted for variations in ambient conditions.

Virtually any class of polyhydroxy-functional resin can be used. Hydroxy-terminated polyester and hydroxy-substituted acrylic resins are the most common coreactants. In general terms, polyesters permit higher solids, greater solvent resistance, and better adhesion to metals while acrylics provide faster dry, lower cost (since the equivalent weight is usually higher than that of polyesters resulting in the need for less of the more expensive isocyanate), and better exterior durability resulting from superior hydrolytic and photochemical stability.

A variety of other coreactants can be cross-linked with isocyanates. Since all alkyd resins have unreacted hydroxyl groups, their rate of dry can be accelerated by adding a polyisocyanate like the isocyanurate trimer of IPDI just before application. Nitrocellulose (shipped wet with xylene instead of ethyl or isopropyl alcohol) is used in formulating cross-linking furniture lacquers. While hydroxy-terminated polyethers are widely used in urethane foams, they are not widely used in coatings applications since the resulting coatings show high moisture vapor permeability, relatively poor exterior durability, and are soft as a result of the low T_g of the polyethers. Bisphenol A epoxy resins are used; the cross-linking reaction is between the hydroxyl groups and the isocyanate. Ketimine derivatives of polyamines can be used in moisture cure coatings, but they afford better properties in baking coatings, possibly due to the formation (see Section 12.1) of cyclic ureas [1].

Polyisocyanate–polyol coatings for maintenance paint applications are generally cured at ambient temperatures, while those for automobile refinishing and aircraft applications are cured at ambient or modestly elevated temperatures. Since one wishes in many cases to have coatings with a T_g somewhat above the curing temperature, selection of the combination of polyisocyanate and polyol that will provide the appropriate final T_g becomes critical. If a polyisocyanate with relatively flexible aliphatic chains such as the isocyanurate of HDI is used, it should be used with a relatively high T_g acrylic or polyester. On the other hand, if a polyisocyanate that yields relatively rigid cross-linked sections is used, such as the trimethylolpropane–TMXDI prepolymer, a significantly lower T_g acrylic or polyester is required.

If the T_g of the partially reacted system is near the temperature at which the cross-linking must occur, the rate of urethane formation will become mobility controlled and the cure rate may be very slow. If the T_g of the fully reacted system is somewhat above the cure temperature, the reaction will virtually stop before the cross-linking reaction has gone to completion [10]. This problem is, of course, reduced or eliminated by baking at temperatures above the ultimate T_g of the fully reacted coating.

An important variable in formulating two-package polyisocyanate–polyol coatings is the ratio of $N{=}C{=}O/OH$ used. In ambient cure systems, it is often found that a ratio of the order of 1.1:1 gives better film performance than a 1:1 ratio. A probable reason is that part of the $N{=}C{=}O$ groups react with water from the solvent, pigment, or air to give urea cross-links. To the extent that this happens, two hydroxyl groups would be left unreacted for each water molecule if the ratio were exactly 1:1; the use of excess $N{=}C{=}O$ minimizes residual unreacted hydroxyl groups. Pot life can also be affected by the reactant ratios.

In the case of aircraft finishes it is common to use $N{=}C{=}O/OH$ ratios as high as 2:1. The resulting reduced hydroxyl concentration gives a longer pot life. It is also possible that the high mobility of water and the high reactivity of the amine groups resulting from the water–isocyanate reaction favors faster reaction rates at higher T_g values than occurs with only the hydroxyl–isocyanate reaction.

There is always a compromise between pot life and curing time because reaction rates are governed by the laws of chemical kinetics (see Section 3.2.2). Several formulating variables are available to increase pot life with little, or at least small, effect on cure time. The concentrations of reactive groups should be kept as low as possible. Of course, this becomes more difficult as one formulates to higher and higher solids. As noted in the preceding paragraph, using mixed hydroxyl–moisture curing systems will help.

Since isocyanate–alcohol reactions proceed slowest in media with high levels of hydrogen-bond accepting groups, solvents should be selected to the extent possible, which are strong hydrogen-bond acceptors and resins should be designed, if possible, with low levels of hydrogen-bond accepting groups. After application, as the solvent evaporates, the medium will become less hydrogen-bond accepting and the reaction rate will increase accordingly. This strategy serves the purpose of both extending the pot life and promoting reactivity after application.

Assuming the system is to be catalyzed, organotin catalysts are generally preferable. As discussed in Section 12.2.2, reaction rates are reported to be dependent on alcohol concentration to the one-half power with organotin catalysts and to the first power with amine catalysts, which tends to reduce the pot life and extend the cure time in the latter case. Furthermore, since the effectiveness of tin catalysts is reduced by the presence of carboxylic acids, one can add a volatile acid such as acetic acid to the formula. It will inhibit the reaction during the pot life stage but will evaporate during application, and the inhibiting effect will disappear. Although carboxylic acids reduce the effectiveness of tin catalysts, they do catalyze the reaction of aliphatic isocyanates with alcohols [11].

If very fast cure at relatively low temperature is needed, reactive coreactants and/or high catalyst levels are used, and pot life will be short. Such formulations can be applied using special spray equipment in which the two packages are fed to the spray gun by separate pumps and mixed inside the gun just before they are sprayed. Care is required to assure that the proper ratios are fed and thoroughly mixed, and, of course, such spray equipment is expensive.

12.4.4. Blocked Isocyanates

It was recognized in early work on utilizing isocyanates in coatings that there would be major advantages to developing systems that could be supplied as one-package coatings having minimal toxic hazard. These objectives led to work on "blocked isocyanates," an approach taking advantage of the fact that many of the addition reactions that isocyanates undergo are reversible. In this approach a di- or polyisocyanate is reacted with certain monofunctional active hydrogen compounds and the adduct (a blocked isocyanate) is formulated into a coating with a polyfunctional hydroxy-substituted resin. The coating is stable at ambient temperature, but when it is baked the monofunctional reactant is released (and usually volatilized), and the polyfunctional resin reacts with the polyisocyanate to form cross-links. Amine-functional resins, which are more reactive than hydroxy-functional resins, are also used in certain applications.

Synthesis of a blocked isocyanate may be accomplished as shown, where B—H represents the blocking group:

$$R-N=C=O \ + \ B-H \ \longrightarrow \ R-NH-\overset{\overset{\displaystyle O}{\|}}{C}-B$$

A large number of blocking agents have been studied [12,13]. The agents most widely used commercially are phenols, oximes, alcohols, ε-caprolactam, and diethyl malonate [14]. Their adducts are stable enough that one-package coatings can be formulated, and the toxic hazard is reduced. Many blocked isocyanates are sufficiently stable in the presence of water at storage temperatures that they can be used in water-borne coatings. However, the use of blocked isocyanates has at least two drawbacks: relatively high temperatures are required for cross-linking with polyol reactants, negating one of the major advantages of isocyanate cross-linked coatings, and, in some cases, release of the blocking agent causes a pollution problem.

It is common to see tables of *unblocking temperatures*. The implication is that there is some "magic temperature" above which the various blocked isocyanates abruptly unblock. There is not. Rates of unblocking reactions are governed by Arrhenius constraints (see Section 3.2.2). Published *unblocking temperature* data are based on some arbitrary measurement of extent of reaction under a specific set of conditions. Examples of arbitrary criteria are (1) the lowest temperature at which free isocyanate can be detected spectroscopically and (2) the temperature required to achieve a specified degree of solvent resistance in cross-linking a specific coreactant resin for a specified time. Provder [15] showed how instrumentation can be used to determine the kinetics of unblocking reactions and the rate of development of cure. In using such data one must always remember that the values depend on the rate of heating as well as the time at some temperature.

SCHEME 4

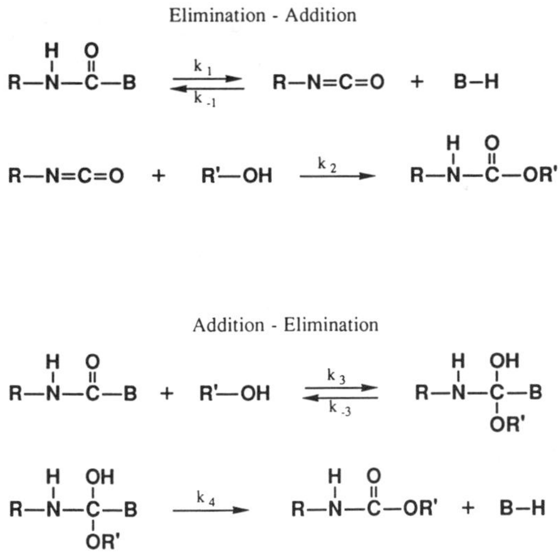

Elimination - Addition

$$R-\underset{\underset{H}{|}}{N}-\underset{\underset{O}{\parallel}}{C}-B \underset{k_{-1}}{\overset{k_1}{\rightleftharpoons}} R-N=C=O + B-H$$

$$R-N=C=O + R'-OH \xrightarrow{k_2} R-\underset{\underset{H}{|}}{N}-\underset{\underset{O}{\parallel}}{C}-OR'$$

Addition - Elimination

$$R-\underset{\underset{H}{|}}{N}-\underset{\underset{O}{\parallel}}{C}-B + R'-OH \underset{k_{-3}}{\overset{k_3}{\rightleftharpoons}} R-\underset{\underset{H}{|}}{N}-\underset{\underset{OR'}{|}}{\overset{\overset{OH}{|}}{C}}-B$$

$$R-\underset{\underset{H}{|}}{N}-\underset{\underset{OR'}{|}}{\overset{\overset{OH}{|}}{C}}-B \xrightarrow{k_4} R-\underset{\underset{H}{|}}{N}-\underset{\underset{O}{\parallel}}{C}-OR' + B-H$$

The reaction pathway for cross-linking is often written as the dissociation of the blocked isocyanate to give the blocking agent and the free isocyanate, which then reacts with an alcoholic hydroxyl group (i.e., elimination followed by addition). Alternatively, the reaction could proceed by addition of the alcohol to the blocked isocyanate to yield a tetrahedral intermediate followed by elimination of the blocking agent (i.e., addition followed by elimination). Both pathways are shown in Scheme 4, in which BH represents the blocking group and NuH represents a nucleophilic coreactant, either an alcohol or an amine. Evidence has been presented for both pathways with elimination–addition favored by increasing temperature. In many cases, including oxime blocked isocyanates, k_2 is small compared to k_{-1}, but volatilization of the blocking agent displaces the equilibria toward the cross-linked product. The volatile, monofunctional blocking agent can escape from the film, but the polyfunctional, nonvolatile coreactant (represented by NuH) cannot.

Frequently, catalysts are used in blocked isocyanate coatings. A wide variety of catalysts is reported: organotin compounds, DABCO, and zinc octoate are among them. In primers, where discoloration is not a problem, copper and cobalt naph-thenates have been found effective. It is not usually known which reaction is being catalyzed. It could be the decomposition of the blocked isocyanate to give the isocyanate. But, it also could be catalysis of the reaction of the nucleophile with the isocyanate so that k_2 increases relative to k_{-1}. And, it could also be catalysis of some stage of the addition–elimination pathway. Research on catalysis could well make a valuable contribution to the utilization of blocked isocyanates.

Phenols were the first blocking agents used commercially despite their environmental problems. Blocked isocyanates made with phenol or alkyl-substituted phenols are less thermally stable than those blocked with alcohols. Furthermore, blocked aromatic isocyanates are more reactive than blocked aliphatic isocyanates; thus, phenol blocked aromatic isocyanates are relatively reactive. A typical curing schedule with a phenol blocked polyisocyanate and a hydroxy-functional resin is 30 min at 160°C. Phenol blocked isocyanates are particularly used for wire coatings, applied by running the wire through a coating bath and then through an exit die appropriately larger than the diameter of the wire. This process requires a coating with one package stability. Since wire is relatively small and can be heated to high temperatures without major energy cost, high cure temperature is not a major disadvantage. Phenols had been used for many years as coatings solvents in the wire coating industry, and ovens were already engineered to permit recovery of the phenols, minimizing the pollution problem. The excellent abrasion resistance of polyurethane wire coatings justifies the extra cost.

Oximes, such as methyl ethyl ketone (MEK) oxime, form less thermally stable blocked isocyanates. Their reactivity is such that reasonably stable one-package coatings can be formulated with hydroxy-functional resins that can be cured in 30 min at 130°C or somewhat lower.

$$R-N=C=O \; + \; \begin{matrix} R' \\ \diagdown \\ C=N-OH \\ \diagup \\ R'' \end{matrix} \; \rightleftharpoons \; \begin{matrix} H & O & & R' \\ | & || & & \diagup \\ R-N-C-O-N=C \\ & & & \diagdown \\ & & & R'' \end{matrix}$$

Oximes find use as blocking agents in magnetic metal oxide coatings for application to oriented polyester tape. Such coatings must be cured in a short time at temperatures no higher than 80°C to avoid distortion of the plastic tape and yet must have sufficient pot life to permit application with a roller coater (at least a few hours). A two-package polyisocyanate–polyol system that would cure rapidly enough would not have sufficient pot life. A conventional polyol–blocked isocyanate system would have adequate pot life but would cure too slowly. An oxime blocked diisocyanate with an amine-terminated polyamide coreactant gives sufficient pot life but will still cure rapidly at 80°C. The final cured coatings undoubtedly contain some unreacted amine groups that could discolor in time, but this is not a disadvantage in magnetic tapes for audio, video, or computer uses. The excellent abrasion resistance and flexibility of polyurethane coatings are major advantages in this application.

It is common that amines react more rapidly with blocked isocyanates than do alcohols. This observation has been interpreted as evidence for the addition–elimination pathway. The intuitive reasoning is that the more nucleophilic amines are expected to be more reactive than alcohols in addition to blocked isocyanates, but are not expected to promote the elimination reaction. However, the higher reactivity of amines can also be rationalized in terms of the elimination–addition pathway when one takes into account that the elimination step is reversible and that amines are better competitors than alcohols for the intermediate isocyanate relative to the blocking group. In terms of Scheme 4, k_2 is much higher for amines than for alcohols.

Strong evidence for the elimination–addition pathway in reactions of amines and alcohols with oxime-blocked aromatic and aliphatic isocyanates has been obtained [16]. With amines the rate of reaction was found to be independent of amine concentration, indicating that essentially all of the intermediate isocyanate proceeds to product (a urea). (Bear in mind that the addition–elimination pathway requires dependence of the reaction rate on amine concentration.) In contrast, the rate of reaction with alcohols was substantially slower than with amines and was dependent on the concentration of alcohol. An exception occurred when there was a very large excess of alcohol; under these conditions the rates with alcohols and amines were almost equivalent. These and other results indicated that, with alcohols, a substantial fraction of intermediate isocyanate reverts to starting materials, resulting in an overall slower rate of conversion (to urethane). The conclusions were supported by demonstrating the order of reactivity with isocyanates: amines >> oximes >> alcohols.

The rate expression for elimination–addition (refer to Scheme 4) is provided in Eq. 12.7, where BI, BH, and NuH represent the blocked isocyanate, blocking group, and nucleophile, respectively.

$$\text{Rate} = k_1[\text{BI}] \frac{k_2[\text{NuH}]}{k_2[\text{NuH}] + k_{-1}[\text{BH}]} \tag{12.7}$$

When $k_2[\text{NuH}] >> k_{-1}[\text{BH}]$, the ratio in Eq. 12.7 approaches 1 and the equation simplifies to Eq. 12.8.

$$\text{Rate} = k_1[\text{BI}] \tag{12.8}$$

In this situation, the rate is independent of the nucleophile concentration, as observed in the reaction of oxime-blocked isocyanates with amines.

On the other hand, when $k_{-1}[\text{BH}]$ is greater than or equivalent to $k_2[\text{NuH}]$, the ratio in Eq. 12.7 becomes less than 1 resulting in a slower reaction rate. When $k_{-1}[\text{BH}]$ is substantially larger than $k_2[\text{NuH}]$, Eq. 12.7 simplifies to Eq. 12.9.

$$\text{Rate} = k_1[\text{BI}] \frac{k_2[\text{NuH}]}{k_{-1}[\text{BH}]} \tag{12.9}$$

This is the situation with alcohols; apparently oximes react faster than alcohols with isocyanates ($k_{-1} >> k_1$).

Steric crowding was found to enhance the dissociation rates of oxime blocked isocyanates, both from the standpoint of the oxime and the isocyanate. Thus, isocyanates blocked with diisopropylketoxime dissociated faster than the same isocyanates blocked with dimethylketoxime (acetone oxime). Oxime blocked TMXDI dissociated faster than other less sterically crowded aliphatic isocyanates and with comparable rates relative to aromatic isocyanates. Generally, blocked aromatic isocyanates are substantially more reactive than correspondingly blocked aliphatic isocyanates. On the other hand, coatings derived from aliphatic isocyanates exhibit substantially better exterior durability. Thus, the high reactivity of blocked TMXDI (and TMI) derivatives, characteristic of aromatic isocyanates, and their exterior durability, characteristic of aliphatic isocyanates, are potentially important advantages.

The largest volume coatings with blocked isocyanates are cationic electrodeposition primers (see Chapter 30). They must be stable in water indefinitely; therefore, blocked isocyanates with low reactivity are used (e.g., *n*-butyl alcohol blocked isocyanates) and the resin has secondary amine groups that are neutralized with weak acids (e.g., lactic or formic acids). During electrodeposition of the coating, the protonated amine groups are converted into free amines at the cathode, causing the polymer to form a uniform coating on the cathode. The coatings are subsequently cured at relatively high temperatures, during which the alcohol blocked isocyanates react with the amine.

Another large scale use of blocked isocyanates is in powder coatings (see Chapter 31). Here the principal blocking agent is ε-caprolactam, the cyclic lactam of $H_2N(CH_2)_5COOH$. Oximes are gaining popularity in powder coatings because of their reactivity at lower temperature.

Diethyl malonate blocked diisocyanates undergo cross-linking with polyols at lower temperatures than other commercially used blocked isocyanates, a typical curing schedule being 120°C for 30 min. However, reaction of diethyl malonate blocked isocyanates with alcohols and amines does not proceed by either of the pathways shown in Scheme 4. Reaction with alcohols does not yield urethanes, but rather transesterification occurs, as shown [17].

Correspondingly, reaction with amines yields amides, not ureas.

Although this transesterification goes fairly rapidly at 120°C, storage stable coatings can be formulated by using monofunctional alcohol in the solvent mixture [14]. The monofunctional alcohol extends the storage stability by competing with the polyols for reaction with the blocked isocyanate. After the coating is applied, the monofunctional alcohol evaporates and reaction with polyol proceeds to produce the cross-linked coating.

12.4.5. Linear Thermoplastic Polyurethanes

Thermoplastic polyurethanes are made by polymerization of oligomeric diols and diisocyanates to form high molecular weight, *linear* polymers. They are used as lacquers (see Chapter 4) without cross-linking. Typical uses are abrasion-resistant coatings for flexible, heat-sensitive substrates (e.g., vinyl fabrics, leather, elastomers, textiles, and foams). The combination of high molecular weight and strong intermolecular hydrogen bonding gives good film properties but also means that large amounts of hydrogen-bond acceptor solvents are needed to achieve application viscosity; as a result the coatings have to be applied at very low solids. Dispersions of thermoplastic polyurethanes in water have been developed to permit reduction in VOC emission (see Section 12.4.7).

12.4.6. Hydroxy-Terminated Polyurethanes

Diisocyanates can be reacted with diols and triols at an $N\!=\!C\!=\!O$/OH ratio of less than 1 to make hydroxy-terminated polyurethanes, which can then be cross-linked with MF resins. Compared to polyesters, the hydrolytic stability of the coatings is probably superior and the coatings can be expected to provide the toughness and abrasion resistance associated with urethane coatings. However, the solids at equal molecular weight and viscosity would be lower because of the stronger intermolecular hydrogen bonding.

Compared to hydroxy-functional acrylic resins, they offer the advantage of abrasion resistance and the possibility of using lower molecular weight resins, at least partly offsetting the viscosity disadvantage of urethanes. As discussed in Section 7.2, molecular weight reduction of acrylic resins made by conventional free radical initiated polymerization is limited by the problem of ensuring that at least two hydroxyl groups are present on each oligomer molecule. In the case of hydroxy-terminated polyurethanes (as with polyesters), all of the molecules will have two (or more) terminal hydroxyl groups even at very low molecular weight.

12.4.7. Water-Borne Urethane Systems

As with all other resin systems used in coatings, increasingly stringent VOC emission control regulations have motivated development of water-borne polyurethane systems. A wide variety of approaches has been investigated.

Perhaps the most obvious approach is the use of blocked isocyanate cross-linkers. This is done on a large scale with alcohol blocked isocyanates in aqueous cationic electrodeposition coatings as mentioned earlier and discussed further in Chapter 30. Blocked isocyanates can also be used with water-reducible anionic acrylic or polyester resins. In effect, one uses a blocked diisocyanate instead of an MF resin with water-reducible coreactant resins (see Section 7.3). Oxime blocked isocyanates provide the most reactive systems. Oxime blocked aliphatic isocyanates are used because the oxime blocked aromatic isocyanates hydrolyze too readily [18].

Another approach is to prepare water-reducible urethanes. For example, a hydroxy-functional polyurethane can be made in which one of the diols used is 2,2-dimethylolpropionic acid, $HOOCC(CH_2OH)_2CH_3$. The sterically hindered carboxylic acid group on the dimethylolpropionic acid is so low in reactivity with isocyanate that few amide bonds are formed. After neutralization of the carboxylic acid groups on the resin with an amine, the system can be diluted with water to yield a dispersion analogous to those discussed in Section 7.3. Class I MF resins can be used as cross-linkers. Such resins do not have the hydrolytic stability problems of water-reducible polyesters; also, the probability of formation of low molecular weight, cyclic, nonfunctional molecules is low.

A third class of water-borne urethane resins is predominantly linear, relatively high molecular weight polyurethane polymers dispersed in water. Such materials might be considered latexes, but they are almost universally called aqueous dispersions. A way of preparing such dispersions is by the so-called *acetone process* [19]. Initially an isocyanate-terminated polymer is prepared in acetone solution from a diisocyanate and a diol (or mixture of diols). This resin is then *chain extended* by reacting with a sulfonate substituted diamine, $H_2NCH_2CH_2NHCH_2CH_2SO_3^-Na^+$. The reaction of the amine groups leads to chain extension giving the desired high molecular weight. The acetone solution is diluted with water, and the acetone is removed by distillation. The sulfonate salt substituent acts as a charge-repulsion stabilizer for the aqueous dispersion. The average number of sulfonate groups per molecule can be adjusted by use of mixtures of unsubstituted diamines and sulfonate substituted diamines. One can incorporate a small amount of trifunctional isocyanate in making the initial isocyanate-terminated polymer. The result is dispersed polymer having a low degree of cross-linking—low enough that the particles can still coalesce to a continuous film but high enough that film properties are superior to those obtained with linear polymers [18].

One can also make combined acrylic–urethane aqueous dispersions [18]. Acrylic monomers are emulsion polymerized in the presence of an aqueous dispersion of a hydroxyl-terminated polyurethane (e.g., one made with dimethylpropionic acid) as mentioned earlier. The polyurethane acts as the stabilizer for the final aqueous dispersion, minimizing the need for surfactant. Coalescence requires balance of T_gs of both the urethane and acrylic parts of the system. Compositions based on an IPDI–polypropylene glycol–dimethylolpropionic acid urethane with styrene–methyl methacrylate–butyl acrylate are reported to form films at low temperatures.

In anticipation of further tightening of VOC limits during the 1990s, intensive research is underway to find commercially feasible water-borne polyurethane coatings with very low VOC [19]. Details of actual developmental products are proprietary.

REFERENCES

1. M. Bock and R. Halpaap, *J. Coat. Technol.*, **59** (755), 131 (1987).

2. M. Sato, *J. Am. Chem. Soc.*, **82**, 3893 (1960).

3. S. W. Wong and K. C. Frisch, *J. Polym. Sci. A Polym. Chem.*, **24**, 2867; 2877 (1986).

4. I. Yilgor and J. E. McGrath, *J. Appl. Polym Sci.*, **30**, 1733 (1985).

5. F. W. Van der Weij, *J. Polym. Sci. A Polym. Chem.*, **19**, 381 (1981).

6. K. Hatada, K. Ute, and S. P. Pappas, *J. Polym. Sci. C Polym. Lett.*, **25**, 477 (1987).

7. H.-K. Ono, F. N. Jones, and S. P. Pappas, *J. Polym. Sci. C Polym. Lett.*, **23**, 509 (1985).

8. K. Hatada, K. Ute, K.-I. Oka, and S. P. Pappas, *J. Polym. Sci. Part A Polym. Chem.*, **28**, 3019 (1990).

9. T. M. Chapman, *J. Polym. Chem. A Polym. Chem.*, **27**, 1983 (1989).

10. D. E. Fiori and R. W. Dexter, *Proc. Water-Borne Higher-Solids Coat. Symp.*, New Orleans, LA, 1986, pp. 186–199.

11. T. Nakamichi and M. Ishidoya, *J. Coat. Technol.*, **60** (766), 33 (1988).

12. Z. W. Wicks, Jr., *Prog. Org. Coat.*, **3**, 73 (1975).

13. Z. W. Wicks, Jr., *Prog. Org. Coat.*, **9**, 3 (1981).

14. T. A. Potter, J. W. Rosthauser, H. G. Schmelzer, *Proc. Water-Borne Higher-Solids Coat. Symp.*, New Orleans, LA, 1986, pp. 162–185.

15. T. Provder, *J. Coat. Technol.*, **61** (770), 33 (1989).

16. S. P. Pappas and E. H. Urruti, *Proc. Water-Borne Higher-Solids Coat. Symp.*, New Orleans, LA, 1986, pp. 146–161.

17. Z. W. Wicks, Jr., and B. W. Kostyk, *J. Coat. Technol.*, **49** (634), 77 (1977).

18. P. L. Jansse, *J. Oil Colour Chem. Assoc.*, **89**, 478 (1989).

19. J. W. Rosthauser and K. Nachtkamp, "Water-Borne Polyurethanes," in K. C. Frisch and D. Klempner, Eds., *Adv. Urethane Sci. Technol.*, Vol. 10, 1987, pp. 121–162.

Other Thermosetting Resins

The thermosetting resins described in Chapters 6 through 12 are used in the majority of current thermosetting coatings. Many other thermosetting resin systems have been studied over the last few decades. This chapter describes several types of thermosetting resin systems that have achieved significant commercial use along with examples of newer types now under development.

Many companies have active research and development programs seeking new cross-linking systems. A significant aspect of the motivation for this research is the increasing concern that formaldehyde, oxiranes, and isocyanates may present environmental and toxic hazards. However, in some cases the potential cross-linking reactions are designed to cross-link hydroxyl, amine, and/or carboxylic acid groups on resins at room temperature or modestly elevated temperatures. Since the human body contains polymers with hydroxy, amine, and carboxylic acid functional groups, it seems unlikely that such new cross-linking systems will be free of toxic hazards.

13.1. UNSATURATED POLYESTER RESINS

Unsaturated polyester resins are used on a large scale in glass-reinforced plastics for boats, tanks, automobile bodies, and many other applications. The plastics are commonly called just *fiberglass* plastics. The resins are used on a smaller scale in some coatings applications.

The unsaturated polyesters, which contain unsaturated maleate (*Z*-configuration) and fumarate (*E*-configuration) groups in their backbone, are used as solutions in a monomer, most commonly styrene. With a free radical initiator, the solution is converted into a cross-linked polymer. The product is a complex mixture derived from copolymerization of monomer with the maleate or fumarate double bonds, graft copolymerization, and probably homopolymerization of the monomer.

A variety of diacids and diols can be used to make unsaturated polyesters. The most common materials are copolymers of phthalic anhydride (PA), maleic anhydride, and propylene glycol or a mixture of propylene glycol and propylene oxide. These are the lowest cost raw materials, the initial reaction can be carried out at moderate temperatures, and a minimum amount of water is formed. The cross-link density of the cured product is affected by the ratio of maleic/PA; mole percentages in the range of 15 to 40% of maleic anhydride are commonly used. During esterification, some of the maleate ester groups formed isomerize to fumarate ester groups. The fraction isomerized depends on processing time and temperature and can also vary depending on the other components of the polyester. Fumarate groups are more reactive during cross-linking then maleate groups. Reaction conditions during esterification must be carefully controlled to ensure that the extent of isomerization is reproducible from batch to batch. A portion of a maleic anhydride–PA–propylene glycol polyester is represented by the following structure:

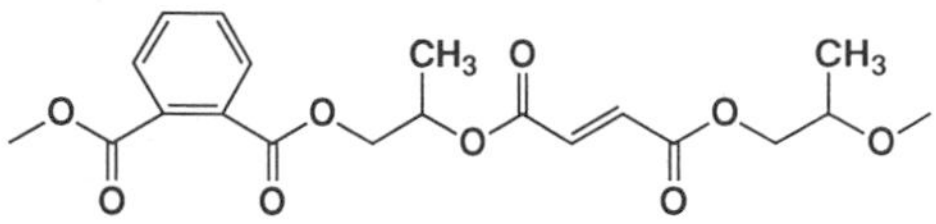

Typical segment of an Unsaturated Polyester Molecule

Use of fumaric acid instead of maleic anhydride results in a higher fumarate diester content; such polyesters generally give harder plastics than those made with maleic anhydride. More flexible plastics can be obtained by substituting an aliphatic diacid (e.g., adipic acid) for part of the PA to reduce T_g. While other glycols can be used, economics usually dictate the use of propylene glycol (or propylene oxide). The properties of the cured materials are also, obviously, controlled by the ratio of polyester to styrene; typically a 70:30 ratio is used.

Peroxides, such as benzoyl peroxide, are used as initiators at temperatures of the order of 70 to 100°C. In many applications of styrene–polyesters it is desirable to initiate polymerization at ambient temperatures. An initiator, such as methyl ethyl ketone (MEK) peroxide, with promoters, such as a mixture of dimethylaniline and cobalt naphthenate, is added to the resin–styrene solution just before use. The combination reacts rapidly at room temperature to generate free radicals that initiate polymerization.

Methyl ethyl ketone peroxide is a complex mixture in which the major components contain both peroxide and hydroperoxide groups, as shown [1]. The cobalt salt acts as a redox catalyst for the decomposition of the peroxide and hydroperoxide groups into free radicals and the dimethylaniline further promotes that reaction.

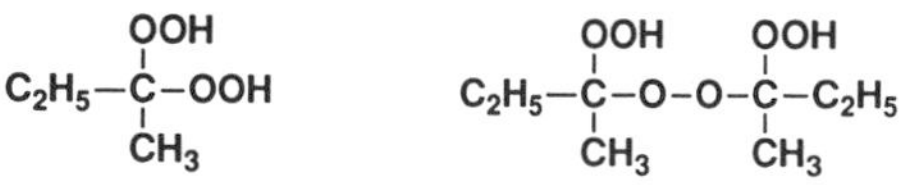

MEK Hydroperoxide (major components)

As with most free radical polymerizations, this polymerization is oxygen inhibited. This is not a problem for many plastics applications since the reaction is commonly carried out in a mold so the resin surface is not exposed to air.

One plastics application is sometimes classified as a coating: namely, what are called *gel coats*. Gel coats are pigmented unsaturated polyester–styrene coatings (with initiator and promoter); they are sprayed on the inside of a mold surface. The exposed (back) surface of the gel coat is then sprayed with a glass fiber loaded unsaturated polyester–styrene compound and the back surface is covered with plastic film. After the entire composition cross-links and is taken from the mold, the surface is the gel coat that was cured in the absence of oxygen. A wide variety of glass-reinforced plastic objects ranging from prefabricated shower stalls to boat hulls are made in this manner. In the case of boat hulls, hydrolytic stability is particularly important so the surface will retain gloss after outdoor exposure and immersion in water. Unsaturated polyesters made using neopentyl glycol and isophthalic acid provide better gloss retention than those made from propylene glycol and PA; they are used in gel coats despite their higher cost.

In most coatings applications, the top surface is exposed to the air and will remain sticky after the polymerization is complete below the surface. A further problem is the volatility of styrene. It is common to minimize these problems by incorporating insoluble semicrystalline paraffin wax in the formula. After the coating is applied, the low surface tension wax particles preferentially come to the surface. The wax layer reduces the rate of styrene loss and also substantially reduces the oxygen concentration in direct contact with the polymerizing mass, thereby minimizing the inhibition of surface cure. The wax layer, however, results in a relatively uneven low gloss surface, which is unsuitable for many applications but is acceptable for others.

Another approach to using unsaturated polyesters in coatings is UV curing. A photoinitiator is used that will generate free radicals on exposure to UV radiation. High intensity radiation sources are used, which generate very large numbers of free radicals at a sufficiently rapid rate at the surface, so that the oxygen in the air at the surface is depleted (by reaction with radicals), and polymerization can proceed. This approach requires that the rate of free radical generation be faster than the rate of diffusion of oxygen to the surface. Acrylate systems (see Section 13.2) are more widely used than the styrene–unsaturated polyesters. UV curing is discussed in Chapter 32.

Still another avenue to minimizing oxygen inhibition has been the addition of a wide variety of materials with allyl groups. For example, allyl ether coreactants reduce the oxygen inhibition effect with styrene–unsaturated polyester systems [2]. It is suggested that a free radical abstracts a hydrogen atom from the methylene group activated by both the adjacent allylic double bond and the ether oxygen. The free radicals from this reaction can then react with oxygen to form peroxy free radicals, which in turn abstract hydrogen atoms from the activated methylene groups to form hydroperoxides. The reactions consume some of the oxygen at the surface and generate new hydroperoxide in a chain reaction; as a result, oxygen inhibition is reduced.

The idea of using allyl ether reactions has recently been extended to water-reducible, unsaturated polyester resins. A low molecular weight ester is prepared by reacting 2 mol of maleic anhydride with 1 mol of a mixture of low molecular weight diols and polyalkylene glycols. The resulting partial ester is further esterified with 2 mol of trimethylolpropane diallyl ether [3]. The final unsaturated polyester can be emulsified in water for application. Films can be cured either with hydroperoxide–cobalt initiators or by photoinitiators and UV radiation.

13.2. (METH)ACRYLATED OLIGOMERS

Thermosetting resins can be made with acrylate or methacrylate double bonds as cross-linkable groups. Such resins are widely used in radiation cured coatings (see

Chapter 32). They can also be used in ambient cure and at somewhat elevated temperatures (force-dry) systems using initiators with appropriate reactivity. Both acrylate and methacrylate polymerizations are air inhibited, particularly methacrylate systems that generally require higher cure temperatures to minimize the effect of oxygen inhibition.

Acrylated oligomers can also be cross-linked with polyfunctional primary amines, where they undergo a Michael addition reaction, as illustrated for monofunctional reactants. This reaction is fast so that practical systems are based on blocked amines, commonly ketimines [4]. When a coating film is exposed to atmospheric moisture, the ketimine reacts with water to release the free primary amine.

$$RN{=}C\!\!\begin{smallmatrix}R'\\[2pt]R''\end{smallmatrix} \;+\; H_2O \;\longrightarrow\; RNH_2 \;+\; \begin{smallmatrix}R'\\[2pt]R''\end{smallmatrix}\!\!C{=}O$$

$$RNH_2 \;+\; H_2C{=}CH{-}\overset{O}{\overset{\|}{C}}{-}OR \;\longrightarrow\; RNH{-}CH_2{-}CH_2{-}\overset{O}{\overset{\|}{C}}{-}OR$$

(Meth)acrylated oligomers are prepared by a variety of routes from a variety of starting oligomers. The properties of films made from the (meth)acrylated oligomers will be affected by the average number of acrylic double bonds per molecule and by the structure of the "core" oligomer. For example, acrylated urethane oligomers tend to give coatings with a good combination of hardness and elasticity, and epoxy resin derivatives tend to give coatings with good toughness, chemical resistance, and adhesion.

Any polyol or hydroxy-terminated oligomer (HO—R—OH) can be reacted with a diisocyanate (OCN—R'—NCO) to yield an isocyanate-terminated oligomer (**1**). This oligomer can be reacted with hydroxyethyl acrylate at ambient or moderately elevated temperature to yield an acrylated urethane oligomer. Hydroxyethyl methacrylate yields the analogous methacrylated urethane oligomers. Due to the wide variety of hydroxy-functional oligomers available, this procedure is very versatile.

$$HO{-}R{-}OH \;+\; 2\,O{=}N{=}C{-}R'{-}N{=}C{=}O \;\longrightarrow\; \left[O{=}N{=}C{-}R'{-}\overset{O}{\overset{\|}{C}}{-}O\right]_2\!\!R$$
1

$$\mathbf{1} \;+\; 2\,HO{-}CH_2{-}CH_2{-}O{-}\overset{O}{\overset{\|}{C}}{-}CH{=}CH_2 \;\longrightarrow$$

$$H_2C{=}CH{-}\overset{O}{\overset{\|}{C}}{-}O{-}CH_2{-}CH_2{-}O{-}\overset{O}{\overset{\|}{C}}{-}NH{-}R'{-}NH{-}\overset{O}{\overset{\|}{C}}{-}O{-}R{-}O{-}\overset{O}{\overset{\|}{C}}{-}NH{-}R'{-}NH{-}$$

$$-\overset{O}{\overset{\|}{C}}{-}O{-}CH_2{-}CH_2{-}O{-}\overset{O}{\overset{\|}{C}}{-}CH{=}CH_2$$

Acrylated urethane oligomer

Another route is to react oxirane groups of epoxy resins with (meth)acrylic acid. The ring-opening reaction yields the acrylic ester and a hydroxyl group as shown.

$$CH_2{=}CH{-}\overset{O}{\overset{\|}{C}}{-}O{-}CH_2{-}\overset{OH}{\overset{|}{CH}}{-}CH_2{-}O{-}\!\!\bigcirc\!\!{-}\underset{CH_3}{\overset{CH_3}{\overset{|}{\underset{|}{C}}}}{-}\!\!\bigcirc\!\!{-}O{-}CH_2{-}\overset{OH}{\overset{|}{CH}}{-}CH_2{-}O{-}\overset{O}{\overset{\|}{C}}{-}CH{=}CH_2$$

Acrylated epoxy oligomer

Various catalysts (e.g., triphenyl phosphine) are used so that the reaction is carried out at as low a temperature as possible. Even with a catalyst, temperatures above 100°C are required. Care is required to avoid chain-growth polymerization of the acrylic acid or esters, which is readily initiated by free radicals. Inhibitors are added to trap any free radicals. It should be noted that some inhibitors, notably phenolic antioxidants, are effective only in the presence of oxygen, so the reaction is commonly carried out under a mixed inert gas, air atmosphere. Variations in reaction conditions and catalyst composition can result in significant differences in the product. The most widely used epoxy resin is the standard liquid BPA epoxy resin (n = 0.13) to yield predominantly the acrylated diglycidyl ether of BPA as shown.

Theoretically, the hydroxyl group of the acrylated resins could be further esterified with acrylic acid to give an oligomer having increased functionality. However, at the high temperatures required for this reaction, polymerization is hard to avoid even with inhibitors. Even small degrees of polymerization increase resin viscosity, and the extent of polymerization is likely to vary from batch to batch. Epoxidized soybean oil or epoxidized linseed oil can also be reacted with acrylic acid to give acrylated oligomers with low T_g and high functionality.

Hydroxy-terminated polyesters can be esterified with acrylic acid to form acrylate-terminated polyesters. Such resins would be relatively low in cost except for the difficulty of making them reproducibly for the reasons noted above.

Acrylated melamine–formaldehyde (MF) resins have been prepared by reacting etherified MF resins with acrylamide [5]. These resins have the potential to cure in two ways: by UV curing through the acrylate double bonds and thermally by residual alkoxymethylol groups on the MF resin. In this way, it is reported that UV curing followed by thermal curing will provide films with increased hardness, improved stain resistance, and durability.

13.3. SILICONE RESINS

Silicones, more properly called polysiloxanes, are polymers with inorganic backbones consisting of $Si(R)_2$—O repeating units. Polysiloxanes are prepared from chlorosilanes. The principal commercial monomers are the following methyl and phenyl substituted silanes.

Me_3SiCl	Trimethylchlorosilane
Ph_2SiCl_2	Diphenyldichlorosilane
$PhSi(Me)Cl_2$	Phenylmethyldichlorosilane
Me_2SiCl_2	Dimethyldichlorosilane
$PhSiCl_3$	Phenyltrichlorosilane
$MeSiCl_3$	Methyltrichlorosilane

Chlorosilanes react with water to form silanols that in turn can condense to form siloxanes. For example, dimethyldichlorosilane reacts with water to form a polysiloxane, often represented as a linear polymer:

$$Me_2SiCl_2 \ + \ H_2O \ \longrightarrow \ \left[\begin{array}{c} CH_3 \\ | \\ Si-O \\ | \\ CH_3 \end{array} \right]_n$$

Structures of the polymers can be more complex than represented above. Siloxanes are more apt than hydrocarbons to form large-ring cyclic structures such as $(Me_2SiO)_4$. Such *cyclics* are intermediates in the polymerization and are present to varying degrees in the products [6].

Monochlorosilanes function as chain-terminating groups, reducing molecular weight in proportion to the ratio of dichlorosilane/monochlorosilane. Oligomeric products are called silicone fluids or silicone oils. As discussed in Chapter 23, small amounts of silicone fluids are sometimes used as additives in coatings. Poly(dimethylsiloxane) fluids reduce surface tension because the solubility of the polymer is somewhat limited and the great ease of rotation around the Si—O—Si bonds leads to rapid orientation of methyl groups at the surface. The methyl groups give a surface with very low surface tension.

$$(CH_3)_3Si\left[\!-O-\underset{\underset{CH_3}{|}}{\overset{\overset{CH_3}{|}}{Si}}\!-\!\right]_{\!n}\!\!O-Si(CH_3)$$

A silicone fluid

Commercial poly(dimethylsiloxanes) have bimodal molecular weight distributions, and the higher molecular weight fractions have limited compatibility with many coatings resins. Chemically modified silicone fluids with broader ranges of compatibility have been described [7]. Examples are polysiloxane–polyether block copolymers, polyethercopolymers that also have some longer alkyl chains on the siloxane groups, and poly(dimethylsiloxanes) modified with esters and aryl-substituted alkyl groups.

Trichlorosilanes are used to impart chain branching. They can be copolymerized with dichlorosilanes to make silicone rubbers. While silicone rubbers are expensive, they have unique and useful properties; they exhibit excellent resistance to oxidation accompanied by very good retention of flexibility at very low temperatures. They have low T_gs as a result of the great ease of rotation around the Si—O—Si bonds.

However, chain branching is not the only possible outcome when trichlorosilanes are copolymerized. They can react with water to form three-dimensional cluster compounds (silsesquioxanes) exemplified by the cubic octamer $(C_6H_5)_8Si_8O_{12}$. Such clusters form to a variable extent when trichlorsilanes are copolymerized with dichlorosilanes, reducing the degree of branching [6].

13.3.1. Silicone Resins for Coatings

If a mixture of mono-, di-, and trichlorosilanes is polymerized, a so-called *silicone resin* is produced. A silicone resin is a branched polysiloxane that has some unreacted hydroxy groups. The initial reaction is carried out by hydrolyzing a solution of the monomers in a water-immiscible solvent. The resulting silicone resin solution is separated from the acidic water phase, and residual acid is carefully removed. A stable solution of a silicone resin is obtained if the last water is removed from the reaction mixture at relatively low temperatures in the absence of a catalyst. The $\overline{M}_n$ of such resins is generally in the range 700–5000. Because the chemistry is complex, different copolymerization processes can be expected to yield different proportions of linear, branched, cyclic, and three-dimensional cluster structures, which affect properties.

Such silicone resins can be formulated into a coating, applied to a substrate, and baked. Cross-linking and film formation will result. The cross-linking process is reversible. While formation of linear siloxane backbones $(R_2Si—O)_x$ is not readily reversible, the presence of a third Si—O bond on a single Si atom at a cross-linked site enhances reversibility.

The cross-linking reaction is slow at room temperature. A typical cure schedule is 1 h at 225°C. The time and temperature can be reduced somewhat by catalysis. A widely used catalyst is zinc octanoate. Since the cross-linking reaction is reversible, silicone films are sensitive to water, especially under basic conditions. Aqueous ammonia is especially destructive to such films. Evidence has been presented that aminolysis occurs selectively at Si groups with three oxygen atoms (i.e., at cross-linking sites), probably resulting from the greater electrophilicity of such Si groups as a consequence of attachment to three electronegative oxygen atoms [8].

The molecular weights and viscosities of silicone resins depend on the proportions of monochloro-, dichloro-, and trichlorosilane monomers and on the process, which affects the proportions of the possible structures (linear, branched, cyclic, and cluster) in the resin. The ultimate cross-link density depends on the fraction of trichlorosilane monomer in the recipe and on the fraction of this monomer that forms branched, rather than cyclic or cluster structures, during the process.

Most silicone resins are copolymers of methyl and phenyl substituted monomers; other alkyl monomers are sometimes included. Properties depend strongly on the phenyl/methyl ratio. The general effect on properties is given in Table 13.1, adapted from Ref. [6].

The rate of the cross-linking reaction is faster with the high-methyl substituted silicone resins. Consequently, there is a larger amount of material lost during heat curing of high-phenyl resins since low molecular weight components evaporate in the oven to a larger degree before polymerizing. Obviously, the faster cure rate of the high-methyl substituted silicones and the longer package stability of the high-phenyl substituted silicones follow from the same factor.

The greater UV resistance of high-methyl relative to high-phenyl substituted silicone resins gives greater gloss retention during exterior exposure. The exterior

Table 13.1. Properties of High-Methyl versus High-Phenyl Silicone Resins

High Methyl	High Phenyl
Lower weight loss in curing	
Faster cure rate	Longer package stability
Greater UV stability	Greater heat stability
Lower temperature flexibility	

Source: Adapted from Ref. [6].

durability of well-formulated silicone coatings is significantly better than that of acrylic–MF, polyester–MF, or urethane coatings and approaches that of highly fluorinated polymers.

On the other hand, high-phenyl silicones are superior to high-methyl silicones for applications requiring high temperature resistance and are far superior to any other organic coatings except certain fluoropolymers. The thermal stability of high-methyl silicones is much greater than that of silicones with longer alkyl groups. The half-lives at 250°C, based on weight loss studies, are reported to exceed 100,000 h for phenylsilicone films compared with over 10,000 h for methylsilicone films and only 2 h for propylsilicone films [6]. These data were obtained in studies of completely condensed silsesquioxanes ($RSiO_{3/2}$). The *useful* life of a silicone film at 350°C is given as 1000 h; for comparison a *polyester* film (presumably cross-linked with an MF resin) has a useful life of 1000 h at only 223°C.

When silicones are thermally decomposed, the ultimate product is silicon dioxide which, though brittle, can serve as a temperature-resistant coating binder. For example, chimney paints are made from silicone resins pigmented with aluminum flake for use at over 500°C for years. At the high service temperature, the organic substituents burn off leaving behind a film of the aluminum pigment in a matrix of silicon dioxide with perhaps some aluminum silicate—essentially glass. While it is brittle, this film continues to provide protection if it is not mechanically damaged.

As expected from the difference noted earlier in the effects of aromatic and aliphatic substituents, coatings from methylsilicone resins have low temperature flexibility superior to those from phenylsilicones and superior to most other organic coatings. They also combine, to an unusual degree, the properties of being repellant to liquid water but highly permeable to water vapor.

13.3.2. Silicone-Modified Resins for Coatings

The high cost of pure silicone coatings as well as their requirement for long, high temperature cures led to the use of silicones as modifiers for other coating resins. The earliest approach was simply to add a silicone resin to an alkyd resin in the reactor at the end of the alkyd cook. While some covalent bonds between silicone resin and alkyd might form, probably most of the silicone resin simply dissolves in the alkyd. Exterior durability of silicone-modified alkyd coatings is significantly better than unmodified alkyds. The extent of improvement in durability increases with increased amounts of added silicone resin; 30% silicone resin is a common degree of modification. Silicone resins designed for this purpose may contain higher alkyl as well as methyl and phenyl groups to improve compatibility.

Alkyd coatings modified with predominately phenyl silicone resins are reported to have greater thermoplasticity, faster air-drying, and higher solubility than predominately methylsilicone-modified alkyds [6]. These results can be understood when we consider that the higher rigidity of the aromatic rings leads to a "solid" film at an earlier stage of cross-linking. Less cross-linking in the phenylsilicone-modified coatings makes them more thermoplastic and soluble.

Further improvements in exterior durability are obtained by coreacting a *silicone intermediate*. The silicone intermediates are designed to react with free hydroxyl groups on the alkyd.

Silicone-modified alkyds are used mainly in air-dry coatings (e.g., on the exteriors of steel petroleum storage tanks). Silicone-modified polyester and acrylic resins find wide use in baked coatings, especially in coil coatings for metal siding. Such modified resins give films with significantly better exterior durability than the corresponding silicone-free polyester or acrylic films.

The resins are made by reacting silicone intermediates with hydroxy-terminated polyester or hydroxy-functional acrylic resins. In some cases, the silicone intermediates have silanol (Si—OH) groups. These can undergo cocondensation with hydroxyl groups on the resin to be modified as well as self-condensation with other Si—OH groups. The ratio of these two reactions is controlled by catalyst choice. By far the best catalysts for promoting the desirable cocondensation between the resin and the silicone intermediate are titanates such as tetraisopropyl- or tetra-isobutyltitanate [6]. In model compound studies, tetraisopropyltitanate was shown to favor cocondensation over self-condensation by $3.4:1$ as compared to a $0.23:1$ ratio with no catalyst. Cocondensation between a hydroxy-functional silicone intermediate and acrylic polyol is shown below. Self-condensation between silicone intermediates leads to Si—O—Si cross-links, without participation of the acrylic (alkyd or polyester).

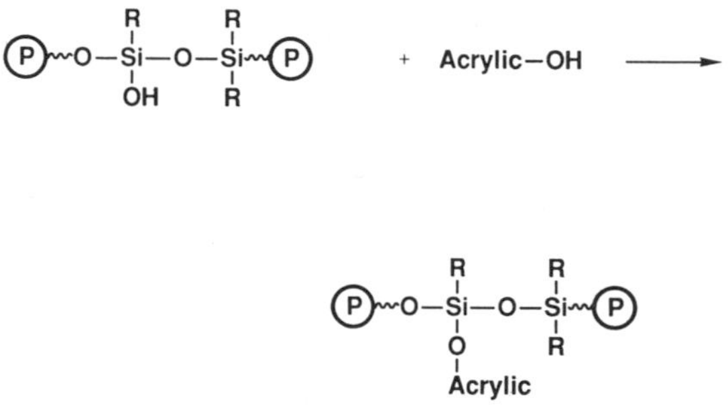

Excessive cocondensation of a polyfunctional silicone intermediate and a polyfunctional resin can result in gelation. It has been found easier to control the reaction when the reactive functional groups on the silicone intermediates are silylmethoxy (Si—OMe) rather than silanol (Si—OH) groups. A methoxylated silicone intermediate has been described, which has an average molecular weight (presumably number average) of 750 and an equivalent weight of 210, corresponding to an average of about 3.5 Si—OMe groups per molecule [9]. A solution of the polyester or acrylic is heated with the desired amount of silicone intermediate and the titanate catalyst at 140°C until a predetermined viscosity is reached. The reaction is relatively slow at 140°C; higher temperature would reduce reaction time and hence cost but would increase the risk of overshooting the desired viscosity as well as the risk of gelation.

The same reactions occur during cross-linking after the coating is applied. Usually zinc octanoate is used as the catalyst for cross-linking the coating since titanate esters are hydrolyzed by the water brought into the coating by pigments. The principal applications are in coil coatings (see Section 33.4), where a typical cure schedule is 90 s to peak metal temperatures of 300°C. Such coatings tend to soften when exposed to high humidity for prolonged periods, probably resulting from hydrolysis of cross-links which, as discussed above, have three oxygen atoms bound to a single silicone atom. The softening is called *reversion*. The film will harden again if the ambient humidity decreases, but the film is subject to physical damage if scraped while it is soft. To minimize this problem it is common to use a small amount of MF resin as a secondary cross-linker in silicone–polyester and silicone–acrylic coatings.

Typical silicone-modified polyesters and acrylics have 30–50 wt% silicone. Modification with less than about 25% silicone provides little improvement in exterior durability. Increasing the silicone content above 30% improves exterior durability but increases cost.

13.3.3. Moisture Cure Silicone Resins for Coatings

Tetraethylorthosilicate, $Si(OEt)_4$, although not a silicone, is used in making binders for zinc-rich primers, which are widely used primers for corrosion protection of steel (see Sections 27.4.3 and 36.3). When exposed to water vapor in the atmosphere, $Si(OEt)_4$ hydrolyzes to give a complex polysilicic acid network, which forms the coating binder.

$$Si(OC_2H_5)_4 \; + \; H_2O \; \rightleftharpoons \; (C_2H_5O)_3Si\!-\!O\!-\!Si(OC_2H_5)_3 \; + \; C_2H_5OH$$

To make a coating an ethyl alcohol solution of $Si(OEt)_4$ is partially polymerized by the addition of a small amount of water—just enough to increase the molecular weight sufficiently to give the desired viscosity, and the zinc powder pigment is dispersed in the resulting oligomer solution. A high proportion of zinc is used. When the coating is applied the alcohol evaporates, water is absorbed from the atmosphere, and the cross-linking reaction continues to completion at ambient temperatures. Note that the reaction is reversible; the ethyl alcohol solvent retards polymerization by small amounts of water in the coating formulation during storage.

While the cross-linked binder is predominantly polysilicic acid, zinc salts form by reaction with the zinc hydroxide–carbonate always present in zinc metal pigments. In fact, some people refer to the coatings as "zinc silicate" coatings. Ferrous and ferric ions from the steel surface may also be incorporated in the matrix.

Similar reactions can be used to cross-link polymers with terminal trialkoxysilyl groups. For example, an isocyanate-terminated resin can be reacted with 3-aminopropyltriethoxysilane to give a resin with terminal triethoxysilyl groups, as shown.

$$RNCO \; + \; H_2N\!-\!(CH_2)_3\!-\!Si(OEt)_3 \; \longrightarrow \; RNH\!-\!\overset{\overset{\displaystyle O}{\|}}{C}\!-\!NH\!-\!(CH_2)_3\!-\!Si(OEt)_3$$

$$\xrightarrow{\;H_2O\;} \; RNH\!-\!\overset{\overset{\displaystyle O}{\|}}{C}\!-\!NH\!-\!(CH_2)_3\!-\!\underset{\underset{\displaystyle O}{|}}{\overset{\overset{\displaystyle O}{|}}{Si}}\!-\!O\!-\!\underset{\underset{\displaystyle O}{|}}{\overset{\overset{\displaystyle O}{|}}{Si}}\!-\!(CH_2)_3\!-\!NH\!-\!\overset{\overset{\displaystyle O}{\|}}{C}\!-\!NHR$$

Coatings made using such resins will cross-link to a polymer network after application and exposure to humid air. Multifunctional isocyanates may be used to prepare resins with increased cross-link density. The original isocyanate-terminated resin would also moisture cure, but an important advantage of the triethoxysilane resin is that part of the solvent can be ethyl alcohol. The ethyl alcohol will permit a reasonable pot life in the presence of some water such as will come into a coating from water on pigment surfaces. If free isocyanate groups were present, the coatings would be unstable in the presence of either ethyl alcohol or water. The trialkoxysilane approach also avoids formation of CO_2, which can lead to film imperfections such as pinholing.

As with other types of resins, water-borne and high solids silicone and silicone-modified resins have been introduced in recent years to reduce solvent emissions [10,11]. The basic chemistry is similar to that of the materials described above.

13.4. OTHER APPROACHES TO CROSS-LINKABLE RESINS

Over many years, a wide variety of potential cross-linking systems have been investigated. This section includes brief discussions of the more promising approaches that have been published.

13.4.1. 2-Hydroxyalkylamide Cross-Linkers

Esterification of most alcohols with carboxylic acids is too slow for practical use as a cross-linking reaction. However, 2-hydroxyalkylamides undergo esterification reactions more rapidly than simple alcohols (e.g., *n*-octyl alcohol). Polyfunctional 2-hydroxyalkylamides (e.g., the tetrafunctional hydroxyalkylamide derived from aminolysis of dimethyl adipate with diisopropanolamine) can serve as cross-linkers for carboxylic acid-functional acrylic or polyester resins [12].

Hydroxyalkylamide Cross-linker

The properties of coatings obtained by cross-linking carboxylic acid-functional acrylic resins with hydroxyalkylamides compare favorably with those obtained using MF resins as cross-linkers with the same resins. An advantage relative to MF cross-linkers is the absence of formaldehyde, which is emitted in low concentrations when MF based coatings are baked. A disadvantage is that higher baking temperatures are required; the lowest reported satisfactory bake is 150°C for 30 min. Unfortunately, no catalyst for the reaction has been reported.

Such hydroxyalkylamides are soluble in both water and in common coatings solvents and, hence, are useful as cross-linkers in either water-borne or solvent-borne coatings. Furthermore, tetra-*N,N,N',N'*-(2-hydroxyethyl)adipamide is a solid well suited for use in powder coatings [13] (see Chapter 31).

The cross-linking reaction exhibits several unique features, compared to general esterification reactions. For one thing, it is not catalyzed by acid. For another, 2-hydroxyalkylamides, in which the alcohol group is secondary, esterify more rapidly than a corresponding primary alcohol derivative. Furthermore, aromatic carboxylic acids esterify with 2-hydroxyalkylamides at a faster rate than aliphatic carboxylic acids, the reverse of the rates observed with other alcohols. Presumably, therefore, cure temperatures with carboxylic acid-terminated polyesters could be lower than with carboxylic acid-functional acrylic resins, since in the former the residual acid groups are predominantly aromatic acid groups and in the later case they are all aliphatic acid groups. Although the 2-hydroxyalkylamides react more rapidly with aromatic acids, the saponification resistance of the products, esters of aromatic acids, is said to be greater than that of the corresponding esters of aliphatic acids.

Mechanistic studies of the hydroxyalkylamide–carboxylic acid esterification reaction, as well as related reactions, have been reported [14]. An explanation for the results, involving participation of the amide groups and intermediate formation of oxazolinium groups, has been advanced and is consistent with experimental observations.

It was also shown that 2-hydroxyalkylamides will react with epoxies. At elevated temperatures, the amide rearranges to the aminoester and the resulting amine group then reacts with the epoxy [15]:

$$R-\overset{\overset{O}{\|}}{C}-\overset{\overset{H}{|}}{N}-CH_2-CH_2-OH \ + \ H_2C\overset{O}{\overset{\diagdown\diagup}{-}}CH-CH_2-OR' \longrightarrow$$

$$R-\overset{\overset{O}{\|}}{C}-O-CH_2-CH_2-NH-CH_2-\overset{\overset{OH}{|}}{C}H-CH_2-OR'$$

13.4.2. Acetoacetate Cross-Linking Systems

The chemistry of β-keto esters such as acetoacetic esters has been widely studied. These esters exist to a significant degree in the tautomeric enol form.

$$H_3C-\overset{\overset{O}{\|}}{C}-CH_2-\overset{\overset{O}{\|}}{C}-OR \longleftrightarrow H_3C-\overset{\overset{OH}{|}}{C}=CH-\overset{\overset{O}{\|}}{C}-OR$$

They undergo a wide range of reactions, some of which can be adapted for coatings purposes.

Several methods are available to prepare resins bearing acetoacetate groups. Acetoacetoxy-functional acrylic resins can be made by copolymerizing acetoacetoxyethyl methacrylate with other acrylate monomers [16]. Hydroxy-functional polyesters and other hydroxy-functional resins can be reacted with diketene or transesterified with methyl acetoacetate to form acetoacetylated resins. The replacement of the hydroxyl group with the less polar acetoacetate group leads to a reduction in viscosity and, in turn, somewhat higher solids at a given application viscosity.

A variety of coreactants with acetoacetylated resins have been investigated. The MF resins react with acetoacetate groups in the presence of an acid catalyst, although somewhat less rapidly than with hydroxyl groups [16]. Film properties of the resulting coatings using acetoacetylated resins are reported to be comparable with those made from hydroxy-functional resins. There are indications of improved wet adhesion, perhaps resulting from chelating interactions with the surface of the steel panels. Low molecular weight acetoacetic esters are generally more readily hydrolyzed than other aliphatic esters. However, there are no published reports of difficulties attributable to chemical lability of the cross-links.

$$\begin{array}{c}
\text{R'} \qquad\qquad\qquad \text{R} \\
\text{C-O} \qquad\quad \text{O=C} \\
\text{HC} \qquad \text{Fe} \qquad \text{CH} \\
\text{C=O} \qquad\quad \text{O-C} \\
\text{R} \qquad\qquad\qquad \text{R'}
\end{array}$$

Isocyanates also react with acetoacetate groups; again the reaction is slower than with hydroxyl groups. This results in systems with somewhat longer pot life but correspondingly slower cure rate. Film properties are reported to be similar to those of conventional polyurethanes [16]. Note that the initial product is an *acetoacetate blocked isocyanate*; presumably, therefore, the cross-links may be thermally labile (see Section 12.4.4). No studies of the consequences of the presence of such a structure in the film have been published.

$$2\ \text{(P)}\text{--O--}\overset{O}{\overset{\|}{C}}\text{--CH}_2\text{--}\overset{O}{\overset{\|}{C}}\text{--CH}_3\ +\ \text{O=C=N--R--N=C=O} \xrightarrow{25^\circ\text{C}}$$

$$\text{(P)}\text{--O--}\overset{O}{\overset{\|}{C}}\text{--}\overset{\overset{\text{CH}_3}{|}}{\underset{}{\text{CH}}}\text{--}\overset{\overset{\text{C=O}}{}}{\underset{}{C}}\text{--NH--R--NH--}\overset{}{\underset{}{C}}\text{--}\overset{\overset{\text{O=C}}{}}{\underset{}{\text{CH}}}\overset{\overset{\text{CH}_3}{|}}{}\text{--}\overset{O}{\overset{\|}{C}}\text{--O--}\text{(P)}$$

A

Polyacrylates (see Section 13.2) undergo Michael reactions with acetoacetate groups at ambient temperatures in the presence of strongly basic organic catalysts such as tetramethylguanidine (TMG) [17].

$$2\ \text{(P)}\text{--O--}\overset{O}{\overset{\|}{C}}\text{--CH}_2\text{--}\overset{O}{\overset{\|}{C}}\text{--CH}_3\ +\ \text{CH}_2\text{=CH--}\overset{O}{\overset{\|}{C}}\text{--O--R--O--}\overset{O}{\overset{\|}{C}}\text{--CH=CH}_2 \xrightarrow{\text{Base}}$$

$$\text{(P)}\text{--O--}\overset{O}{\overset{\|}{C}}\text{--}\overset{\overset{\text{CH}_3}{|}}{\text{CH}}\text{--CH}_2\text{--CH}_2\text{--}\overset{O}{\overset{\|}{C}}\text{--O--R--O--}\overset{O}{\overset{\|}{C}}\text{--CH}_2\text{--CH}_2\text{--}\overset{\overset{\text{CH}_3}{|}}{\text{CH}}\text{--}\overset{O}{\overset{\|}{C}}\text{--O--}\text{(P)}$$

B

Amines react rapidly with the carbonyl group of an acetoacetic ester; hence polyamines can be used as cross-linkers for acetoacetate functionalized resins. The reaction is so fast at ambient temperatures that pot life is limited. This problem can be minimized by blocking the amine groups with ketones. The resultant ketimine will hydrolyze in the presence of water, permitting cross-linking to proceed [17].

13.4.3. Thermosetting Fluorinated Resins

As discussed in Section 4.2.3, fluorinated polymers show outstanding exterior durability. However, the thermoplastic materials are not easily applied and, in the case of organosol applied vinylidene fluoride polymers, only low gloss coatings can be made. Considerable research has been carried out over the years on fluorinated copolymers with functional groups such as hydroxyl groups that can be cross-linked after application. For example, perfluoroacrylate esters have been copolymerized with HEMA to form solvent-soluble cross-linkable resins; however, the monomers are very expensive. Attempts to copolymerize monomers such as $CF_2{=}CF_2$ and $CF_2{=}CH_2$ with functional acrylic monomers have been unsuccessful.

Solvent-soluble halofluoroethylene–vinyl ether copolymers have been introduced [18]. They are used in coatings for steel building panels and to a limited extent in clear coats for automobiles. The $CF_2{=}CFX$ and vinyl ether monomers form alternating polymers, as shown in Figure 13.1. Functional groups can be introduced by copolymerizing hydroxy- and/or carboxy-substituted vinyl ether co-monomers. The T_g is controlled by the ratio of vinylidene fluoride derivative/vinyl ether monomers and by the chain length of the alkyl group on the vinyl ether.

Copolymers with hydroxy groups can be cross-linked with MF resins, isocyanates, or blocked isocyanates. Ambient cure systems are based on cross-linking with aliphatic polyisocyanates. Gloss retention of the isocyanate cross-linked coatings in accelerated tests for exterior durability is reported to be excellent, however, as discussed in Chapter 25, such tests are unreliable predictors of actual outdoor exposure results. Thus far exterior durability results have only been reported for limited time periods. Gloss retention after 3 to 5 years of outdoor exposure in Japan is reported to be superior to an (unidentified) acrylic–MF coating but some-

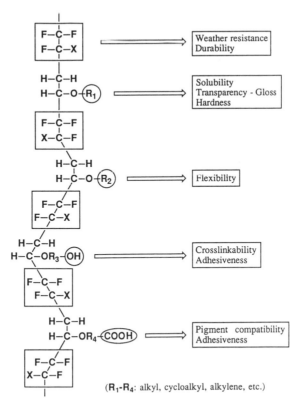

Figure 13.1. Molecular structure of fluoroethylene–vinyl ether copolymers. (From Ref. [18], with permission.)

what inferior to a poly(vinylidene fluoride) one. The reduced gloss retention as compared to poly(vinylidene fluoride) is not surprising in view of the presence of large numbers of methylene groups alpha to ether oxygen atom. Longer exposure data will be required to assess the merit of the system, but it would now appear to provide exterior durability properties intermediate between acrylics and fluorocarbons with the advantages of easier application and high gloss. Comparisons with silicone-modified polyester and acrylic coatings have not yet been reported.

13.4.4. Acrylamidoglycolate Ester Ethers

Methyl acrylamidoglycolate methyl ether (MAGME), a monomer with three different reactive sites, is available [19]. Being an acrylic ester, it can be copolymerized with other acrylic monomers forming polymers having side chains with two different reactive groups.

$$H_2C=CH-\overset{\overset{\displaystyle O}{\|}}{C}-NH-CH-\overset{\overset{\displaystyle O}{\|}}{C}-OCH_3$$
$$\underset{OCH_3}{|}$$

MAGME

The methyl ester is activated by the two heteroatoms alpha to the carbonyl group and will react under mild conditions with primary amines to form amides. Acrylic copolymers of MAGME can be cross-linked with polyfunctional primary amines at ambient temperatures.

The methyl ether is substituted on a carbon alpha to an amide nitrogen; that is, it is an alkoxymethyl derivative of acrylamide. Therefore, it will undergo cross-linking reactions like amino resins derived from acrylamide itself (see Section 6.4.4). In this manner the MAGME derived acrylic resins can also be used with hydroxy-functional resins in coatings that will cure at temperatures somewhat over 100°C with acid catalysis.

13.4.5. Polyaziridine Cross-Linkers

Aziridine, the nitrogen three-membered ring counterpart of oxirane, and its derivatives have been studied for many years; in some cases polyaziridines have been used as cross-linkers. The common name of aziridine is ethylene imine. Ethylene imine is a highly toxic material and may be carcinogenic. Concern about toxic hazards of aziridine derivatives has partly inhibited their commercial use.

With acids ethylene imine is even more reactive than ethylene oxide. In the presence of relatively strong acids it polymerizes very rapidly to yield poly(ethylene imine).

$$\underset{H_2C-CH_2}{\overset{\overset{\displaystyle H}{\underset{\displaystyle N}{|}}}{}} \quad \xrightarrow{H^+} \quad -(CH_2-CH_2-NH-CH_2-CH_2-NH)_n$$

Among the many other reactions of aziridines, probably the one of greatest interest in coatings applications is the reaction of an aziridine with a carboxylic acid. The initial product is the 2-aminoester but most 2-aminoesters spontaneously rearrange to yield the corresponding 2-hydroxyamides. In the presence of an acid such as hydrochloric acid, the 2-hydroxyamide will rearrange back to the hydrochloride salt of the 2-aminoester.

$$\underset{\underset{H}{\overset{|}{N}}}{\overset{H_2C-CH_2}{\diagdown\diagup}} + \ RCOOH \ \longrightarrow \ R-\overset{\overset{O}{\parallel}}{C}-O-CH_2-CH_2-NH_2 \ \longrightarrow$$

$$R-\overset{\overset{O}{\parallel}}{C}-NH-CH_2-CH_2-OH$$

A variety of polyfunctional aziridines has been investigated. One example is the trifunctional product made by a Michael addition of 3 mol of aziridine to 1 mol of trimethylolpropane triacrylate.

$$H_3C-CH_2-C\left(CH_2-O-\overset{\overset{O}{\parallel}}{C}-CH_2-CH_2-N\overset{\diagup CH_2}{\underset{\diagdown CH_2}{|}}\right)_3$$

Such polyaziridines have been most commonly used in coatings to cross-link carboxylic acid-functional latex polymers. The reaction with the carboxylic acid probably is the same as the reaction of the unsubstituted aziridine shown earlier. It is much faster than the reaction of the aziridine groups with water, but the reaction with water is fast enough that the cross-linker must be added shortly before application. Pot life can be extended to a few hours by adjusting the pH to 8.5–9 with aqueous ammonia.

In view of the potential toxic hazards, manufacturer's recommendations for safe handling should be carefully followed.

13.4.6. Polycarbodiimide Cross-Linkers

Carbodiimides also react with carboxylic acids and react slowly enough with water so that they can be used in water-borne systems. The product of the reaction with a carboxylic acid is an N-acylurea [20].

$$RN=C=NR \ + \ R'COOH \ \longrightarrow \ \underset{\underset{\underset{R'}{\overset{|}{C=O}}}{\overset{|}{}}}{RN}-\overset{\overset{O}{\parallel}}{C}-\overset{\overset{H}{\overset{|}{}}}{NR}$$

Multifunctional carbodiimides are available to serve as cross-linkers for carboxylic acid-functional resins such as carboxylic acid-substituted latexes, carboxylic acid-functional aqueous polyurethane dispersions, and presumably any other carboxylic acid-functional resin. Curing schedules of 30 min at 85°C are recommended.

REFERENCES

1. C. S. Sheppard, "Peroxy Compounds," in *Encyclopedia of Polymer Science and Engineering*, 2nd ed., Vol. 11, Wiley, New York, 1988, pp. 1–21.

2. H.-J. Traenckner and H. U. Pohl, *Angew. Makromol. Chem.*, **108**, 61 (1982).

3. M. J. Dvorchak and B. H. Riberi, *Proc. 16th Intl. Conf. Org. Coat. Sci. Tech.*, 1 (1990).

4. A. Noomen, *Prog. Org. Coat.*, **17**, 27 (1989).

5. J. J. Gummeson, *J. Coat. Technol.*, **62** (785), 43 (1990).

6. L. H. Brown, "Silicones in Protective Coatings," in *Treatise on Coatings*, Vol. I, Part III, R. R. Myers and J. S. Long, Eds., Marcel Dekker, New York, 1972, pp. 513–563.

7. F. Fink, W. Heilen, R. Berger, and J. Adams, *J. Coat. Technol.*, **62** (791), 47 (1990).

8. Y-C. Hsiao, L. W. Hill, and S. P. Pappas, *J. Appl. Polym. Sci.*, **19**, 2817, 1975. See also S. P. Pappas and R. L. Just, *J. Polym. Sci. Polym. Chem. Ed.*, **18**, 527 (1980).

9. Dow Corning Corp. *Bulletin 03-032, Dow Corning Z-6188 Intermediate*, Midland, MI, 1967.

10. W. Sittenthaler, *Proc. FATIPEC Congr.* II-329 (1988).

11. W. A. Finzel, *Polym. Mater. Sci. Eng.*, **63**, 777 (1990).

12. J. Lomax and G. F. Swift, *J. Coat. Technol.*, **50** (643), 49 (1978).

13. A. Mercurio, *Proc. 16th Intl. Conf. Org. Coat. Sci. Tech.*, Athens, Greece (1990), p. 235.

14. Z. W. Wicks, Jr., M. R. Appelt, and J. C. Soleim, *J. Coat. Technol.*, **57** (726), 51 (1985).

15. F. N. Jones and I-C. Lin, *J. Appl. Polym. Sci.*, **29**, 3213 (1984).

16. F. D. Rector, W. W. Blount, and D. R. Leonard, *J. Coat. Technol.*, **61** (771), 31 (1989).

17. R. J. Clemens and F. D. Rector, *J. Coat. Technol.*, **61** (770), 83 (1989).

18. S. Munekata, *Prog. Org. Coat.*, **16**, 113 (1988).

19. H. R. Lucas, *J. Coat. Technol.*, **57** (731), 49 (1985).

20. Anonymous, "Ucarlink Crosslinkers," Union Carbide Product Bulletin F-60756 (1989).

CHAPTER XIV

Solvents and Solubility

By far the majority of coatings contain volatile material that evaporates during application and film formation. The volatile components are there primarily to reduce the viscosity of the coating low enough for application (usually in the range of 0.05 to 1 Pa·s) and to control viscosity changes during the application and film formation processes. Selection of volatile components affects the occurrence of such film defects as popping, sagging, and leveling, and can affect such properties as adhesion, corrosion protection, and exterior durability. Too often formulators do not consider the broad effects that solvent choice can have in coating performance; especially, it is common for solvent substitutions to be made with inadequate thought about the possible effects of the substitutions. The wide range of factors that must be considered in selecting volatile components for a particular formulation will be covered in this and Chapter 15. These factors include solubility criteria, volatility, viscosity effects, surface tension, flammability, toxicity, odor, air pollution, and cost.

In some cases, it is necessary that the volatile material be a solvent for the resins in the formulation; in other cases, nonsolvent volatile components are desired. Unfortunately, words are used sloppily and, commonly, any volatile organic material in the coating is called a solvent whether it dissolves the resin or not. The word solvent is generally not used to include water, only organic materials. For example, coatings are commonly classified as *solvent-borne* and *water-borne*. While, in most cases, the resins in solvent-borne coatings are dissolved in the solvents, this is not always the case. Furthermore, most water-borne coatings have at least some organic solvent in their formulations; usually, the resins are not truly soluble in water or the water–solvent blend.

As discussed in previous chapters, a major objective of coatings research and development is to reduce solvent usage, so as to reduce air pollution. While ozone is naturally present in the atmosphere at low concentrations, volatile organic compound (VOC) emissions from automobiles, coatings, and other sources cause ozone concentrations to be at dangerous levels in many cities much of the year. Massive research and development efforts have led to decreased VOC from coatings, but current levels still cause problems. It has been estimated that in 1987, even after over 20 years of technology-forcing regulations, about 180 million kilograms of solvent was still used in coatings in California. Reduction of air pollution from coatings, while retaining, and often improving, coating quality, remains one of the great challenges in coatings science.

14.1. SOLVENT COMPOSITION

A wide variety of organic compounds and mixtures of compounds are used as solvents in coatings. They can be classified in three broad categories: weakly hydrogen-bonding solvents (hydrocarbons and chlorinated solvents), hydrogen-bond acceptor solvents (esters and ketones), and hydrogen-bond donor–acceptor solvents (alcohols). Table 14.1 lists some of the more widely used solvents together with their boiling ranges, relative evaporation rates (see Chapter 15), and densities (g mL^{-1}). The data given are for commercial grades of the solvents, not for the pure solvents; there are variations from source to source and lot to lot.

14.1.1. Weak Hydrogen-Bond Solvents

Hydrocarbon solvents are divided into two classes: aliphatic and aromatic. Commercial aliphatic solvents are mixtures of straight-chain and branched-chain aliphatic hydrocarbons, alicyclic hydrocarbons, and usually some aromatic hydrocarbons. Commercial mixtures vary in volatility and solvency; data for a few typical solvents are given in Table 14.1. Varnish makers and painters (VM & P) naphthas are examples of aliphatic solvents having high volatility. Mineral spirits is the name generally given to slower evaporating aliphatic hydrocarbon solvents. Special grades of mineral spirits with low aromatic compound content are widely used because of their less pungent odor; somewhat optimistically, they are called *odorless mineral spirits*. A primary advantage of aliphatic solvents is low cost, especially on a volume basis since the density as well as the price per unit weight is low. Turpentine, a mixture of alicyclic hydrocarbons derived from pine trees, was formerly used in large volumes but has been almost completely supplanted by mineral spirits, which costs less.

Aromatic hydrocarbon solvents are generally more expensive than aliphatic solvents but can dissolve a broader range of resins. Use of benzene is prohibited because of its toxicity. Toluene and "xylene" are used on a large scale. In coatings "xylene" is a mixture of the three isomeric xylenes and ethyl benzene; ratios vary from supplier to supplier and from time to time. The xylene listed in Table 14.1 is typical. Higher boiling mixtures of aromatic hydrocarbons (high flash naphthas) are also used. They are predominantly mixed alkyl and dialkyl benzenes with three to five carbon atoms in the substituent groups.

Chlorinated solvents are weak hydrogen-bond donors with higher dipole interactions than hydrocarbons. Their use has been increasingly limited by toxic hazard concerns. The major chlorinated solvent now in use in coatings is 1,1,1–trichloroethane (methyl chloroform). As mentioned in Chapter 15, in some states this solvent is excluded from VOC for air pollution regulatory purposes. It must be used with adequate ventilation and is usually restricted to air drying coatings due to potential corrosion problems in baking ovens. Manufacture of 1,1,1–trichloroethane is expected to be phased out during the 1990s since it is implicated in reduction of ozone in the stratosphere. Its high density makes it expensive on a volume basis.

Table 14.1. Representative Organic Solvents[a]

Name	Boiling Range (°C)	Relative Evaporation Rate (25°C)	Density (25°C)
Low Hydrogen-Bonding Solvents			
VM & P naphtha	119–129	1.4	0.742
Mineral spirits	158–197	0.1	0.772
Toluene	110–111	2.0	0.865
Xylene	138–140	0.6	0.865
High flash naphtha	181–201	<0.1	0.881
1,1,1-Trichloroethane	73–75	6.0	1.325
Hydrogen-Bond Acceptor Solvents			
Methyl ethyl ketone	80	3.8	0.802
Methyl isobutyl ketone	116	1.6	0.799
Methyl *n*-amyl ketone	147–153	0.46	0.814
Isophorone	215–220	0.02	0.919
Ethyl acetate	75–78	3.9	0.894
Isopropyl acetate	85–90	3.4	0.866
n-Butyl acetate	118–128	1.0	0.872
1-Methoxy-2-propyl acetate	140–150	0.4	0.966
2-Butoxyethyl acetate	186–194	0.03	0.938
1-Nitropropane–nitroethane blend	112–133	1.0	0.987
Hydrogen-Bond Donor–Acceptor Solvents			
Methyl alcohol	64–65	3.5	0.789
Ethyl alcohol	74–82	1.4	0.809
Isopropyl alcohol	80–84	1.4	0.783
n-Butyl alcohol	116–119	0.62	0.808
sec-Butyl alcohol	98–101	0.81	0.805
1-Propoxypropan-2-ol	149–153	0.21	0.890
2-Butoxyethanol	169–173	0.07	0.901
Monobutyl ether of diethylene glycol	230–235	<0.01	0.956
Ethylene glycol	196–198	<0.01	1.114
Propylene glycol	185–190	0.01	1.035

[a]Data from technical literature of Angus Chemical Co., Arco Chemical Co., Dow Chemical Co., Eastman Kodak Co., Exxon Chemical Co., Shell Chemical Co., and Union Carbide Corp.

14.1.2. Hydrogen-Bond Acceptor Solvents

Esters and ketones are widely used when hydrogen-bond acceptor solvents are needed. Ketones are generally less expensive than esters with corresponding vapor pressures. The cost differential is particularly marked on a volume basis since the density of ketones is lower than that of esters with comparable volatility. Ethyl, isopropyl, and butyl acetates and methyl ethyl ketone (MEK), methyl isobutyl ketone (MIBK), and methyl *n*-amyl ketone (MAK) are the most widely used. When slower evaporating solvents are needed, esters such as 1–methoxy–2–propyl acetate and 2–butoxyethyl acetate are often preferred over slow evaporating ketones such as isophorone because of odor. In very general terms, esters have more pleasant odors than ketones. However, change of odor is commonly more important than odor itself. People notice changes from a familiar odor, and users may suspect that coatings performance is changed when the odor changes. Esters of 2–ethoxy-ethanol were widely used at one time but are now suspected of imparting long-term adverse health effects.

Esters should not be used as solvents for resins bearing primary or secondary amine groups because of the potential for aminolysis reactions, which would convert the amines into amides.

Nitroparaffins such as a nitroethane–nitropropane blend are used to a degree. A principal advantage is that their high polarity results in increased electrical conductivity. This property is useful in adjusting solvent combinations for electrostatic spraying (see Chapter 22).

14.1.3. Hydrogen-Bond Donor–Acceptor Solvents

Alcohols are used when strong hydrogen-bond donor–acceptor solvents are needed. The most widely used relatively volatile alcohols are methyl, ethyl, isopropyl, *n*-butyl, *sec*-butyl, and isobutyl alcohols. Most latex paints contain a slow evaporating, water-soluble solvent such as propylene or ethylene glycol; one purpose is to suppress the freezing point—they do not dissolve the latex polymer. In water-reducible acrylic and polyester resin systems, ether–alcohols such as 1–propoxy-propan-2–ol, 2–butoxyethanol, or the monobutyl ether of diethylene glycol are used.

When using alcohols as solvents, one must remember that they will react with some resins. Obviously, one does not use alcohols as solvents with isocyanates. In fact, there are *urethane grade* esters and ketones that contain no more than traces of water or alcohols. MF resins undergo reversible transetherification reactions with alcohols in the solvent during storage of the coating (see Section 6.3.1.3). The presence of the monofunctional alcohol stabilizes MF containing coatings by reducing the probability of cross-linking reactions with the hydroxy-functional polymer. If the alcohols in the solvent are different than those used to prepare the MF resin, there will be also be changes in curing rates of the MF resin. For polyesters, transesterification can occur with alcohols in the solvent leading to molecular weight reduction. This problem can be particularly severe with water-reducible polyesters where transesterification can lead to loss of the "solubilizing" carboxylic acid end groups. Such reactions can be minimized by using less reactive secondary, or better still, tertiary alcohols rather than primary alcohols. Epoxy resins can react slowly with alcohols, especially primary alcohols, at storage temperatures.

14.2. SOLUBILITY

The basic requirement for the solvents in many coatings is that they dissolve the resins. In the early days of the paint industry, there was little problem selecting solvents—almost all the resins used dissolved in hydrocarbon solvents. The exception was shellac and that dissolved in ethyl alcohol.

Early in the twentieth century, when nitrocellulose was introduced, solvent selection became a much greater challenge. RS type nitrocellulose is soluble in most esters and ketones but not in hydrocarbons or in most alcohols. But esters, ketones, and alcohols are more expensive than hydrocarbons. It was soon found that a solution of nitrocellulose in esters or ketones could be diluted with some hydrocarbon solvent to reduce cost. Usually, somewhat more aromatic solvent than aliphatic solvent could be tolerated without precipitation. Often, a mixture of aliphatic and aromatic hydrocarbon solvents was used. Then it was discovered that if ethyl alcohol was included in the solvent mixture, a higher ratio of hydrocarbon solvents could be used. Esters and ketones were classified as *true solvents*, hydrocarbons as *diluents*, and alcohols as *latent solvents*. This terminology remains in use.

In formulating a solvent mixture for a nitrocellulose lacquer, it is necessary to select solvents so that the resins will stay in solution throughout the evaporation of the solvents. If the slowest evaporating solvent is, for example, a hydrocarbon, the nitrocellulose could precipitate before all the solvent is gone. An uneven film with poor appearance and physical properties would result. In order to control the rate of drying under various conditions, mixtures of solvents are required. At a minimum, nitrocellulose lacquer solvent blends contain two esters or ketones, two hydrocarbons, and one alcohol. As many as 10 solvents are blended to provide control of drying and solubility throughout the drying cycle at the lowest cost per unit volume. The solvent combination required depends not only on the nitrocellulose but also on the other components of the lacquer, such as hard resins and alkyds. Empirical formulation of such a lacquer solvent combination is a time consuming job even for an experienced formulator.

After 1930 more and more types of resins were adopted for use in coatings, and the problems of empirical selection of solvents and solvent mixtures became more and more complex. The general rule of "like dissolves like" was broadened by the experience with nitrocellulose, notably that mixtures of weak hydrogen-bonding hydrocarbon solvents with strong hydrogen-bonding alcohols gave solvency properties similar to medium hydrogen-bonding esters and ketones.

14.2.1. Solubility Parameters

In the 1950s, Harry Burrell [1] initiated studies to develop a more scientific basis for selecting solvents and formulating solvent mixtures. He turned to the work of a physical chemist, Joel Hildebrand [2], who had been studying miscibility (mutual solubility) of small molecule organic compounds since 1916, analyzing miscibility in terms of basic thermodynamics. Hildebrand showed that the tendency of any pair of chemicals to mix spontaneously can be described in terms of Gibbs' free energy equation.

$$\Delta G_m = \Delta H_m - T\Delta S_m$$

Mixing is thermodynamically favored when the change in free energy of mixing (ΔG_m) is negative. Mixing may be very fast or very slow, depending on the physical circumstances, but if ΔG_m is negative, the mixture, once formed, will not separate spontaneously. Hildebrand pointed out that the change in entropy (ΔS_m) would be expected to be positive since the solution is usually less ordered after mixing than before. (There are exceptions, usually where relatively strong interactions among the different kinds of molecules increase the order of mixtures relative to pure materials resulting in a negative entropy change.) For the majority of cases, ΔS_m is positive and the entropy factor will favor mixing. Hildebrand focused attention on ΔH_m, the enthalpy of mixing, since it is generally the main factor that determines whether ΔG_m will be positive or negative.

The change in enthalpy of mixing is in turn related to the change in energy of mixing, ΔE_m. R is the gas constant and T is temperature in kelvin.

$$\Delta E_m = \Delta H_m - RT$$

Liquids have intramolecular attractive forces strong enough to hold the molecules together at that temperature; otherwise, they would be gases, not liquids. The forces can be measured by determining the energy needed to vaporize the liquid at that temperature. The results are expressed in terms of the molar energy of vaporization divided by the molar volume (V) $\Delta E_v/V$, called the *cohesive energy density*. Hildebrand reasoned that the energy required to separate molecules during mixing must be related to the cohesive energy densities. He expressed the change in energy of mixing of ideal solvents by the following equation where V_m is the average molar volume with ϕ_1 and ϕ_2 being the volume fractions of the two components.

$$\Delta E_m = V_m\phi_1\phi_2[(\Delta E_v/V)_1^{1/2} - (\Delta E_v/V)_2^{1/2}]^2$$

Hildebrand defined the square root of the cohesive energy density of a solvent as the *solubility parameter*, δ. He proposed that a pair of solvents would be miscible, if the square of the differences in solubility parameters approached zero. Then the ΔE_m of the pair would approach zero and the ΔH_m for that pair would be small; as a result, the free energy of mixing would be controlled by ΔS_m. When ΔS_m is positive, as it usually is, the solvents would be miscible. It should be emphasized that ΔE_v and V_m vary with temperature and that δ, therefore, varies with temperature. Most tables of solubility parameter give the values at 25°C although, commonly, no temperature is specified.

There is risk of confusion about units. The older units for solubility parameter were $(cal\ cm^{-3})^{1/2}$, sometimes designated as hildebrands, h. In the SI system the proper units are $(MPa)^{1/2}$; $1(MPa)^{1/2} = 0.488\ (cal\ cm^{-3})^{1/2}$. Unfortunately, solubility parameters often are published without stating the units. The SI units have not been widely adopted and $(cal\ cm^{-3})^{1/2}$ are still used, although frequently not specified.

When data are not available for heat of vaporization, solubility parameters can be estimated by use of empirical equations from boiling points, vapor pressure data, or surface tension data [3]. They can also be estimated by summation of Small's molar attraction constants (G) using the following equation where ρ is density and M is molecular weight.

$$\delta = \left(\frac{\rho}{M}\right) \sum G = \left(\frac{1}{V}\right) \sum G$$

Hoy [4] calculated Small's constants using data from 640 compounds. Several of the more important of these constants are listed in Table 14.2 in SI units, that is, $(MPa)^{1/2} cm^3 mol^{-1}$. As with solubility parameters, the units for Small's constants are commonly not specified; often they are given in the obsolete units, $(cal\ cm^3)^{1/2} mol^{-1}$.

Both Small's constants and the solubility parameters of hydrogens-bond donor–acceptor molecules vary with the environment. The value for the alcohol hydroxyl group in Table 14.2 depends on the composition of other groups in the solvent with which the hydroxyl group might hydrogen-bond and on the polarity of other components in mixed systems. Water is the most extreme case. While specific values of "the solubility parameter of water" are sometimes encountered, the values vary widely since they are very dependent on the medium. Their usefulness is limited to comparing systems with very similar compositions.

It had been confirmed experimentally that many small organic molecules are miscible in all proportions when the difference in solubility parameter is small. Burrell [1] tried to apply this idea to the prediction of solubility of resins used in coatings. Obviously, the volatility of the resins is so low that one cannot measure ΔE_v of resins at 25°C. Burrell got around this problem by determining the solubility of a resin in a series of solvents of known solubility parameters. The solubility parameter range of a resin was then taken as the range of solubility parameters of the solvents that would dissolve the resin. He soon found, however, that in many cases, resins were not soluble in all the solvents having solubility parameters that fell within the solubility parameter range determined for the resin.

As Burrell analyzed his data, it became apparent that the cases in which solubility parameters did not predict miscibility were usually ones in which the solvents and the resins differed markedly in degrees of hydrogen bonding. He then divided all solvents into three classes: poor hydrogen-bonding solvents, primarily hydrocarbons; medium hydrogen-bonding solvents, primarily esters and ketones; and high hydrogen-bonding solvents, primarily alcohols. He then proceeded to determine solubility of the resins in a series of each of these three classes of solvents. The resulting ranges of solubility parameters for a resin in poor, medium, and high hydrogen-bonding solvents permitted fairly good predictions of solubility.

More importantly, Burrell found that one can predict with some confidence whether some solvent mixture would dissolve a resin. One can calculate the weighted

Table 14.2. Molar Attraction Constants $(MPa)^{1/2} cm^3 mol^{-1}$ at 25°C

Hydrocarbon Groups	G	Other Groups	G
—CH$_3$	284	O (ethers)	236
—CH$_2$—	270	O (oxiranes)	361
—CH—	176	Cl	420
		CO (ketones)	539
=CH—	249	COO (esters)	668
=CH$_2$	259	OH (see text)	463
Phenyl	1400		
Phenylene	1370		

average solubility parameter of a solvent mixture that was known to dissolve the resin using the following relationship, where the x values are the mole fractions of solvents in the mixture.

$$\delta_{(mix)} = \frac{x_1 v_1 \delta_1 + x_2 v_2 \delta_2 + x_3 v_3 \delta_3 + \cdots}{x_1 v_1 + x_2 v_2 + x_3 v_3 + \cdots}$$

Since molar volumes for most solvents are fairly similar, it is a reasonable, simplifying approximation to use their volume fractions for calculating average solubility parameter.

$$\delta_{(mix)} = \phi_1 \delta_1 + \phi_2 \delta_2 + \phi_3 \delta_3 + \cdots$$

One does a similar calculation to determine the average degree of hydrogen bonding. Using this procedure, it was possible to predict reasonably often whether or not an alternative solvent mixture would also dissolve the resin. The procedure has been refined further, as we will see, but Burrell's relatively simple procedure proved quite successful.

Next, Burrell attempted to apply the same idea to the prediction of solvents or solvent mixtures for new resins. To minimize the experimental work he used Small's molar constants to estimate the solubility parameter of the resin. Gram equivalent weight of the average repeating unit in the resin was used for molecular weight in Small's equation. Results were fair; in a majority of cases, but not all cases, the resins were soluble in solvents and solvent mixtures with solubility parameters similar to those calculated for the resin, provided the need for similar levels of hydrogen bonding was taken into consideration. However, the question remains as to whether the predictions were better than could be made following the old truism that *like dissolves like*.

A variety of refinements and extensions of Burrell's two-variable, or two dimensional techniques have been proposed. For example, Lieberman [5] mapped solubility of polymers in a grid in which the axes are *solubility parameter* and *hydrogen-bonding index*. The plot predicts that any solvent or solvent blend having solubility parameter and hydrogen-bonding index, which fall within the boundaries of the plot, will dissolve this resin. Such plots can be useful, but their preparation requires a substantial amount of experimental effort. Nonetheless, this and similar two dimensional systems have found some uses [6].

14.2.2. Three-Dimensional Solubility Parameters

Other investigators considered that the two dimensional approach was too simplistic and proposed three-variable systems. The three-dimensional system of Hansen [7] came to be most widely accepted. Hansen reasoned that since there are three types of interactive forces between molecules there should be three types of solubility parameters: dispersion (nonpolar), δ_d; polar, δ_p; and hydrogen bond δ_h. The total solubility parameter was set equal to the square root of the sum of the squares of the partial solubility parameters.

$$\delta = (\delta_d^2 + \delta_p^2 + \delta_h^2)^{1/2}$$

Note that Hansen arbitrarily assigned equal weight to the three parameters.

In the case of mixed solvents, one can calculate the weighted average of the three partial solubility parameters analogously with the calculation of the weighted average of total solubility parameters given earlier.

$$\delta_d(\text{blend}) = (\phi\delta_d)_1 + (\phi\delta_d)_2 + \cdots + (\phi\delta_d)_n$$

$$\delta_p(\text{blend}) = (\phi\delta_p)_1 + (\phi\delta_p)_2 + \cdots + (\phi\delta_p)_n$$

$$\delta_h(\text{blend}) = (\phi\delta_h)_1 + (\phi\delta_h)_2 + \cdots + (\phi\delta_h)_n$$

Three-dimensional solubility parameters can be determined or calculated by a variety of methods. Although these values are commonly given to three figures, it should not be inferred that all three figures are significant. Table 14.3 gives solubility parameters for some representative solvents, selected from tables in the Polymer Handbook [8].

Hansen [7] determined three-dimensional solubility parameters for a group of representative resins by experimentally testing the solubilities of some 34 polymers in some 90 solvents. As an alternative to this laborious procedure, Hoy [9] calculated three-dimensional solubility parameters of resins by a method analogous to Small's. The values determined or calculated by various methods are not exactly the same; tables of (presumably) self-consistent values have been published and are available on computer data bases.

14.2.3. Uses and Limitations of Solubility Parameters

Three-dimensional solubility parameters are based on thermodynamic laws, but several assumptions and arbitrary choices are involved in their derivation and use.

Table 14.3. Three-Dimensional Solubility Parameters[a]

Solvent	Total δ	δ_d	δ_p	δ_h
n-Hexane	14.9	14.9	0	0
Toluene	18.2	18.0	1.4	2.0
o-Xylene	18.0	17.8	1.0	3.1
Methyl ethyl ketone	19.0	16.0	9.0	5.1
Methyl isobutyl ketone	17.0	15.3	6.1	4.1
Isophorone	19.8	16.6	8.2	7.4
Ethyl acetate	18.2	15.8	5.3	7.2
Isobutyl acetate	16.8	15.1	3.7	6.3
n-Butyl acetate	17.4	15.8	3.7	6.3
Methyl alcohol	29.7	15.1	12.3	22.3
Ethyl alcohol	26.6	15.8	8.8	19.4
Isopropyl alcohol	23.5	16.4	6.1	16.4
n-Butyl alcohol	23.1	16.0	5.7	15.8
2-Butoxyethanol	20.9	16.0	5.1	12.3
1,1,1-Trichloroethane	17.6	17.0	4.3	2.0

[a]Units are in $(\text{MPa})^{1/2}$.

Thus assertions that they are "theoretically sound" [3] can be misleading. They are most safely regarded as an empirical method that has proven useful for finding alternative solvent mixtures with similar solubility characteristics.

Attempts have been made to apply solubility parameters to many other problems, but the theoretical foundations are shaky and the actual results have been erratic. For example, they do not accurately predict the solubility of all polymers in all solvent combinations. Furthermore, use of solubility parameters to predict the solubility of one polymer in another polymer (compatibility) gives erroneous predictions more often than it gives correct ones. The idea was even taken to the extreme of *determining* three-dimensional solubility parameters for insoluble pigments and using these to predict the stability of pigment dispersions. Essentially, experimental data were used to estimate parameters that were then used to predict the results of experiments very similar to those that provided the parameters. The satisfactory nature of these predictions was taken as evidence that the technique was valid. While three-dimensional solubility parameters are not appropriate to use in considering dispersions of solid pigments, this is not to say that solvent–resin–pigment interactions are not important; they are, as discussed in Chapter 20.

It has been increasingly recognized that the use of three-dimensional solubility parameters is a gross oversimplification of the complexity of the factors involved in solubility. A major fraction of the difficulties flow from two probably interrelated factors. First, as noted in Section 14.2.1, Hildebrand's theory applies to situations in which entropy can be neglected and miscibility is controlled by changes in enthalpy. A second source of problems lies in oversimplifying hydrogen-bond effects. A hydrogen-bond results from the interaction between two sites of opposite nature: a proton donor and a proton acceptor. Some molecules (e.g., chlorinated solvents) are only proton donors; others (e.g., ketones and ethers) are primarily proton acceptors; and others (e.g., alcohols and carboxylic acids) are both. Hansen's hydrogen-bond solubility parameter combines these differences into a single parameter, δ_h. Important effects of entropy changes are particularly likely to occur in systems where hydrogen bonding is relatively important.

The difficulties increase as molecular weight increases. It has been well known since the early studies of polymers that the range of solvents that will dissolve a resin decreases as molecular weight increases. As the molecules get larger, the interaction between the solvent and the polymer molecules must be greater to overcome intermolecular polymer–polymer molecule interactions. Intramolecular interactions may also play a role. Estimates of solubility parameters of resins and polymers by use of group attraction constants do not take molecular weight into consideration.

People first working with polymers may be surprised by some of the solubility effects. They may be accustomed to situations where there is an upper limit of solubility; for example, 36.1 g of NaCl will dissolve in 100 g of water at 25°C. In general, there is no upper limit to the solubility of a polymer in solvent(s). If a small amount of a relatively homogeneous polymer is entirely soluble, it is quite safe to conclude that any larger amount will be soluble. On the other hand, it is very common to have a lower limit to solubility. A high concentration of a polymer may be soluble in some solvent(s) but on dilution, part of the polymer may precipitate. This phenomenon can be utilized to fractionate polymers by dilution. The fractions precipitating first are, generally, the highest molecular weight components. As dilution is continued further, lower and lower molecular weight fractions are precipitated. In some cases, the fractionation is based on the polarity of parts of the resin. Many alkyds are soluble at high concentrations in aliphatic solvents but partially precipitate on dilution. The first fractions precipitating are high molecular weight molecules with a larger than average number of hydroxyl and/or carboxylic acid groups. In considering these common phenomena it is useful to think of the situation in reverse: Some solvent is soluble in the polymer (resin) and all of the polymer is soluble in the combination of solvent and polymer. As more solvent is added, the solvency of the system changes, and parts of the polymer become insoluble in the more dilute polymer solution. The precipitate is highly swollen with solvent.

It is probable that a major limitation on the use of three-dimensional solubility parameters is the effects of inter- versus intramolecular hydrogen-bonding interactions. In the case of low molecular weight materials, Hoy [4] showed that some alcohols can change apparent polarity to assume the polar nature of their surrounding environment. In polar solvents, such substances behave as polar solutes. However, in nonpolar solvents, they undergo intramolecular or intermolecular hydrogen bonding, forming dimers in the latter case. Hoy particularly studied glycol ether solvents. More extreme examples are seen with carboxylic acids. Low molecular weight carboxylic acids (e.g., acetic acid) exist as dimers even in dilute solution in low hydrogen-bond accepting solvents like benzene and toluene [10]. Acrylic oligomers primarily substituted with single carboxylic acid groups behave like dimers in xylene and like single molecules in methyl isobutyl ketone [11]. In view of these observations it is not surprising that it is difficult to predict precisely the interactions of many coatings resins with various solvents because the resins are often polydisperse, polyfunctional materials with multiple hydrogen-bond donors and acceptor sites of different types.

14.2.4. Other Solubility Theories

Numerous additional refinements to Hansen's three dimensional solubility parameters have been proposed. Some have involved adding a fourth parameter, for example, by dividing δ_h into hydrogen-bond donor and acceptor terms. Entirely different approaches have been investigated, such as rationalization of all solubility data on the basis of acid–base interactions.

Huyskens and Haulait-Pirson [12] proposed equations that reflect the changes in entropy and attempt to take the differences resulting from both hydrogen-bond acceptor and donor groups into consideration. They provide somewhat better predictions of solubilities of a limited number of resin–solvent combinations than obtained with Hansen's method. However, their model equation still does not provide perfect correlations, and only the relatively simple cases of solubility versus nonsolubility of poly(vinyl acetate), poly(methyl methacrylate), and poly(ethyl methacrylate) were studied. Their paper explains quite well the shortcomings of the three-dimensional solubility parameter system but falls short of providing a broadly applicable alternative.

A major series of studies involving cooperative work in several laboratories in the United States, England, and Spain has lead to the development of what a technical news story called a "universal solubility equation" [13]. The researchers themselves "believe this to be somewhat of an overstatement" [14]. The complexity can be seen because this *general solubility equation* has 13 factors in a 5 term equation and still does not take into consideration several important situations. Among the limitations, critical to the coatings field, is that the equation has thus far been applied mainly to monofunctional solutes. It is evident that the problems are complex and no overall solubility theory is yet available.

Meanwhile, what can practical formulators dealing with coatings problems do? In the coatings industry, solvent changes are frequent—motivated by new toxic hazard information, changes in relative cost, temporary shortages, and so on. This is the type of problem for which solubility parameters are most useful, recognizing, of course, that they are not *theoretically sound* and that the results may be imperfect. When one must change a solvent blend that is a satisfactory solvent for some resin system, one can use solvent computer programs to calculate several alternative solvent combinations that could be useful starting points for reformulation. The data banks include not only solubility parameters but also evaporation rate data, cost, density, and so on. Thus, for example, such programs permit calculation of alternative solvent mixtures whose solubility characteristics will not change substantially during evaporation while minimizing cost per unit volume. In the 1960s, air pollution regulations took effect that permitted unlimited use of some solvents but restricted the use of others. Most solvent combinations had to be reformulated and solubility parameters proved to be of considerable value.

Many solvent suppliers maintain computer programs for calculating alternative solvent mixtures and provide them on line as a service to customers. An example of such a calculation is given in Table 14.4 [15]. [Solubility parameter units have been converted to $(MPa)^{1/2}$ and rounded-off.] Solvent blend 1 contained 2-ethoxyethyl acetate, which has been found to be a teratogen and, therefore, had to be removed from the formula. Two of the formulas, 2 and 3, calculated to have close to equal average solubility parameters and evaporation rates are shown. Formula 2 proved to be unsatisfactory since the slowest evaporating solvent, xylene, was not a solvent for the resins. Blend 3 was a better starting point because the slowest evaporating solvent, 2-methoxypropyl acetate, is a true solvent for the resins. Relative evaporation rates for the solvent mixtures are also given in Table 14.4. Relative evaporation rates of mixed solvents must be viewed with some caution (see Section 15.1.3). It must also be emphasized again that solvent composition can affect many properties of coatings (e.g., flow characteristics and corrosion protection), so consideration of the possible effects of changes in solvent composition must be considered more broadly than just achieving solubility.

A second type of problem faced by practical chemists is to predict solvents and solvent combinations for some new resin. In this case, solubility parameters may be useful, but in most cases, a qualitative application of the idea that like dissolves like will be as effective in finding a starting point for formulation and will take less time. If one does not easily find a solvent or solvent mixture in a few initial attempts, then it may be worth trying calculations of solubility parameters. Fortunately, the problem of selecting solvents that will dissolve resins has become somewhat easier since, in order to increase solids, lower and lower molecular weight resins are being used. As noted earlier, a wider and wider range of organic chemicals will act as true solvents as the molecular weight gets lower.

Table 14.4. Solvent Blends, Weight Percents

Component	1	2	3
MEK	9.9	6.1	14.7
MIBK	29.7	32.8	19.7
Xylene	21.9	25.2	24.3
Toluene	20.1	18.4	18.2
n-Butyl alcohol	13.5	17.5	17.3
Ethoxyethyl acetate	4.9		
Methoxypropyl acetate			5.7
Property	**1**	**2**	**3**
Solubility parameter, δ^a	19	19	19
δ_d	18	18	18
δ_p	5.7	5.7	5.3
δ_h	5.7	5.1	5.3
Relative evaporation rate	1.16	1.17	1.14
Cost^b	0.70	0.68	0.68

$^a\delta$ is in $(MPa)^{1/2}$.
^{b}At time of study in dollars per kilogram ($/kg).

REFERENCES

1. H. Burrell, *Off. Dig.*, **27**, 726 (1955).

2. J. Hildebrand, *J. Am. Chem. Soc.*, **38**, 1452 (1916).

3. T. C. Patton, *Paint Flow and Pigment Dispersion*, 2nd ed., Wiley-Interscience, New York, 1979, pp. 306–310.

4. K. L. Hoy, *J. Paint Technol.*, **42** (541), 76 (1970).

5. E. P. Lieberman, *Off. Dig.*, **34** (444), 30 (1962).

6. W. H. Ellis, *Solvents*, Federation Societies Coatings Technology, Blue Bell, PA, 1986.

7. C. M. Hansen, *J. Paint Technol.*, **39**, 505 (1967).

8. J. Brandrup and E. H. Immergut, *Polymer Handbook*, 3rd ed., Wiley, New York, 1989, pp. VII/540–543.

9. K. L. Hoy, *Tables of Solubility Parameters*, Union Carbide Corp., Chemicals & Plastics, R & D Dept. (1969).

10. M. Carmichael and F. Cantwell, *Can. J. Chem.*, **60**, 1286 (1982).

11. Z. W. Wicks, Jr., and L. G. Fitzgerald, *J. Coat. Technol.*, **57** (730), 45 (1985).

12. P. L. Huyskens and M. C. Haulait-Pirson, *J. Coat. Technol.*, **57** (724), 57 (1985).

13. R. Rawls, *Chem. Eng. News*, March 18, 20 (1985).

14. M. J. Kamlet, R. M. Doherty, J.-L. M. Abboud, M. H. Abraham, and R. W. Taft, *Chemtech*, September, 566 (1986).

15. ARCO Chemical Co., *ARCOCOMP Solvent Selector Computer Program*, Newton Square, PA (1987).

CHAPTER **XV**

Other Solvent Properties

Many properties of solvents besides solvency are critical in formulating coatings. This chapter discusses the other important factors involved in solvent selection.

15.1. SOLVENT EVAPORATION RATES

During application and film formation, the volatile material must evaporate out of the coating. The rate at which evaporation occurs can affect not only the time required to convert the coating to a dry film but also the appearance and physical properties of the final film. Like so many topics in the coatings field, evaporation rate would appear to be simple, but it turns out to be very complex and only partly understood. First the complexity of evaporation of single solvents and solvent mixtures are discussed followed by the even more complex matter of solvent evaporation from coatings.

15.1.1. Evaporation of Single Solvents

The rate of evaporation of a single solvent is affected by four variables: temperature, vapor pressure, surface/volume ratio, and rate of air flow over the surface; the rate of evaporation of water is also affected by relative humidity.

The critical *temperature* is the temperature of the liquid, specifically of its surface at the air interface. While this temperature may initially be that of the surrounding air, it decreases as the solvent evaporates. While the surface is cooled by evaporation, it is warmed by thermal diffusion from within the sample and its surroundings. Depending on the circumstances, thermal diffusion may occur rapidly, so that the surface temperature does not fall much during evaporation, or slowly, resulting in a sharp drop in surface temperature. The cooling effect is largest for those solvents and situations in which solvent is evaporating most rapidly. Also, the higher the heat of vaporization of the solvent, the greater the extent of temperature drop if other variables are equal.

Vapor pressure of the solvent is an important variable, but it must be emphasized that the critical vapor pressure is that at the temperature(s) at which evaporation is occurring. A common error is to assume that boiling points, the temperatures at which solvents have a vapor pressure of 10^2 kPa (1 mmHg = 0.1333 kPa; 1 atm = 101.3 kPa), are directly related to vapor pressures at other temperatures. However, boiling point is a poor indicator of vapor pressures at other temperatures. For example, benzene has a boiling point of 80°C and ethyl alcohol has a boiling point of 78°C, but at 25°C their respective vapor pressures are 1.3 and 0.79 kPa. Consequently, benzene will evaporate more rapidly at 25°C than ethyl alcohol under the same physical circumstances. Similarly, n-butyl acetate (bp 126°C) evaporates more rapidly at 25°C than n-butyl alcohol (bp 118°C).

The effect of the *ratio of surface area/volume* results because solvent evaporation occurs at the solvent–air interface. If 10 g of solvent are spread out over an area of 100 cm^2, it will evaporate much more rapidly than if the surface area were only 1 cm^2. Solvent will evaporate more slowly from paint in an open can than after application as a film. When paint is applied by a spray gun, it is atomized to small particles as it comes out of the orifice of the gun. The ratio of surface area/volume is very high and the rate of solvent loss is very high compared to the situation in the applied film. Thus, it is common for a major fraction of solvent to evaporate during the brief time after the spray droplets leave the orifice of the spray gun and before they arrive at the surface to which the coating is being applied. As coating films dry, the initial rate of solvent evaporation is independent of film thickness; but, because the ratio of surface/volume affects evaporation rate, the fraction of solvent present in a 50-μm film after a given time will be greater than that remaining in a 25-μm film. The concentration of the resin solution and the viscosity will increase more slowly when a coating is applied in thicker films.

The rate of *air flow over the surface* of the solvent is critical because the rate of evaporation depends on the partial pressure of the solvent vapor in the air at the air–solvent interface. If the vaporized solvent molecules are not carried away from the surface quickly, the partial pressure of solvent will build up and evaporation will be suppressed. Air flow rates vary substantially depending on the application method and, therefore, the solvents used in a coating must be selected for the particular application conditions. For example, spraying a coating with a compressed air spray gun results in significantly more loss of solvent than when the coating is sprayed from an airless gun because of the greater air flow over the surface of the droplets by the former method. The rate of solvent loss from the freshly painted surface will depend on the rate of air flow through the spray booth. If the same coating is applied to the outside and the inside of a pipe, the solvent would probably evaporate much more rapidly from the outside unless the inside is ventilated. Air flow effects often cause nonuniform evaporation from painted objects— solvent will evaporate more rapidly near the edges of a painted panel than from its center.

Relative humidity has little effect on the evaporation rates of most solvents, however, it has a major effect on the evaporation rate of water. The higher the relative humidity, the more slowly water evaporates when all other conditions are equal. As relative humidity approaches 100%, the rate of evaporation of water approaches zero. It should be remembered that relative humidity decreases as the air is warmed. Sometimes it is feasible to partly compensate for the effects of humidity on drying by modestly increasing the air temperature, taking advantage of the fact that warming the air both increases the vapor pressure of the water and decreases the relative humidity.

15.1.2. Relative Evaporation Rates of Single Solvents

As mentioned in Chapter 14, the introduction of nitrocellulose lacquers led to the need to formulate complex solvent combinations. One of the important criteria involved in solvent selection was evaporation rate. From the standpoint of solvent selection, vapor pressure is an important variable affecting evaporation rate. However, it is difficult to look at a set of vapor pressure data and judge how much more rapidly one solvent will evaporate than another. Procedures for measuring absolute evaporation rates were developed, but because of the difficulties of controlling air and heat flows in different types of apparatus it became common to determine "relative evaporation rates." Rates of evaporation of other solvents were related to the evaporation rate of n-butyl acetate, which at the time was the "standard" nitrocellulose solvent. Relative evaporation rate E is defined by Eq. 15.1, where t_{90} is the time for 90% of a sample to evaporate in a given type of apparatus under strictly controlled conditions.

$$E = \frac{t_{90}(n\text{-butyl acetate})}{t_{90}(\text{test solvent})} \tag{15.1}$$

Using Eq. 15.1, the relative evaporation rate of n-butyl acetate is 1 by definition. Faster relative evaporation rate solvents have larger E values. Some authors express E as a percentage figure, corresponding to 100 for n-butyl acetate. In either system, the higher the E value, the faster the relative evaporation rate. One should be very sure which reference point is being used; there is at least one paper in the literature where an author used data from two different tables without realizing that he/she was mixing data that differed by a factor of 100; the conclusions were absurd.

The determination of relative evaporation rate requires measurements under carefully controlled, standardized conditions. The Shell Thin-Film Evaporometer is an example of an apparatus that is used. A top loading balance is placed in an enclosure designed to minimize variations in air flow over the surface of the pan of the balance. A flow of 25°C air with a relative humidity of less than 5% at a rate of 21 L min^{-1} is maintained. A sample of 0.70-mL solvent is dispensed onto a piece of filter paper on the balance pan, and the time required to lose 90% of the sample weight is determined. Rates determined this way are *volume-based relative evaporation rates*; some examples are provided in Table 14.1.

While volume-based evaporation rates are the most common in the literature, some experimental procedures use a fixed weight, rather than a fixed volume of solvent. Care must be taken not to mix data obtained by the different methods. One should also avoid comparing relative evaporation rates determined using different instruments because different evaporation conditions can change not only the absolute evaporation rates but also the relative evaporation rates.

A study of Rocklin [1] illustrates the effects of changes in conditions on relative evaporation rates. He compared the relative evaporation rates of 66 solvents measured by the standard procedure (evaporation from filter paper) with rates measured by evaporation directly from the flat aluminum pan of the balance. Table 15.1 gives data obtained with several of the more widely used solvents. Note that these filter paper data are not exactly identical with those in Table 14.1, presumably reflecting different experimental conditions or solvent compositions. Note also that

Table 15.1. Volume-Based Relative Evaporation Rates at 25°C

Solvent	E_{paper}	E_{metal}	Ratio
n-Pentane	12	38	0.32
Acetone	5.7	10	0.55
Ethyl acetate	4.0	6.0	0.67
Methyl ethyl ketone	3.9	5.3	0.74
n-Heptane	3.6	4.3	0.83
Toluene	2.0	2.1	0.92
Ethyl alcohol	1.7	2.6	0.65
Methyl isobutyl ketone	1.7	1.7	1.0
Isobutyl acetate	1.5	1.5	1.0
n-Butyl acetate	1	1	1
sec-Butyl alcohol	0.93	1.2	0.81
m-Xylene	0.71	0.71	1.0
n-Butyl alcohol	0.44	0.48	0.92
2-Ethoxyethanol	0.37	0.38	0.98
Methyl *n*-amyl ketone	0.34	0.35	0.96
Water	0.31	0.56	0.56
2-Ethoxyethyl acetate	0.20	0.19	1.1
n-Decane	0.18	0.16	1.1
2-Butoxyethanol	0.077	0.073	1.1
Isophorone	0.023	0.026	1.0
Diethyleneglycol monoethyl ether	0.013	0.014	0.99

both sets of data are relative to *n*-butyl acetate. As can be seen by comparing the E_{paper}/E_{metal} ratios, significant differences exist between relative evaporation rates when the solvents evaporate from filter paper and the corresponding rates when those same solvents evaporate from a smooth metal surface in the same instrument at the same settings.

The most notable differences between E_{paper} and E_{metal} occur with the faster evaporating solvents and with water and alcohols. In the case of evaporation from filter paper, the ratio of surface area/volume is much higher than in the case of the evaporation from the metal. Therefore, solvents evaporate more rapidly from filter paper at first, resulting in a sharp drop in temperature, leading to a decrease in vapor pressure that slows down evaporation. In the case of evaporation from a metal surface, the surface area is much smaller and the thermal conductivity is much higher. Both factors presumably minimize the temperature drop leading to less decrease in vapor pressure and evaporation rate. The difference is greater with fast evaporating solvents such as *n*-pentane and acetone than with the slower evaporating *n*-butyl acetate resulting in their low ratios of E_{paper}/E_{metal}. The experimental temperature was 25°C but that is the air temperature. What controls the rates of evaporation, and hence the relative rates, is the actual temperature(s) of the surfaces of the evaporating solvents.

The E_{paper}/E_{metal} ratios are also low in the cases of water and alcohols. This may result from the greater extent of hydrogen-bond interaction of the hydroxyl groups with the very large surface area of the cellulose of the paper as compared with the smooth aluminum surface. This effect tends to retard the evaporation of water and alcohols relative to *n*-butyl acetate. Note that relative to *n*-butyl acetate, *sec*-butyl alcohol evaporates more rapidly from the smooth metal surface and more slowly from the filter paper.

In formulating baking coatings for spray application, it is common to use a mixture of fast and very slow evaporating solvents. A significant fraction of the fast evaporating solvent evaporates before the spray droplets reach the object being painted, raising viscosity and reducing the tendency of the coating to "sag," while the slow evaporating solvent keeps the viscosity low enough to promote "leveling" and to minimize the probability of "popping" when the coated object is put into the baking oven (see Chapter 23 for a further discussion of sagging, leveling, and popping).

In selecting slow evaporating solvents, formulators have generally used tables of relative evaporation rates at 25°C and boiling points because they have been the only available data. Now, evaporation rate data have been published for a limited number of slow evaporating solvents over the range of 75 to 150°C [2]. The rates were determined using a thermogravimetric analyzer (TGA) isothermally at a series of furnace temperatures. The rate–temperature relationships can be quite different at higher than at lower temperatures. For example, at 25°C, the evaporation rate for a commercial mixture of the dimethyl esters of succinic, glutaric, and adipic acids is about five times slower than for isophorone. However, at 150°C, their evaporation rates are approximately equal.

Which evaporation rate data are "correct"? All of them are "correct" because they depend on the particular set of circumstances under which they were determined. However, we do not apply coatings to filter paper (or to aluminum balance pans or to TGA pans) nor do we dry them in a Shell evaporometer (or a TGA furnace). The rates at which solvents evaporate in actual use depend on the particular situation. Solvents evaporate more rapidly from coatings applied by a spray gun than if they are applied by roller coating. As pointed out earlier, the type of spray gun can make a substantial difference. Other significant variables include the rate of air flow through the spray booth, the shape and mass of the object coated, the film thickness being applied, the flash-off time before entering an oven, and the way the heat in the oven is zoned.

This description makes it sound almost impossible to formulate a satisfactory solvent mixture for a coating. However, formulation does not start in a vacuum of knowledge. Experience with other formulas that proved useful under somewhat similar circumstances can provide valuable guidance. Using such formulas together with any reasonable table of relative evaporation rates, one can make a first attempt at a new formula. The coating is then applied under the particular circumstances and adjustments made as needed. The final adjustment is almost always made by an experienced person in the user's factory with production spray guns and operating conditions. The use of relative evaporation rate tables assists development of a formula that is in the "ball park" as to what is needed; because final, in-factory adjustments are needed anyway, it may not matter much which relative evaporation rate table was used. It follows that the limited precision of measurement of evaporation rates may not be as significant a drawback as might be expected.

15.1.3. Evaporation of Mixed Solvents

Evaporation of solvent blends rather than pure solvents adds further complications to this already complex situation. In *ideal* homogeneous solutions, vapor pressure is governed by Raoult's law, which predicts that the vapor pressure P_i of the ith component of the solution is reduced from the vapor pressure of the pure liquid P_i° in proportion to its mole fraction x_i.

$$P_i = x_i P_i^\circ$$

Since the vapor pressures of the different solvents are different, the composition of the solvent that evaporates is different than the composition of the solvent blend. Therefore, the partial vapor pressures change continuously as solvent evaporates from a mixture.

Raoult's law provides a good approximation for many combinations of solvents, especially when structures are similar and intermolecular interactions are minimal. However, many other solvent mixtures are nonideal as a result of interaction effects. These effects change as ratios of solvents change. The vapor pressure P_{total} of any mixture of miscible solvents can be calculated by the following equation, where χ represents an empirical adjustment factor for the interaction effects—usually given the more dignified sounding title of an activity coefficient.

$$P_{total} = P_1 + P_2 + \cdots + P_i = \chi_1 P_1^0 x_1 + \chi_2 P_2^0 x_2 + \cdots + \chi_i P_i^0 x_i$$

Activity coefficients for many solvents have been evaluated and incorporated into computer programs, which can calculate the partial vapor pressures of each solvent in a mixture throughout its evaporation. An example is called UNIFAC [3]. Results of such calculations are often presented as the partial vapor pressures at each 10% interval through the evaporation. Several assumptions are involved in calculating vapor pressures, and in order to relate vapor pressure to evaporation rate it may be assumed that all other factors (temperature, surface/volume ratio, and air flow) affecting evaporation rate are fixed. Thus the results of such calculations are only approximations. However, in view of the major uncertainties in relative evaporation rates, which depend on application conditions, a high level of accuracy would be wasted. An extensive review paper discusses the evaporation of solvent blends [4].

Relative evaporation rates of mixed-solvent compositions can be determined experimentally in an evaporometer. Relative evaporation rates at a temperature of mixed solvents (E_T) have been calculated from volume fractions (c), activity coefficients (a), and relative evaporation rates (E), for the individual solvents [5].

$$E_T = (caE)_1 + (caE)_2 + \cdots + (caE)_i)$$

Such calculated E_T values are of dubious accuracy since the composition changes over time so that the value of E_T changes. The difference between experimental and calculated relative evaporation rates would be expected to be especially large in comparing a solvent mixture with a narrow range of E values with another mixture having a wide range of E values. For example, one could calculate E_T for a mixture of two solvents with high and low E values that would be equal to the E value of a single solvent with an intermediate E value. However, it is evident that the mixed solvent would actually have a smaller relative evaporation rate than the single solvent since after the fast evaporating solvent was gone, the slow evaporating solvent would evaporate more slowly than the single intermediate E value solvent.

Evaporation rates of water–organic solvent mixtures require special treatment for at least four reasons. First, strong interactions often cause substantial deviations from Raoult's law. Second, relative humidity (RH) strongly affects the evaporation rate of water but has little or no effect on the evaporation rates of organic solvents. Third, azeotropic effects may occur. Fourth, the heat capacity of water is unusually high.

The substantial humidity effect occurs because the relative evaporation rate E of water at 0 to 5% RH and an air temperature of 25°C is 0.31 but at 100% RH it is 0. If a solution of 2-butoxyethanol ($E = 0.077$) in water evaporates at low RH, water evaporates more rapidly and the remaining solution becomes enriched in 2-butoxyethanol. At high humidity, 2-butoxyethanol evaporates more rapidly, and the remaining solution becomes enriched in water. At some intermediate RH, the relative evaporation rates of water and 2-butoxyethanol will be equal. When an aqueous solution evaporates at this RH, the composition of the remaining solution will be constant. This RH has been called the critical relative humidity (CRH) [6]. The CRH for 2-butoxyethanol solutions in water is estimated at about 80%. Of course, if the relative evaporation rate of the solvent were larger than that of water even at 0 to 5% RH, there would be no CRH since the solvent would evaporate more rapidly than water at any RH. At the other end of the scale, if the relative evaporation rate of the solvent were very low, the CRH would approach 100%.

Azeotropic behavior is particularly likely to occur with water solutions. Rocklin [7] studied the role of azeotropy in speeding up water–solvent evaporation in humid air. He reports development of a computer model, the AQUEVAP program, that permits calculation of the fastest evaporating water–solvent blends at various RH values. For example, at 40% RH, the time required for evaporation of 90% of a 20 wt% solution of 2-butoxyethanol in water is 1820 s compared with 2290 s for water alone. Thus codistillation of water and 2-butoxyethanol accelerates evaporation.

The high heat capacity of water can also affect the evaporation rates of water and water–solvent blends in an oven. For example, the time for 99% weight loss of 2-butoxyethanol (bp 171°C), water, and a 26:74 blend of 2-butoxyethanol–water in a TGA, when room temperature samples were put into the furnace at 150°C, were 2, 2.6, and 2.5 min, respectively [8]. While the air temperature in the furnace was 150°C, the samples took some time to heat up. The higher heat of vaporization of water (2260 J g^{-1} at its boiling point) compared to 2-butoxyethanol (373 J g^{-1} at its boiling point) apparently slowed the rate of heating of the water and water–solvent blends enough to more than offset the expected evaporation rates based on boiling points or E values.

15.1.4. Evaporation of Solvents from Coatings Films

Except in high solids coatings (see Section 15.1.5) there is little effect of the resin or other coatings components on the initial rates of organic solvent evaporation when coatings films are applied. It has been shown that the initial rates of evaporation of solvent from resin solutions are, within experimental error, the same as, or at least close to, the rates of evaporation of the solvents alone under the same conditions. Such observations are not inconsistent with Raoult's law, which predicts that dissolved resins will have little effect on vapor pressure because of their high molecular weights.

However, as solvent loss from a coating film continues, a stage is reached at which the rate of evaporation slows sharply. This observation is generally attributed to diffusion effects. As viscosity of the remaining film increases, free volume availability decreases and the rate of solvent loss becomes dependent on the rate of diffusion of the solvent through the film to the surface rather than on the rate of evaporation from the surface. The solids level at which the transition from evaporation rate control to diffusion rate control occurs varies widely, but it is often in the 40–60 NVV range.

Hansen [9] described the situation as "first stage" and "second stage" losses of solvent. In the first stage, the rate is governed by the same factors that govern evaporation of solvent mixtures—vapor pressure, surface temperature, air flow over the surface, and surface/volume ratio; at this stage one would predict theoretically a first-power dependence on wet film thickness, which is observed. After a transitional stage, evaporation slows and the rate of solvent loss becomes dependent on the rate of diffusion of the solvent molecules through the film. During this stage the evaporation rate depends on the square of film thickness. In line with the changes in viscosity, the first stage is sometimes called the "wet stage" and the second stage the "dry stage." A typical example of a plot of weight of solution remaining as a function of time in the evaporation of MIBK from a soluble vinyl chloride copolymer resin solution (Vinylite VYHH, Union Carbide Corporation) is shown in Figure 15.1 [10].

During the second stage, diffusion rate is primarily controlled by free volume availability. That is, the solvent molecules move through the film by jumping from free volume hole to free volume hole. The most important factor controlling free

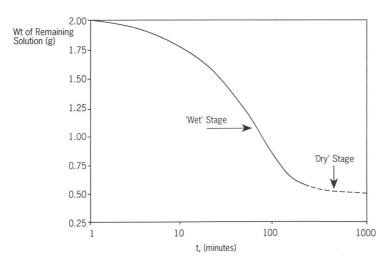

Figure 15.1. Two-stage release of solvent; MIBK in Vinylite VYHH at 23°C, initially at 20% w/w polymer–solvent. (From Ref. [10], with permission.)

volume availability is $(T - T_g)$. If solvent evaporation is occurring at a temperature well above the T_g of the solvent-free resin, diffusion rate will not restrict the rate of solvent evaporation at any stage in drying. If the T_g of the resin is above the temperature at which drying is occurring, solvent evaporation will become diffusion rate controlled when $(T - T_g)$ becomes small. No experimental data are available as to the values of $(T - T_g)$ when diffusion rate control becomes important; it is probable that such values are system dependent.

When second stage is reached, any solvent molecule reaching the surface evaporates and the rate of evaporation becomes dependent on how fast the solvent molecules get to the surface, not on their vapor pressure. As solvent loss continues, concentration of the remaining resin solution increases and, therefore, T_g increases and the rate of diffusion slows further. If the T_g of the resin is sufficiently higher than the temperature of the film, the rate of solvent loss will in time approach zero. Years after films have been formed, there will still be residual solvent left in the film. If solvent must be essentially completely removed in a reasonable time, the film must be baked at a temperature above the T_g of the solvent-free resin. It is a common mistake for new polymer chemists to think that the rate of solvent loss from polymer samples at temperatures that are below the T_g of the blend can be increased by drying under a vacuum. Since the rate of solvent loss has reached a stage where it is independent of the vapor pressure of the solvent, it will also be independent of the atmospheric pressure above the sample.

In keeping with this relationship of diffusion rate and solvent evaporation, solvent evaporation is affected by the addition of a plasticizer. The concentration of solvent at which solvent loss will become diffusion controlled decreases as plasticizer concentration is increased. The rate at which solvent will be lost after reaching the diffusion control stage increases as plasticizer concentration increases. The amount of solvent retained after some time interval decreases as plasticizer concentration increases. Perhaps, less obviously, in the case of mixed-solvent systems the ratio of slow/fast evaporating solvent retained in the film will increase as plasticizer concentration increases.

There have been many attempts to quantify the diffusion of solvents from coating films during solvent evaporation [11, 12] but success has been limited. A recent paper summarizes some of the problems and suggests possible approaches to modeling the diffusion through scaling analysis [13].

It has been shown repeatedly that the amount of solvent retained by a film does not depend directly on the volatility of the solvent during the second, diffusion controlled stage. Other factors are involved. One is molecular size. Since the molecules must jump from hole to hole, the smaller the size of the solvent molecule the greater the chance of finding sufficiently large holes. Even though its relative evaporation rate is higher, cyclohexane is retained in films to a greater degree than toluene presumably because the cyclohexane is bulkier. Figure 15.2 shows the results of an interesting experiment to illustrate the effect of cross-sectional size. Nitrocellulose and acrylic lacquers were made at 20% w/w solids in 60:40 isobutyl acetate (IBAc)/n-butyl acetate (BAc) and the ratios of the two solvents remaining in the films were monitored. In common with other branched–linear isomeric pairs, the relative evaporation rate of the branched IBAc is higher than that of the linear BAc. Thus the ratio of IBAc/BAc remaining in the film decreases during the first or wet stage of solvent loss, reaching a minimum of about 35:65. Then, during the

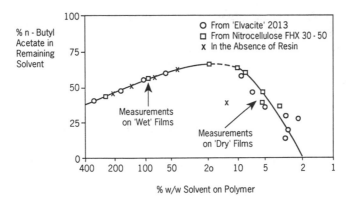

Figure 15.2. Changes in remaining solvent concentration during wet and dry stages of solvent evaporation at 23°C from films of an acrylic resin (Elvacite 2013) and a nitrocellulose resin, each initially in 60/40 IBAc/BAc. (From Ref. [10], with permission.)

second or dry stage, the linear compound diffuses more rapidly so that the ratio of IBAc/BAc remaining in the film increases and approaches 90:10 [10].

As a broad generalization it is true that small molecules tend to evaporate faster than large molecules in both the first and second stages, but this is a coincidental result because small molecules tend to be more volatile in the first case and to diffuse faster in the second case. Generally, evaporation rate is a poor indicator of second-stage behavior—quantitative correlation with diffusion rates is poor.

In air dry coatings the T_g quite often increases to a level above ambient temperature while significant amounts of solvent remain in the film. When this happens, solvent evaporation becomes extremely slow, and detectable amounts of volatile solvents may remain in the film for years. Residual solvent has been shown to affect film properties such as corrosion resistance (see Section 27.3.1.1) and moisture resistance, generally more or less adversely. Thus selection of solvents that will have minimal detrimental effect may be a significant factor in optimizing long-term film performance. When it is necessary to change a solvent mixture, care must be exercized that properties such as corrosion protection are not adversely affected.

15.1.5. Evaporation of Solvents from High Solids Coatings

Generally, it is more difficult to control sagging of spray applied high solids coatings than was the case with their earlier lower solids counterparts. While other factors may be involved, there is considerable evidence that high solids formulations lose much less solvent than conventional coatings during the brief time that the atomized droplets are traveling between the spray gun and the object being coated [14, 15]. As a result, there is less increase in viscosity and, therefore, a greater tendency to sag.

There has not yet been an adequate elucidation of the reasons for this lower solvent loss. One factor must be the colligative effect of the lower mole fractions of solvent(s) in the high solids coatings. A sample calculation illustrates this difference. If the vehicle of a conventional coating consists of a 30% by weight solids solution of a resin with an $\overline{M}_n$ of 15,000 in 70% of a solvent mixture with an $\overline{M}_n$

of 100, the mole fraction of solvent is 0.997. However, if a high solids coating has 70% solids of a resin with an $\overline{M}_n$ of 1000 and 30% of a solvent with an $\overline{M}_n$ of 100, the mole fraction of solvent is 0.811. While this difference is certainly in the direction of slowing down solvent evaporation from the high solids coating, it would not seem large enough to account for the large differences in solvent loss that have been reported.

It was suggested, without experimental evidence, that high solids coatings may undergo transition from first-stage to second-stage solvent loss with relatively little solvent loss as compared to conventional coatings [16]. The reasoning was that the T_g of the solution in high solids coatings could change much more rapidly with concentration, hence reach a stage of free volume availability limitation of solvent loss after only a little loss of solvent. Consistent with this hypothesis, Ellis [17] found that high solids polyesters were formulated at concentrations that were already above the transition concentration where solvent loss rate become diffusion controlled. Interestingly, he also pointed out that the transition points come at higher solids with linear molecules such as *n*-octane versus isooctane and *n*-butyl acetate as compared to isobutyl acetate. Further research is needed to fully understand the reasons for slower loss of solvents from high solids coatings; meanwhile the slower solvent loss can be a serious problem for the application of high solids coatings. More detailed discussion of sagging and sag control in high solids coatings appears in Chapters 23 and 28.

15.1.6. Volatile Loss from Water-Borne Coatings

For solvent-borne coatings, the formulator has available a variety of solvents having a very wide range of properties and volatility. In contrast, for water-borne coatings, the formulator is limited to water that comes with only one vapor pressure–temperature curve, one heat of vaporization, and so forth. There is only one kind of water. Furthermore, the relative humidity at the time of application and drying of the coatings can have a major effect on the rates of volatile losses from water-borne coatings. Formulators do use limited levels of organic solvents to modify evaporation rates; however, future regulations can be expected to reduce the levels permitted.

A major class of water-borne coatings is based on water-reducible resins. As described in Section 7.3, these resins are not actually soluble in water. A solution of a salt of the resin in an alcohol or an ether alcohol is diluted with water. During dilution, the resin forms into aggregates, which are swollen with solvent and water, dispersed in a continuous aqueous phase that also contains some solvent. The relative evaporation rate of water is affected by the relative humidity, so that with solvents like 2-butoxyethanol, there will be a CRH as described above in Section 15.1.3. Interestingly, it has been found that the CRH is different in coatings than it is in water-solvent blends without resin. For example, CRH is 65% for 10.6 wt% (based on volatile components) 2-butoxyethanol in a coating, in contrast to 80% calculated for this concentration 2-butoxyethanol–water blend without resin [18]. A possible explanation of the difference is that the concentration of 2-butoxyethanol in the continuous water phase is below the average and the solvent/water ratio inside the aggregates is above the average. This type of distribution and its effect on composition of vapor above a model aggregate system using a more volatile solvent (*t*-butyl alcohol) has been demonstrated [19]. Such effects can be critical in controlling sagging and particularly popping of such water-borne coatings (see Chapters 23 and 29).

The other major class of water-borne coatings uses latexes as the principal binder. Evaporation of water from a drying latex paint film resembles first-stage drying throughout most of the process; it is controlled by the temperature, humidity, evaporative cooling, and rate of air flow over the surface [20]. After most of the water has left, evaporation slows as a result of coalescence of a surface layer through which water must diffuse. In latex paints that are to be applied by brush or roller, it is desirable to retard the development of a surface coalesced layer to permit lapping of wet paint on wet paint. This generally requires the presence of some slow evaporating solvent such as ethylene or propylene glycol. The presence of these solvents does not affect the initial rate of water loss but does slow the development of a surface skin [21]. The presence of such a water-soluble solvent also facilitates the loss of coalescing solvent. Use of both types of solvents are discussed further in Chapter 35.

15.2. DENSITY

Density can be an important variable. It can have a major effect on cost. Most solvents are sold on a weight basis but the critical factor and, correspondingly, critical cost in almost all cases in the coatings field is the cost per unit volume. For example, the density of toluene is 0.87 g mL^{-1} and that of 1,1,1–trichloroethane is 1.44 g mL^{-1}, causing a 65% difference between costs on a weight and a volume basis.

United States air pollution regulations are based on weight of solvent per unit volume of coating (see Section 15.7), which favors use of low density solvents in formulations.

15.3. VISCOSITY EFFECTS

Solvent selection can have a major effect on the viscosity of resin solutions. This results from two factors: the viscosity of the solvent itself and the effect of solvent–resin interactions. This subject has acquired great importance since the 1970s as high solids coatings are developed. In high solids coatings it is generally critical to attain low viscosity with the lowest possible amount of solvent.

The direct effect of solvent viscosity on solution viscosity can be seen easily by examining equations relating viscosity and resin concentration. Even the simplest relationship, which is valid only over a narrow range of concentration of solutions with viscosities in a range of about 0.1 to 10 Pa·s, shows this dependence on solvent viscosity.

$$\ln \eta_{solution} = \ln \eta_{solvent} + K(conc)$$

For example, a difference of viscosity of 0.2 mPa·s between the viscosities of two solvents (1.0 and 1.2 mPa·s) may appear trivial but can cause a difference of

2000 mPa·s in the viscosity of a 50 wt% solution of a resin (10 vs 12 Pa·s) [22]. The relationship between concentration and viscosity of resin solutions is discussed in more detail in Chapter 19.

One must be very careful in comparing viscosities of solutions. There are data in the literature that make comparisons based on weight relationships, volume relationships, weight of solvent–volume of coating, and weight of resin–volume of solution. It is critical to decide what type of comparison is appropriate for a particular purpose. Most commonly, weight ratios are used in considering the viscosity of resin solutions, but they can be misleading because air pollution regulations are based on weight of solvent–volume of coating. If one compares the viscosity of resin solutions with different solvents on a weight basis, one might make the "wrong" choice in trying to minimize weight of solvent–volume of coating. Since flow is generally related to volume considerations, use of volume fraction comparisons might seem desirable, but polymer solutions are seldom ideal, creating uncertainty about the significance of volume fraction numbers.

An example of the effect of solvent choice on solution viscosity of solutions containing 400 g of the various solvents per liter of solution is given in Table 15.2 [23]. For comparison purposes, the densities and viscosities of the solvents alone are also given. The data in Table 15.2 illustrate the substantial effect that solvent selection can have on the solution viscosity of a given polymer. It should be noted that the weight concentrations of these solutions are not constant; it would have been interesting to compare also the viscosities of equal weight ratio solutions. The data suggest that solvents with low density and low viscosity will, as predicted by theory, tend to give low solution viscosity. However, there are also other factors involved.

Another factor that is sometimes important is the effect of solvent–solvent interactions on the viscosity of solvent mixtures. In general these effects are small except for mixtures that include alcohols or water. When relatively small (generally <40%) amounts of alcohols are mixed with other solvents they do not increase the viscosity of the solution proportionally to their relatively high neat viscosities [24]. This results from reduced multiple hydrogen-bond interactions between the

Table 15.2. Viscosities at 25°C of Solutions of a High Solids Acrylic Solvent Concentration 400-gL^{-1} Solution

Solvent	Solvent Viscosity (mPa · s)	Solvent Density (g mL^{-1})	Solution Viscosity (mpa · s)
Methyl propyl ketone	0.68	0.805	80
Methyl isobutyl ketone	0.55	0.802	110
Ethyl acetate	0.46	0.894	121
Methyl amyl ketone	0.77	0.814	147
n-Butyl acetate	0.71	0.883	202
Toluene	0.55	0.877	290
Isobutyl isobutyrate	0.83	0.851	367
Xylene	0.66	0.877	367

alcohol molecules in the presence of other solvents. The effects are much larger and less obviously predictable in the case of solutions of solvents in water.

Large viscosity effects can result from solvent–resin interactions. In order to distinguish these effects from the effects of solvent viscosity (discussed above) it is common to compare relative viscosities (η/η_s) of solutions. The effects of interactions are complex and not fully understood in even relatively simple systems. At least two factors play a substantial role.

First, most resins used in coatings have polar and hydrogen-bonding substituents such as hydroxyl or carboxylic acid groups, which tend to associate with polar groups on other molecules, often substantially increasing viscosity. To reduce viscosity it is important to choose solvents that can prevent, or at least minimize, such interactions by interacting with the polar groups themselves. Polar solvents with single hydrogen-bond acceptor sites are effective in minimizing relative viscosity; examples are ketones, ethers, and esters.

A second important factor is the effect of a given solvent on the hydrodynamic volume of the individual resin molecules with their closely associated solvent molecules. If the interaction between resin and solvent molecules is very strong, the chains of the resin molecules become extended and their hydrodynamic volume increases. If the interaction is not so strong, the molecules will contract, and the hydrodynamic volume will become smaller. Relative viscosity tends to be directly related to hydrodynamic volume. However, if solvent–resin interaction is so weak that there can be resin–resin interaction rather than only resin–solvent interaction, clusters of resin molecules will form and relative viscosity and absolute viscosity will increase.

In the majority of coatings applications and especially high solids coatings, which utilize resins of low molecular weight, resin–resin interactions are the stronger of the two effects, although hydrodynamic volume effects can be significant. Thus most high solids formulations include hydrogen-bond acceptor solvents.

The situation with mixed solvents can be even more complex. Only one paper has been published, which attempts to study the viscosity effects resulting from using solvent mixtures. Erickson and Garner [25] found in a study involving a relatively limited number of systems that the relative viscosity was dominated by the effect of the solvent that interacted most strongly with the resin.

A further word of caution: Relative viscosities can be very useful in attempts to understand the factors affecting viscosity, but the viscosities that are important in applying coatings are absolute viscosities, not relative viscosities. A solvent giving a relatively high viscosity solution could give a relatively low relative viscosity solution. Viscosity of resin solutions is discussed further in Chapter 19.

15.4. OTHER PHYSICAL PROPERTIES

Conductivity is another factor that can affect solvent choice. As discussed in Chapter 22, use of electrostatic spray guns requires control of conductivity of the coating. In general, formulations having appreciable, but low, conductivity work best. The conductivity of hydrocarbon solvents is too low to permit pickup of adequate electrostatic charge. Alcohols, nitroparaffins, and even small amounts of amines are common solvents or additives to increase conductivity to a desired range.

The conductivity of water-borne coatings poses problems for electrostatic application. Problems include the need to insulate the spray apparatus and relatively fast loss of charge from spray droplets. Additives can minimize the latter problem. For example, inclusion of glycol ethers improves sprayability of water-borne coatings, apparently by reducing surface conductivity. Presumably, the alkyl groups of the ethers orient quickly to the surface of the droplets.

Surface tension can be another important factor influencing solvent selection (see Section 2.4). Solvent can affect the surface tension of coatings, which, in turn, can have important effects on the flow behavior of coatings during application as discussed at length in Chapter 23. Solvent selection can also be a major factor affecting the development of surface tension differentials across the surface of a drying film during application and film formation, which can have substantial effects on flow behavior [26, 27]. Since surface tensions are very dependent on temperature and concentration of resins in solution, solvent volatility can have a particularly great effect on the development of surface tension differentials.

In order for a coating to wet a substrate, the surface tension of the coating must be lower than that of the substrate. In this regard, solvent selection can be an important factor especially in aqueous systems. While surface tension of aqueous systems can be reduced by adding surfactant, it is often more desirable to accomplish the same purpose with a solvent, such as 2-butoxyethanol. Solvent will evaporate, while surfactant residues in the final film may be detrimental to properties such as adhesion and humidity resistance.

Surface tension is a time dependent phenomenon. When a liquid is sheared, its surface is perturbed and surface tension usually increases temporarily. During application of coatings, it has been proposed that dynamic surface tension may well be more important than equilibrium surface tension [28, 29]. Solvent selection may affect dynamic surface tension, but the subject has not yet been adequately studied.

15.5. FLAMMABILITY

Fires and explosions of flammable coatings solvents can inflict terrible burns—their severity can be shocking. Many tragic accidents have occurred in which people were killed or severely burned. Sadly, almost all of these accidents could have been prevented.

Most solvents used in coatings are flammable. Care should be exercised in working with solvents in the laboratory, in the coatings factory, and in the end use environments. Flammability depends on structure and particularly on vapor pressure. It is also very dependent on conditions. Generally, there will be both an upper and a lower level of vapor concentration for flammability or explosion. If the partial pressure is low enough, not enough energy will be released during burning to maintain the vapor–air mixture above the ignition temperature of the system. If the partial pressure of the solvent is high enough, there will be insufficient oxygen for explosion or fire. Tables of upper and lower explosive limits of many solvents are available. It is important to realize that a full container of solvent may present less of a fire risk than a recently emptied container; the vapor phase in the former might have the solvent at concentrations above the upper explosive limit, while the "empty" container may have a concentration in the explosive range. Factories and laboratories should be equipped with explosion proof electrical installations. One should never try to bypass these installations.

The most common cause for fires igniting in coatings factories has been static electricity. For example, solvent flowing out of one tank and into another tank by gravity can pick up enough electrostatic charge to cause a spark, which can set off a fire or explosion. To avoid such charge accumulation, all tanks, pipes, and so forth, used in handling solvents and solvent containing mixtures, should be electrically grounded at all times.

There are two main types of flammability tests: open cup and closed cup; both measure a flash point, the minimum temperature at which solvent can be ignited by a hot wire under conditions of a particular test. Generally, the open-cup testers give results that are more appropriate in indicating the possible degree of hazard of a mixture when exposed to air as during a spill. The closed-cup flash point more nearly describes the fire hazard of a liquid enclosed in a container. The U.S. Department of Transportation regulations for shipment of flammable liquids base their limits on closed-cup tests. A closed-cup flash point will always be lower than an open-cup flash point. Solvents used in architectural paints should have closed-cup flash points over 38°C. Transportation costs can be substantially affected by flash points of the material being shipped. Extensive tables of flash points of solvents in various testers are available. A few representative closed-cup flash points are given in Table 15.3. Note that different sources sometimes disagree on flash points. Discrepancies are predictable in the cases of solvents that are variable mixtures of chemical species, such as naphthas and mineral spirits, but are harder to rationalize for one-species solvents such as n-butyl acetate. The American Society for Testing and Materials (ASTM) method is said to be accurate to ± 2.5°C, although its reproducibility suffers at temperatures below 0°C.

Water, of course, is not flammable. Most, but not all, chlorinated solvents do not show a flash point. Many chlorinated solvents can, however, burn even though they do not show a flash point. One must be very careful in estimating the flash points of mixtures of solvents because the flash point of a mixture can be lower

Table 15.3. Flash Points

Tag Closed Cup	
Solvent	Flash Point (°C)
VM & P naphtha	−1 to 16
Mineral spirits	~ 44
Toluene	5
Xylene	27
Methyl ethyl ketone	−5
Methyl isobutyl ketone	16
n-Butyl acetate	22–29
n-Butyl alcohol	35–37

Sources: "UCAR Solvents Selection Guide for Coatings," Union Carbide Corporation, Danbury, CT, 1988; "Solvents Data," Texaco Chemical Company, Houston, TX, 1990 [*32*].

than the flash point of any component. For example, although the chlorinated solvent, 1,1,1-trichloroethane, shows no closed-cup flash point and that of a mineral spirits solvent is 43°C, a mixture of 65:35 mineral spirits–1,1,1-trichloroethane showed a flash point of 21°C [30]. It is safest to determine flash points of mixtures experimentally. A discussion of the factors affecting flash point, including molecular interactions in blends, is given by Ellis [31]. Recently, it has been reported that good predictions of closed-cup flash points can be made by a UNIFAC computer program requiring only flash points and molecular structures of the pure components [33]. Best results were reported using UNIFAC group interaction parameters derived from flash points of binary solvent mixtures.

Risk of fire or explosion can be eliminated by meeting either of two conditions. If solvent vapor concentrations in air are kept well outside of the concentration range within which ignition is possible *or* if all sources of ignition are eliminated, there can be no fire. Unfortunately, many tragic accidents have occurred because neither of these conditions was satisfied. Because it is difficult to be absolutely certain that either of these conditions can be met in all circumstances, prudent practice dictates that all possible steps should be taken to meet both conditions, providing redundancy. The importance of good ventilation cannot be overemphasized, especially since the vapors of relatively dense solvents may stratify in stagnant air, being higher than the overall average at the lower part of the workspace.

15.6. TOXIC HAZARDS

In considering the risks associated with the use of a coating containing volatile solvents, the extent of exposure must be considered in combination with toxicity data. All solvents are toxic at some level of exposure. Obviously, one should avoid ingesting them. The hazard from skin contact can be controlled by wearing protective clothing. Generally, the greatest potential risk comes from inhalation. The method of application can have a marked effect on the hazard. The evaporation of solvents between the orifice of the spray gun and the surface being painted is much greater with compressed air spray than airless spray which, in turn, is much greater than encountered in roller coating. Commonly, there is a greater hazard when workers are exposed daily to a relatively low level of solvent vapors than when a person is exposed infrequently to a somewhat higher level.

Three general types of toxicity data are important, First, acute toxicity data indicates the level of intake in single doses that can be injurious or lethal—this kind of information can be particularly important in cases of accidental ingestion or spills. A second type of toxicity data concerns the level of exposure considered safe when people are going to be exposed for 8 h a day for long periods of time. This kind of data is used, for example, to set the upper concentration limit for each type of vapor in the air in a spray booth. Third, data is available on the level of exposure over periods of years to certain materials that can increase the risk of cancer, that is, *carcinogenic* materials. When it is found, usually by animal tests, that a solvent is carcinogenic, very low levels of permissible exposure are set. The levels are generally too low to be controlled by economically feasible methods.

The net effect is that carcinogenic solvents are banned. For example, benzene has not been used in coatings for many years for this reason. While most solvents now used in large quantities in coatings have been tested and are thought to be non-carcinogenic, it is only prudent for the user to be aware of current knowledge of the materials he/she is using and to minimize inhalation and contact with all solvents.

Extensive tables of all three types of data are available. The more common difficulty is to know what the level of exposure will be. Not only is the method of application important, but also the design of the work area can be critical. For example, the air flow through a spray booth can be controlled, but it is also important to design the booth so that there are no pockets of relatively still air. Reference [34] describes an approach to assessing possible exposures when retail consumers apply coatings in a room.

In formulating coatings, it is necessary to consider the clientele that will use them. While coatings sold to retail consumers are carefully labeled to include application cautions, one must assume that many people will not read the labels. When selling to a large corporation, it is reasonable to assume that the Material Safety Data Sheets will be read and appropriate practices will be established. But when selling to industrial customers such as automobile repair shops, one cannot assume that the customer will pay attention to precautions. Thus, it may be completely ethical to sell a coating with a somewhat toxic solvent to one class of customer but unethical, although perhaps legal, to sell it to another class of customer.

15.7. ATMOSPHERIC PHOTOCHEMICAL EFFECTS

Since the 1950s it has been increasingly realized that the presence of organic compounds in the atmosphere can lead to serious air pollution problems. Terminology can be confusing; in the older literature, such compounds are referred to as "hydrocarbons" meaning any organic compound not just unsubstituted hydrocarbons. More recently, they are called "volatile organic compounds" (VOC). The terminology "reactive organic gases" (ROG) is also being used. Three end effects of VOC emissions into the atmosphere are important: formation of eye irritants, formation of particulates that reduce visibility, and formation of toxic oxidants, especially ozone.

While all of these factors are important, the most critical has been identified as ozone. While ozone is a naturally occurring component of the atmosphere, it is toxic to plants and animals. When you go up in the pine forests in the Rocky Mountains and smell the wonderful *fresh air*, very probably the odor is ozone. The pine trees emit substantial quantities of VOC into the atmosphere, the UV level increases with altitude, and as a result ozone generation is high. Plants and animals (including humans) evolved in the presence of some ozone and can tolerate its presence up to a point. However, with the rapid growth of VOC emissions from man-made sources since 1900, ozone levels on many days of the year in many parts of the world, especially in and around cities, have exceeded the levels that many plants can withstand and have become human health problems. There is not complete agreement on the level of ozone that is *safe*. The National Ambient Air Quality Standard for ozone allows no more than one time per year (on average over 3 years) exceeding a daily maximum 1 h average ozone concentration of 0.12 ppm by volume. This ozone standard was exceeded (in most cases many times in a year) in 76 urban areas of the United States in the years 1983–1985 [35].

The largest source of man-made VOC emissions is auto and truck tailpipe emissions along with leakage during distribution of fuel. The second largest source of VOC emissions is coatings. In the early investigations of effects of VOC on air pollution, organic compounds were divided by rabbit eye irritation tests into photochemically active compounds of high and low reactivity. It was proposed that if the emission of the highly reactive compounds could be limited, the less reactive ones could dissipate and avoid high local concentration of pollutants. This led to the establishment of a definition of "photochemically reactive solvents" in "Rule 66" of the Los Angeles Air Control District. Any solvent mixture was classified as photoreactive if it contained a combined total of more than 20 vol% of the chemical compounds classified below or which exceeded any of the individual composition limitations:

1. A combination of hydrocarbons, alcohols, aldehydes, esters, ethers, or ketones having an alkene or cycloalkene type of unsaturation: 5%.
2. A combination of aromatic hydrocarbons with eight or more carbon atoms except ethylbenzene: 8%.
3. A combination of toluene, ethylbenzene, ketones having branched hydrocarbon structures, or trichloroethylene: 20%.

Regulations were then established limiting the total photoreactive solvent, using this definition, that could be emitted. Other parts of California and other parts of the country that thought they had less severe air pollution problems than Los Angeles, set up other, generally less restrictive, regulations.

After some years of experience, it was realized that essentially all organic compounds are photoreactive and that the extent of dissipation in the atmosphere after local emission had been overestimated. Changing from highly reactive compounds to less reactive ones might in effect diminish the air pollution effects near the scene of emission in exchange for increasing air pollution downwind from the emission site. Furthermore, the coatings industry objected to having to use different solvent combinations in different parts of the country. This situation led to the conclusion that it would be best to limit the emission of almost all organic compounds into the atmosphere. The only exceptions were a handful of compounds having relatively low photoreactivity, 1,1,1-trichloroethane being the only one still of some importance in coatings. The term "compliance coatings" has been used to designate coatings in which the amount of VOC has been minimized by using 1,1,1-trichloroethane as a major component in the solvent mixture. The terminology may sound strange but 1,1,1-trichloroethane is not classed as a volatile organic compound in most parts of the United States. However, since 1,1,1-trichloroethane has been implicated in the reduction of ozone in the stratosphere, production is expected to be phased out during the 1990s, so that this option may not be available to the coatings formulator in the future.

Volatile organic compound regulations will become even more restrictive in the future. In establishing future regulations there is a difference of opinion as to whether all solvents should be considered as equally undesirable in the atmosphere as they are now. It is simpler to formulate coatings that way and it may be simpler to enforce regulations. However, it may well be that using less reactive solvents to replace more reactive ones would be advantageous by allowing at least some opportunity for dissipation in the atmosphere to minimize the probability of local excess ozone concentrations.

Photochemical reactions in the atmosphere are complex and dependent on many variables in addition to the amount and structure of VOCs, especially on concentrations of various nitrogen oxides. Full discussion of the reactions is beyond the scope of this introductory text. References 35 and 36 provide brief and more detailed reviews, respectively, on the state of the science. Probably the principal pathways leading to the generation of ozone are by way of hydrogen abstraction from the VOC compounds. Some of the reactions that have been proposed to explain ozone generation are shown in Scheme 1, where RH represents a VOC compound.

Scheme 1

$$RH + \cdot OH \rightarrow H_2O + R\cdot$$
$$R\cdot + O_2 \rightarrow ROO\cdot$$
$$ROO\cdot + NO \rightarrow RO\cdot + NO_2$$
$$NO_2 + h\nu \rightarrow NO + O$$
$$O_2 + O \rightarrow O_3$$

An important variable in the amount of ozone generation is the ease of abstraction of hydrogen atoms from reactive organic gases by free radicals such as the hydroxyl free radicals shown in the first equation of Scheme 1. Data are available on the rate constants for a wide variety of organic compounds [36]. In general terms, compounds with hydrogens on carbon atoms alpha to amines or ethers, hydrogens on tertiary carbon atoms, allylic hydrogens, and benzylic hydrogens are examples of easily abstractable hydrogen atoms. The placement of branched-chain ketones in Class 3 of the Rule 66 definition reflects the ease of abstraction of the hydrogen on the tertiary carbon in methyl isobutyl ketone. The absurdity of this generalization is illustrated because Rule 66 would put the same limitation on the use of methyl t-butyl ketone, which has no easily abstractable hydrogen atoms. On the other hand, toluene only has three hydrogens on benzyl carbon atoms and ethylbenzene only two, whereas xylenes have six and were, therefore, appropriately classified as more reactive.

An ingenious approach to the VOC problem is the proposed use of supercritical carbon dioxide as a component in a solvent mixture [37]. The critical temperature and pressure of CO_2 are 31.3°C and 7.4 MPa (72.9 atm), respectively. Below that temperature and above that pressure, CO_2 is a supercritical fluid. It has been found that under these conditions, the solvency properties of CO_2 are similar to aromatic hydrocarbons. The concept, which is now in a pilot plant stage, is to ship the coating in a concentrated form. The high solids coating and supercritical CO_2 are metered into a proportioning spray gun system in such a ratio as to reduce the viscosity to the level needed for proper atomization. Airless spray guns are used; it has been found that the rapid evaporation of CO_2 as the coating leaves the orifice of the spray gun assists atomization (see Section 22.2.5). Volatile organic compound emission reductions of 50% or more are projected.

Current regulations treat all solvents (except water, 1,1,1-trichloroethane, and CO_2) as equally undesirable. Within this context, regulators have recognized that some coatings require higher VOC levels than others for adequate performance. This realization led to the establishment of specific levels of permissible VOC for many individual applications. The levels were set based on feasibility with utilization of the most advanced technology.

The units used are weight of solvent per unit volume of coating as applied excluding water (grams of VOC per liter as applied minus water). This unusual definition follows from the conventions that air pollution data are based on weights of pollutants and that coatings are sold and applied on a volume basis. (Film thickness is a volume measurement since it is equivalent to film thickness per unit area.) Note that the limitations are based on VOC content of the coating "as applied." If, as is very common, the user is to add solvent to the coating as it is shipped by the coating manufacturer, the VOC is based on the coating after this solvent has been added. The exclusion of water is to prevent circumventing regulations by developing coatings with low solids that are diluted with water but still have high VOC contents relative to the dry film thickness that has to be applied.

Table 15.4 gives an illustrative cross section of such limitations. These limits are a combination of those in effect in the South Coast Air Quality Management District of California and in New Jersey at the end of 1990. Other districts and states have different regulations, which span a wide range. These regulations will undoubtedly be made more restrictive. For example, it is projected that by mid-1994 the limit for clear wood furniture top coats in Southern California will be 275 g L^{-1}, one-half the 1990 level.

The problem is further compounded by the desirability of taking into account the *transfer efficiency* of spray applied coatings. When a coating is sprayed, only a part of the coating is actually applied to the object being coated. Transfer efficiency represents the percentage of the coating used that is actually applied to the product. As the transfer efficiency increases, the VOC emissions decrease since

Table 15.4. VOC Limitations for Coatings

Coatings Application	Grams of VOC per Liter as Applied Minus H_2O
Coil coatings	200
Three-piece can sheet exterior base coat	225
Two-piece can exterior base coat	250
Low gloss architectural coatings	250
Traffic coatings (pavement striping)	250
Industrial metal coatings	340
Gloss and semigloss architectural coatings	380
Auto top coats	420
Beverage cans—interior spray	440
Clear wood furniture top coat	550
Semitransparent stains	550
Swimming pool coatings	600
Electric insulating varnish	620
High-performance architectural coatings	750

less coating is used. Transfer efficiency depends on many variables, particularly on the type of spray equipment utilized (see Chapter 22). In some cases, regulations have been established requiring certain kinds of spray equipment or setting a lower limit on transfer efficiency, such as 65%.

There are two other broad approaches to minimizing VOC emissions: solvent recovery and incineration. In some cases, it is feasible to recover the solvent used in coatings. One can pass the air flowing out of a drying chamber through activated carbon beds where the solvent is adsorbed. After the surface of the carbon is saturated with solvent, the material is heated to distill and recover the solvent. Alternatively, the solvent-laden air can be passed through condensers cooled with liquid nitrogen; the nitrogen that is vaporized is fed into the drying chamber reducing the oxygen content and, hence, permitting higher concentrations of solvent without exceeding the lower explosive limit. Solvent recovery is very desirable when it is feasible but feasibility is limited by low solvent concentration in the air stream needed to stay safely below the lower explosive concentration. In the case of effluent air from spray booths, the solvent concentration is generally too low, in order to be safely below concentrations that would have toxic hazards, to permit economic recovery of the solvent.

Volatile organic coating emissions can also be minimized by incineration. The effluent solvent-laden air stream is heated in the presence of a catalyst to a temperature high enough to burn the solvent. As with solvent recovery, this approach is feasible only when solvent concentrations are relatively high. Incineration has been found to be particularly applicable in coil coating (see Section 33.4). In this case, most of the solvent is released in the baking oven. Part of the effluent air from the baking oven is recirculated back into the oven; the amount of such recirculation is limited so that the solvent content does not approach the lower explosive limit. The balance of the effluent air is fed to the natural gas burners that heat the oven. The solvent in the air is burned along with the natural gas; the fuel value of the solvent reduces the natural gas requirement. Thus, VOC emission is minimized and the fuel value of the solvent is used. The VOC limit of 200 g L^{-1} for coil coatings given in Table 15.4 applies when incineration is not done.

In the case of industrially applied coatings, the so-called "bubble concept" is sometimes used for regulating solvent emissions. The VOC emissions from a factory are determined over a year or more and an agreement is made with the regulatory authorities to reduce the total emissions to some fraction of this level—perhaps one-half. Any combination of coatings may be used to achieve this reduction without limitations on the solvent level used in specific applications within the factory.

15.7.1. Determination of VOC

The amount of VOC that will be emitted by a coating is neither obvious nor easily determined. As we saw earlier, there can be solvent retained in films for very long periods of time. In latex paints, coalescing solvents are used that are only slowly released from the coating. In cross-linking coatings, volatile byproducts may be generated by the reaction. For example, MF cross-linking leads to the evolution of a molecule of volatile alcohol for each cocondensation reaction. In the case of self-condensation reactions, there can be emission of alcohol, formaldehyde, and methylal. The amount released will depend on the curing conditions and the amount of catalyst used. On the other hand, when slow evaporating glycol ether solvents are used in an MF cross-linking system, it is probable that some of the glycol ether transetherifies with the MF resin and is not emitted from the film. Amines used in "solubilizing" water-reducible coatings may volatilize to different extents depending on conditions and amine structure [38]. Thus in many cases, only rough approximations of potential VOC emissions can be calculated even when the formulation of a coating is known.

Experimental determination of VOC is not straightforward. The amount of VOC released will depend on conditions under which the coating is used. Time, temperature, film thickness, air flow over the surface, and, in some cases, the amount of catalyst used are among the variables that will affect the results obtained. While it would seem that determination under the conditions of actual use would be most appropriate, this is not easy to do. For air dry coatings, the time required for the determination would be very long. For baking coatings, there could well be differences in conditions for use of the same coating. It is generally agreed that it would be desirable to have a standard method for determining VOC. However, there is not agreement as to what that standard method should be; in view of the effect of application variables, it is doubtful if a single appropriate standard method will ever be developed.

An ASTM publication, MNL 4, was issued in 1989 [39]. Periodic revisions are expected. To date, the most widely used procedure is specified in ASTM D2369-86, which involves heating the sample for 60 min at 110°C. This method may overstate air dry emissions from air dry coatings and understate emissions from coatings that are baked at higher temperatures in actual use, particularly those that cross-link with release of volatile organic byproducts.

The problem is made more complex by the need to determine water content in water-borne coatings. Again general agreement has not been reached on the most suitable method of determination, and, to date, the most widely used method is gas chromatography, as described in ASTM Method D3792. Other methods are being evaluated.

In the case of solvent-borne coatings, VOC is calculated by the following equation, where NVW is the weight solids of the coating determined under specified conditions and ρ_1 is the density of the coating in grams per milliliter. The factor 10 serves to adjust the VOC units to grams of solvent per liter of coating.

$$VOC = 10(100 - NVW)\, \rho_1$$

For units of kilograms per liter, one divides by 100 instead of multiplying by 10. In some U.S. regulations, the units are pounds per gallon; multiplication by 8.345 $\times 10^{-3}$ converts these units into grams per liter.

Because of weight/volume units, VOC cannot be calculated directly from the common formulating parameters, NVW or NVV. The density of both the solvent and the nonvolatile components of the coating must be known. The density of the coating depends particularly strongly on the pigmentation. There can be large variations in the amount of pigment in different coatings and the densities of pigments vary over a wide range.

The VOC of water-borne coatings can be calculated using the following equation where 0.997 is the density of water at 25°C. (The precision of the analyses is not high and, therefore, the correction for the minor variation in the density of the water is really a useless frill.)

$$\text{VOC} = \frac{\rho_1(100 - \text{NVW}) - \rho_1(\%H_2O)}{100 - \rho_1(\%H_2O)/0.997} \times 1000$$

Some regulators would prefer that the units be kilograms per liter of applied coatings solids. This may be a desirable objective but requires determination of the density of the applied cured film. The ASTM Method D 2697 includes a procedure for using Archimedes' liquid volume displacement principle to determine the density of the film cured on a metal disk for 60 min at 110°C. The accuracy of the values is limited by the same factors discussed above. As of 1992, the EPA does not recognize this method but does accept the coatings manufacturer's calculation of the value. Unfortunately, there is no way of making such a calculation accurately either.

GENERAL REFERENCE

W. H. Ellis, *Solvents*, Federation of Societies for Coatings Technology, Blue Bell, PA, 1986.

REFERENCES

1. A. L. Rocklin, *J. Coat. Technol.*, **48** (622), 45 (1976).
2. H. L. Jackson, *J. Coat. Technol.*, **58** (741), 87 (1986).
3. S. Skjold-Jorgenson, B. Kolbe, J. Gmehling, and P. Rasmussen, *Ind. Eng. Chem., Proc. Res. Dev.*, **18**, 714 (1979).
4. T. Yoshida, *Prog. Org. Coat.*, **1**, 72 (1972).
5. T. C. Patton, *Paint Flow and Pigment Dispersion*, 2nd ed., Wiley-Interscience, New York, 1979, p. 340.
6. P. W. Dillon, *J. Coat. Technol.*, **49** (634), 38 (1977).
7. A. L. Rocklin, *J. Coat. Technol.*, **58** (732), 61 (1986).
8. B. C. Watson and Z. W. Wicks, Jr., *J. Coat. Technol.*, **55** (698), 59 (1983).
9. C. M. Hansen, *Ind. Eng. Chem., Prod. Res. Dev.*, **9**, 282 (1970).
10. D. J. Newman and C. J. Nunn, *Prog. Org. Coat.*, **3**, 221 (1975).
11. C. M. Hansen, *J. Oil Colour Chem. Assoc.*, **51**, 27 (1968).
12. R. A. Waggoner and F. D. Blum, *J. Coat. Technol.*, **61** (768), 51 (1989).
13. R. C. Lasky, E. J. Kramer, and C.-Y. Hui, *Polymer*, **29**, 673 (1988).
14. S. H. Wu, *J. Appl. Polym. Sci.*, **22**, 2769 (1978).
15. D. R. Bauer and L. M. Briggs, *J. Coat. Technol.*, **56** (716), 87 (1984).
16. L. W. Hill and Z. W. Wicks, Jr., *Prog. Org. Coat.*, **10**, 55 (1982).
17. W. H. Ellis, *J. Coat. Technol.*, **53** (696), 63 (1983).
18. L. B. Brandenburger and L. W. Hill, *J. Coat. Technol.*, **51** (659), 57 (1979).
19. Z. W. Wicks, Jr., E. A. Anderson, and W. J. Culhane, *J. Coat. Technol.*, **54** (688), 57 (1982).
20. S. G. Croll, *J. Coat. Technol.*, **59** (751), 81 (1987).

21. D. A. Sullivan, *J. Paint Technol.*, **47** (610), 60 (1975).

22. T. C. Patton, *Paint Flow and Pigment Dispersion*, 2nd ed., Wiley-Interscience, New York, 1979, p. 109.

23. G. F. Sprinkle, Jr., *Modern Paint and Coatings*, April, 44 (1983).

24. A. L. Rocklin and G. D. Edwards, *J. Coat. Technol.*, **48** (620), 68 (1976).

25. J. R. Ericson and A. W. Garner, *ACS Org. Coat. Plast. Chem. Prepr.*, **37** (1), 447 (1977).

26. F. J. Hahn, *J. Paint Technol.*, **45** (562), 58 (1971).

27. W. S. Overdiep, *Prog. Org. Coat.*, **14**, 159 (1986).

28. G. P. Bierwagen, *Prog. Org. Coat.*, **3**, 101 (1975).

29. G. P. Bierwagen, *Prog. Org. Coat.*, **19**, 59 (1991).

30. H. A. Wray, *J. Coat. Technol.*, **56** (717), 37 (1984).

31. W. H. Ellis, *J. Coat. Technol.*, **48** (614), 45 (1976).

32. "Solvent Properties," Texaco Chemical Co. (1990); "Solvent Selection Guide," Union Carbide Corp. (1988).

33. D. T. Wu, S. Lonsinger, and J. A. Klein, *FATIPEC 1988 Congress Book*, IV, 227 (1988).

34. R. L. Smith, L. J. Culver, and S. L. Hillman, *J. Coat. Technol.*, **59** (747), 21 (1987).

35. J. H. Seinfeld, *Science*, **243**, 745 (1989).

36. R. Atkinson, *Atmospheric Environment*, **24A**, 1 (1990).

37. K. A. Nielsen, D. C. Busby, C. W. Glancy, K. L. Hoy, A. C. Kuo, and C. Lee, *Polym. Mater. Sci. Eng.*, **63**, 996 (1990).

38. Z. W. Wicks, Jr. and G. F. Chen, *J. Coat. Technol.*, **50** (638), 39 (1978).

39. J. J. Brezinski, Ed., *Manual on Determination of Volatile Compounds in Paints, Inks, and Related Coating Products*, ASTM Manual Series: MNL 4, ASTM, Philadelphia, 1989.

CHAPTER XVI

Color

Color and the interrelated topic of gloss are critical to the decorative aspects of the use of coatings and are sometimes also important in the functional aspects of their use. We have all dealt with color since we were babies, but most people have little understanding of color. Most technical people think of it as an aspect of physics dealing with the distribution of visible light. While that is a factor, it is critical to recognize that color is a psychophysical phenomenon. In a sense, the difficulty of understanding color can be seen by considering the most rigorous definition of color that has been prepared: Color is that characteristic of light by which an observer may distinguish between two structure-free fields of view of the same size and shape. In effect, it says that color is what is left to distinguish between two objects when all the other variables are removed. Not a very satisfying definition.

Color requires that three components be present: an observer, a light source, and an object. (The single exception is when the light source is also the object being viewed.) There is no color on an uninhabited island. This is not just a semantic statement; color requires that an observer be present. There is no color in the absence of light; in a completely darkened room there is no color, not because you cannot see it, because it is not there. There must be an object; if you look out the window of a space ship without looking at a planet, there is no color—there is an observer, there is light, but there is no object. Color results from the interaction of light, object, and observer. In the following sections we discuss each of these aspects and their interactions.

16.1. LIGHT

Light is that range of electromagnetic radiation to which our eyes are sensitive. The range varies somewhat from individual to individual but in most cases includes wavelengths of 380 through 770 nm. The sensitivity of the eye varies as a function of wavelength (see Fig. 16.1) [1]. As can be seen, the eye is relatively insensitive to both the shorter and longer ends of the range. The response of our eyes depends on the distribution of wavelengths of light emitted by the light source. In the case of monochromatic lights, the colors we see range from violet through blue, green,

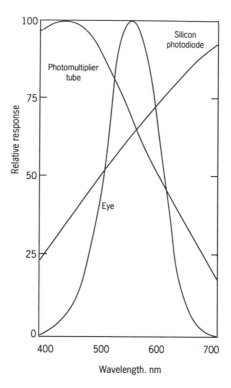

Figure 16.1. Sensitivity of the eye, photomultiplier tube, and silicon photodiode as a function of wavelength. (From Ref. [1], with permission.)

yellow, and red as we look at monochromatic light sources of increasing wavelengths. Figure 16.1 also shows the color response of photomultiplier tubes and silicon photodiodes, which are the other important detectors of light.

We see different colors as the ratios of wavelengths in polychromatic light sources change. If we look at a light source with nearly equal content of all wavelengths, we see white. Sunlight is considered the standard light, but sunlight varies depending on the latitude, season, cloudiness, and so forth. When work on understanding color began, the light from an overcast north sky (in the northern hemisphere) was accepted as standard. Based on many measurements of energy distribution, a light source designated as D_{65} was adopted as a standard related to north sky daylight. A graph of this standard is shown in Figure 16.2 [2]. Tables giving the energy distributions as a function of wavelength for various bandwidths from 1 to 20 nm wide are available [3].

The energy distribution from tungsten lights is quite different. Another standard light source, A, is a carefully specified tungsten light operated under specified conditions. A graph of this standard is also shown in Figure 16.2 and tables of its energy distributions are available.

Fluorescent lights are another important type of light source and many types are available. As shown in Figure 16.3, they exhibit a continuous energy distribution with peaks at a series of wavelengths. Even if the underlying continuous spectrum

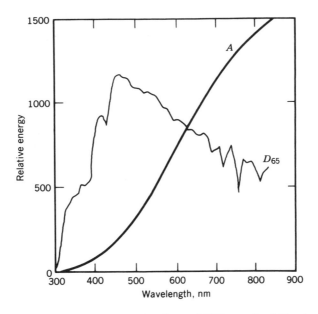

Figure 16.2. Relative spectral power distributions of CIE standard illuminants A and D_{65}. (From Ref. [2], with permission.)

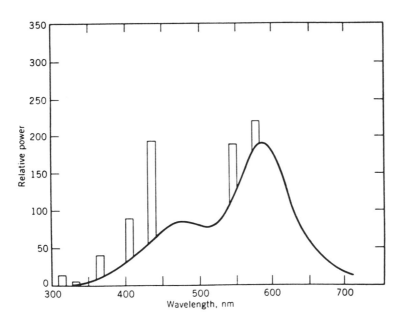

Figure 16.3. Spectral power distribution of a cool white fluorescent lamp (IES 1981). (From Ref. [1], with permission.)

were the same as D_{65}, the peaks would lead to changes in colors when these lights are used as a light source in comparison with overcast north sky light.

16.2. LIGHT–OBJECT INTERACTIONS

An important factor affecting color is the interaction between the light and the object being viewed. In this section we discuss the various modes of interaction between light and objects.

16.2.1. Surface Reflection

When a light beam is directed at a surface, some light will be reflected at that surface and some will pass into the object. As shown in Figure 16.4, if the surface is optically smooth, the light that is reflected will be reflected at the same angle (r) as the angle of incidence (i). This kind of reflectance is called specular (mirrorlike) reflectance. As shown in the figure, the accepted convention is that the angle of incidence normal to the surface is designated as $0°$ and the grazing angle is $90°$.

The fraction of light reflected (R) varies as a function of the angle of incidence and of the difference in refractive index (n) between the two phases. If there is no difference in index of refraction, no light will be reflected at the interface; as the difference increases, the fraction of light reflected increases. For angles of incidence near $0°$, the fraction reflected can be calculated using the following equations, which are given in both the general form of Eq. 16.1a and where the first medium

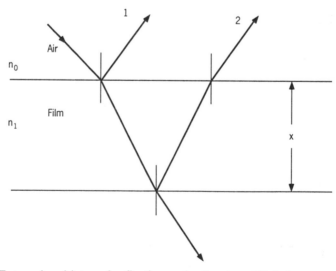

Figure 16.4. External and internal reflection and refraction of light by a nonabsorbing film (refractive index, n_1; thickness, x) with optically smooth parallel surfaces. (From Ref. [4], with permission.)

is air ($n_1 = 1$) Eq. 16.1b.

$$R = \left(\frac{n_2 - n_1}{n_2 + n_1}\right)^2 \qquad (16.1a)$$

$$R = \left(\frac{n - 1}{n + 1}\right)^2 \qquad (16.1b)$$

Since most resins and polymers have refractive indexes of about 1.5, it can be readily calculated from Eq. 16.1b that approximately 4% of incident light will be reflected from a polymer surface when the angle of incidence is near 0°.

The dependence of reflection on the angle of incidence is illustrated in Figure 16.5 [5]. As shown, reflectance approaches 100% as the angle of incidence approaches 90°. The family of curves in Figure 16.5 illustrates the combined effects of refractive index difference and angle of incidence.

Light that is not reflected at the surface enters the object. First, we consider what happens to the light when no light is absorbed by the object. When the light beam enters in the object it is refracted, that is, it is bent. The angle of refraction varies with the ratio of the refractive indexes of the two media as given in the following equation:

$$\sin r = \frac{n_1}{n_2} \sin i$$

Figure 16.4 shows that when light passes from air into a nonabsorbing plastic film with parallel, optically smooth surfaces and an index of refraction of 1.5, the angle of refraction is smaller than the angle of incidence. If the angle of incidence

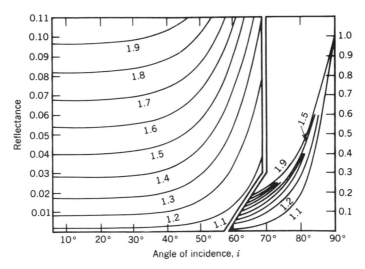

Figure 16.5. Fraction of light reflected from a smooth surface as a function of angle of incidence (i) with various differences in refractive index. (From Ref. [5], with permission.)

is near 0°, the fraction reflected at the first surface will be 0.04, and 0.96 will be transmitted into the film. If there is no absorption, 0.96 will reach the second film—air interface. There 0.96 × 0.96 of the original light will be transmitted into the air on the other side of the film, being refracted so that the angle of refraction is equal to the original angle of incidence. However, there will also be reflection at that second surface and 0.04 × 0.96 will be reflected back towards the first surface. Again, there will be partial reflection of 4% and transmission of 96% at the back of the first surface. As a result $(0.96)^2 \times (0.04)$ of the original light will be transmitted out of the film at an angle equal to the original angle of incidence and $(0.96) \times (0.04)^2$ will be reflected back into the film. This bouncing back and forth will continue, if there is no absorption, until the total transmission equals 0.92 and the total reflectance equals 0.08.

It should be noted that when the light beam passes from a medium of higher refractive index to one of lower refractive index, the angle of the beam increases. It follows that, if the angle of incidence in such a case is high enough, all of the light will be reflected back and none will be transmitted out. In the case of refractive indexes of 1.5 and 1, the critical angle (at which all the light is reflected) is calculated to be 41.8°. As we will see, this increased internal reflection has important effects on color. It also is the basis for optical fibers. If the light enters the end of a nonabsorbing fiber with a refractive index of 1.5 and there are no bends such that the angle of incidence with the surface of the fiber is less than 41.8°, all of the light will follow the fiber even all the way across the ocean.

16.2.2. Absorption Effects

In almost all cases involving coatings, the color we observe is affected by differential absorption of various wavelengths of light. Colorants, dyes, and pigments (and to a degree some resins), absorb some wavelengths of light more strongly than others. These absorptions are controlled by the chemical structures of the colorants.

First, we consider the absorption effects in transparent systems, that is, systems where the colorant is in solution or is so finely divided that it does not significantly reflect light at the colorant—resin interface. The extent of absorption depends on several factors:

Chemical composition and wavelength
Particle size
Optical path length—film thickness
Concentration
Media-colorant interactions

Each colorant has a specific absorption spectrum that controls the absorption of the various wavelengths of light. Chemists generally speak of molar absorptivities ε in units of liter per mole per path length, corresponding to the absorbance per molar concentration of a substance. Physicists generally speak of absorption coefficients K in units of path length^{-1} mass^{-1}, corresponding to the absorbance per unit mass of the system.

In all systems where it is the molecules that absorb the light, which includes almost all systems of importance in coatings, the smaller the particle size, the greater the fraction of light absorbed by the same quantity of a colorant. The highest molar absorptivities will be exhibited by individual molecules in solution. In the case of pigments, the smaller the particle size of the pigment, the greater the absorption.

The longer the path followed by a beam of light through a medium containing absorbing chemicals, the greater the degree of absorption. In the case of transmission of a beam of light at 0°, the optical path equals the film thickness. In any other case, the optical path length will be greater than the film thickness. If, in passing through a unit path length, one-half of the light of a particular wavelength is absorbed and one-half is transmitted (ignoring surface reflection), on passing through two units of path length, three-fourths will be absorbed and one-fourth will be transmitted. Mathematically this relationship is expressed by an exponential equation where X is path length, I is intensity of light transmitted, and I_0 is the original light intensity (chemists commonly use base 10 and physicists base e).

$$\frac{I}{I_0} = 10^{-\varepsilon X} \qquad \frac{I}{I_0} = e^{-KX}$$

Ideally, the same relationship holds when the concentration of the colorant in the medium is changed leading to the corresponding mathematical relationships.

$$\frac{I}{I_0} = 10^{-\varepsilon c X} \qquad \frac{I}{I_0} = e^{-KcX}$$

Actually, the relationship holds only over a limited range of concentrations, and the width of the range is system dependent. In other words, when systems are in solution, there are less likely to be intermolecular interactions between molecules in dilute solutions than in more concentrated solutions. In the case of pigment dispersions (insoluble systems), concentration effects are likely to be even more marked.

Another complication of these systems is their interaction with media. In the case of solutions, change in solvents can sometimes lead to association of molecules effectively increasing particle size and hence reducing absorption. There can also be effects of hydrogen bonding between a soluble dye molecule and the solvent in one solvent and not in another. This, in effect, changes the structure of the compound leading to a change in absorption spectrum. In the case of dispersions of pigments, changes in the medium, such as dilution with solvent, can lead to flocculation (agglomeration) of pigment particles; this causes a larger particle size and hence reduced adsorption.

Figure 16.6 shows transmission spectra of idealized reddish magentas. Spectra *a* and *b* result from transmission of light through the same transparent coating of path length x and $2x$, respectively. (Surface reflection effects are ignored.) Note that the fraction of light of all wavelengths transmitted through the thicker coating, spectrum *b*, is less. But note also that the relative transmission of the blue (B), green (G), and red (R) parts of the spectra are different. The color we would report seeing in case *b* would be a redder purple than what we would report in case *a*. The composition is the same but the shade of the color is strongly affected by path length. The same type of changes would take place if the concentration

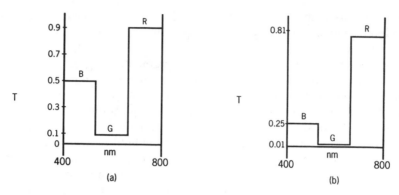

Figure 16.6. Transmission spectra of idealized magentas: (*a*), path length = *x*; (*b*), path length = 2*x*.

were doubled at the same film thickness. This is an idealization of a very real situation. One of the inks in four color process printing that permits printing of multicolored pictures is a reddish magenta. The film thickness of the magenta ink affects the color of that part of the print and hence affects the color of the combined color picture.

16.2.3. Scattering

Scattering is another important phenomenon that can occur during the passage of light through a film. If there are small particles dispersed in the film that have a different refractive index than that of the medium, light will be reflected and diffracted at the interface between the particles and the medium. The physics involved in the scattering process is complex and beyond the scope of this brief presentation of color concepts. Nevertheless, the results can be stated in a straight-forward manner. If one passes a linear beam of light into a film containing a substantial number of nonabsorbing particles, the light will be internally reflected in all directions so that it will change from a beam to diffuse illumination inside the film. Light that reaches the back of the top surface at angles greater than the critical angle will be reflected back into the film, while some of the light that reaches this surface at angles less than the critical angle will leave the film. If the film is thick enough, essentially no light will be able to pass completely through it. All light will be reflected back out of the top surface. The reflection, however, will not be just at the specular angle but at all possible angles—that is, this light will be diffusely reflected even though the top surface of the film is optically smooth. The degree of light scattering by nonabsorbing particles is dependent on several variables:

Refractive index difference between particles and medium
Particle size
Film thickness
Particle concentration

The greater the difference in refractive index, the greater the degree of light scattering. As shown in Figure 16.7, the degree of scattering increases more and more steeply as the difference in refractive index increases. Note that the degree of scattering is the same whether the particles have a higher or lower refractive index than the media. For example, water droplets in air (fog) scatter light as efficiently as air droplets of the same size at the same concentration in water (foam).

An ideal white pigment would absorb no light and have a very high refractive index so that the difference in refractive index from that of binders would be large. Rutile grade TiO_2 comes close to meeting these specifications, it absorbs some light below about 420 nm and has an index of refraction of 2.76. Another crystal type of TiO_2, anatase, absorbs less light, but its refractive index is lower, 2.55, and the smaller index of refraction difference from the average medium leads to substantially less efficient light scattering as compared to the rutile grade.

Scattering is importantly affected by particle size. Figure 16.8 shows the relationship between the scattering coefficient and particle size for 560-nm light for rutile TiO_2 in a resin with a refractive index of about 1.5. As particle size decreases, the scattering coefficient increases until a maximum is reached at 0.19 μm diameter; when the particle size is decreased further, the scattering coefficient drops off steeply [6]. Commercial TiO_2 has a range of particle sizes. Since the efficiency drops off more rapidly on the small diameter size side of the maximum, commercial TiO_2 pigments are produced with an average particle diameter somewhat over 0.2 μm.

You will read in some coatings texts that scattering maximizes for all pigments at a diameter of about one-half the wavelength of light. That is mistaken. The diameter with maximum scattering is dependent on the refractive index difference. An extreme example is calcium carbonate ($n = 1.57$), whose scattering coefficient is also shown in Figure 16.8 (note different scales). The maximum is at about 1.7 μm, and, as would be expected from the small difference in refractive index, the scattering by even the optimum particle size is low.

The extent of light scattering is affected by film thickness. If no absorption occurs, the light that is not reflected back out of the top surface is transmitted

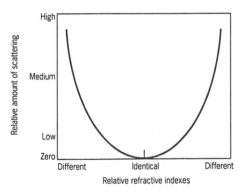

Figure 16.7. Scattering as a function of refractive index difference; particles have higher refractive indices than the media on the right and lower values on the left. (From Ref. [1], with permission.)

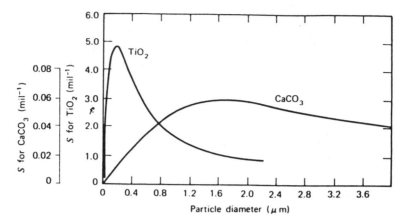

Figure 16.8. Scattering coefficients as a function of particle size for rutile TiO₂ and CaCO₃. (From Ref. [6], with permission.)

unless the film thickness is great enough so that essentially all the light is reflected back. Scattering is affected by concentration. At low concentrations of particles (e.g., rutile TiO_2) transmission through a film (neglecting surface reflection) follows a relationship similar to that given in the case of absorption.

$$\frac{I}{I_0} = e^{-ScX}$$

However, as concentration increases, scattering efficiency decreases and, in the case of rutile TiO_2 pigmented coatings, the efficiency will decrease so much that the fraction of light transmitted will actually increase and that reflected will decrease. The result can be seen in the plot of scattering coefficient (S) as a function

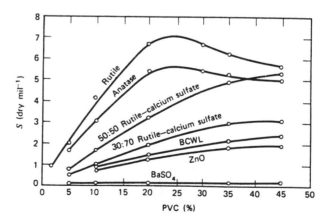

Figure 16.9. Experimental plots of scattering coefficients versus PVC for selected pigments in a dry acrylic lacquer. BCWL signifies basic carbonate of white lead. (From Ref. [6], with permission.)

of pigment volume concentration (PVC) of a rutile TiO_2 in final dry films shown in Figure 16.9 [6]. PVC is defined as the volume percent of pigment in a *dry* film of coating. In commercial practice, the cost effectiveness of pigmenting with TiO_2 drops off sufficiently at PVCs above about 18% that it is not generally economically sound to use a higher PVC of TiO_2 than that. The optimum value will vary somewhat from system to system; it is particularly dependent on the stability of the dispersion and on the TiO_2 content of the particular TiO_2 pigment, which may vary from grade to grade by more than 10% (see Chapter 18).

16.2.4. Multiple Interaction Effects

In the preceding three sections, we discussed the separate effects of surface reflection, absorption, and scattering as independent factors but in real life they are interdependent factors. Light sources generally are not narrow beams but relatively broad beams or diffuse illumination sources. Surfaces are not often optically smooth; in fact, in some cases, we strive for surface roughness. In coatings, we seldom deal with pigments or pigment combinations that only absorb or only scatter light, commonly both are occurring simultaneously. And in coatings, we seldom are interested in free films but rather in films on a substrate of some kind so the internal reflectance at the bottom surface of the film is not the same as that of the top surface of a film. The eye does not distinguish between the light that reaches it that has been reflected from the top surface, from within the film, or from the bottom surface of a film. The combined light from all three sources is integrated when the eye reacts to it.

As shown in Figure 16.5, surface reflectance increases as the angle of incidence increases. If a panel of a high gloss (smooth surface) blue paint, illuminated with a beam of light, is observed at any angle other than the specular angle, one will see a comparatively dark blue color resulting from the diffuse reflectance back out of the paint film. On the other hand, if the angle of incidence is small, and one observes at the specular angle, the color will be a somewhat lighter blue because the observer will see a mixture of that same light plus some "white" light reflected from the surface. If the angle of incidence approaches 90°, the observer will see a very light blue color at the specular angle because the observed light will contain a very high fraction of surface reflected light and only a little light from within the film. If the illumination is diffuse instead of a beam, the color will be a lighter blue from most angles of observation, because one will see a higher fraction of surface reflected light.

If the blue paint is of low gloss (rough surface) with exactly the same composition as in the last example, a lighter blue color will be seen under diffuse illumination at most viewing angles, as compared with the high gloss paint. This results from a greater amount of surface reflected light reaching the eye. The color of low gloss paint will be darkened at most viewing angles by wetting the surface with water since the water fills in the roughness of the surface thereby reducing surface reflection.

As shown earlier, the effects of concentration and film thickness on systems that only absorb or only scatter light are relatively simple and can be modeled by simple equations. However, the interactions resulting when both absorption and scattering occur are complex and the equations to model the resulting reflections from such films are complex. Figure 16.10 gives a set of Kubelka-Munk equations that models reflectance (R_t) and transmittance (T_t), including the effects of absorption (K), scattering (S_i), and reflections from both sides of the top surface (r_e' and r_i') and the inner side of the bottom surface r_i'' [7]. As we discuss later, simplifying assumptions permit use of these equations in color calculations for color matching.

16.3. HIDING

There can be important effects on color resulting from reflectance back through the film of light reaching the substrate to which the coating is applied. It is common to compare the colors of a coating applied over a striped black and white substrate. If one can see the pattern of the stripes through the coating, the coating is said to exhibit poor "hiding." The difference results from the reflectance of light reaching

$$R_t = r_e' + \frac{t_e' t_i' [R_i(1 - r_i'' R_i) + r_i'' T_i^2]}{D}$$

$$T_t = \frac{t_e' t_i'' T_i}{D}$$

$$D = (1 - r_i' R_i)(1 - r_i'' R_i) - r_i' r_i'' T_i^2$$

$$R_i = a - (T_i^2 + b^2)^{1/2}$$

$$T_1 = \frac{b}{a \sinh bS_i X + b \cosh bS_i X}$$

$$a = \frac{K + S_i}{S_i}$$

$$b = (a^2 - 1)^{1/2}$$

Assumptions

1. The equations apply to only one wavelength at a time.
2. S_i and K are the same throughout the entire thickness of the film.
3. Pigment particles are oriented randomly.
4. In those cases where there is scattering, that the internal radiation flux is perfectly diffuse.
5. Edge effects are ignored.
6. The bottom surface of the film is in optical contact with the substrate.

Figure 16.10. Kubelka-Munk equations and assumptions.

the white stripes compared with absorption of light reaching the black stripes. The effect is as if some black pigment was put in the paint above the black stripes. Color changes are even greater if the stripes are red and blue. Obviously, if all light entering the film is absorbed or scattered back out of the film prior to reaching the substrate, there will be no effect of the substrate on the color; the hiding would be complete.

Hiding is a very complex phenomenon and is affected by many factors. Hiding improves as film thickness increases. This means that "coverage" decreases; that is, the area that can be covered (hidden) by a liter of coating will be less and the cost will be higher. Hiding increases as the efficiency of light scattering increases, that is, hiding will be affected by the refractive index differences, particle sizes, and concentrations of scattering pigments present. Hiding increases as the absorption of light increases. Black pigments that have high absorption coefficients for all wavelengths of light are particularly effective for improved hiding (provided you do not mind a black wall). Colorants also improve hiding but not as effectively as black. Surface roughness will improve hiding—a larger part of the light is reflected at the top surface reducing the differences of reflection resulting from differences in the substrate to which the coating is applied.

An important factor affecting hiding that is sometimes forgotten in testing and developing coatings is uniformity of film thickness. The application of coating films commonly results in nonuniform thickness. Coatings are designed to "level," that is, to flow after application to make the film thickness more uniform than it would have been without flow (see Chapter 23). Leveling is often incomplete, however, and there can be serious effects on hiding.

For illustration, let us consider a poor leveling coating where an average film thickness of 50 μm of dry coating has been applied, but there are brushmarks remaining in the film so that adjacent to each other there are lines with film thicknesses of 65 and 35 μm. If the hiding at 50 μm would be just adequate, there will not be good hiding at 35 μm. The difference in color will be emphasized by the thin sections of coating being right next to the thick sections. The contrast resulting will commonly mean that the hiding of such a film will be rated as poorer than that of a uniform 35-μm film of the same coating.

Further complicating the practical problem is the variation of substrates over which the coating might be applied. Hiding of a white coating to be applied over a white surface might be rated as excellent, but hiding of the same coating over a black surface might be rated very poor.

There are quality control tests that satisfactorily compare batches of the same or similar coatings, but no test is available that can provide an absolute measure of hiding [8]. The only way to establish the covering power of a coating is to apply it to an appropriate large surface with a film thickness that gives adequate hiding and then calculate the actual coverage (m^2L^{-1}) value.

In some cases, hiding is not desired; an example is a coating to be applied to metalized plastic Christmas tree ornaments. One wants, for this purpose, transparent red, green, blue, or other color coatings. Preparation of a transparent coating requires that there is no light scattering within the film and, therefore, the particle size of the pigment particles, if any, must be very small. The refractive indexes of the pigment and the binder should also be similar.

16.4. METALLIC COLORS

A large volume class of coatings, called *metallic coatings*, are widely used on automobiles. They are made with transparent colorant systems together with "non-leafing" aluminum flake pigment or, less often, with flake pigments of other metals (see Section 18.2.5). These metallic coatings exhibit unusual shifts in color as a function of viewing angle. As noted in Section 16.2.4, with regular high gloss paints the color seen is dark when the panel is looked at from relatively small viewing (normal) angles and light when the panel is observed from large angles of viewing. In contrast, with metallic coatings the color is lighter when viewed near the normal angle (the face color) and darker when viewed from a larger angle (the flop color). It is desirable to achieve a high degree of "color flop," that is, a large difference between face and flop colors. To do so requires that the surface be very smooth (high gloss), that there be no light scattering from the resin or color pigment dispersion, and that the aluminum flake particles be aligned parallel to the surface of the film.

As seen in the idealized diagram in Figure 16.11, when the observer looks at the film from an angle near the normal, the path length of light through the film is short because it is reflected back by the aluminum. Therefore, there is little chance for absorption by pigment and the color is light. On the other hand, from a greater angle, the observer sees light that has been reflected back and forth within the film so that the path length is longer and the color is darker.

Before the coating is sprayed onto the car surface, the aluminum flakes are randomly oriented. During application and film formation the flakes become oriented parallel to the coated surface if conditions are favorable. No completely satisfactory explanation of the mechanism of orientation has been advanced. Two contending theories are (1) that orientation occurs when the spray droplet first arrives at the surface being painted and (2) that it results from shrinkage of the film by solvent evaporation after application. Both may be partly correct. However, the former mechanism requires that the viscosity remain high enough throughout film formation to prevent substantial motion of the flake particles, but the latter mechanism requires that the viscosity stay low enough for a sufficient time period to allow the particles to move and orient. The viscosity must increase first near the surface. This higher viscosity upper layer keeps one edge of the flake from

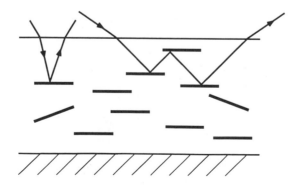

Figure 16.11. Idealized diagram of reflection of light in a metallic coating.

moving much during the shrinkage while the bottom edge of the flake is pushed towards parallel by the force of shrinkage. (See Chapter 33, Section 33.1.3 for further discussion.)

The coating film in which the aluminum flakes are oriented must be transparent. If there is light scattering within the film, the optical path length of the light viewed at large angles will be reduced and flop will be less.

16.5. THE OBSERVER

Another critical aspect of color is the observer, involving the eye and the brain in the case of a human observer. The retina of the eye has two types of photodetectors—rods and cones. These detectors are stimulated by light photons and send signals through the optic nerves to the brain. The rods are sensitive at low levels of illumination to all wavelengths of light and become insensitive as the level of illumination increases. The cones are insensitive at low levels of illumination and are sensitive at higher levels.

The response of the cones is wavelength dependent. The cones have three ranges of sensitivity that overlap—one peaks in the blue region of the spectrum, another in the green, and the third in the red region. These overlapping ranges lead to complex signals being sent to the brain, which integrates the signals so that we can "see" thousands of different colors. The response to the mixture of the wide range of wavelengths reflected from a yellow coating is very similar to the response of the eye to a monochromatic yellow wavelength of light.

At low levels of illumination, we see only with the rods and our vision is limited to shades of gray. At intermediate levels of illumination, one can have responses from both rods and cones, hence one does not see bright colors, only grayish colors. At higher levels of illumination, the rods are inactive and we see the full range of colors from the responses of the cones.

The mechanisms of color vision are complex and only partly understood; they are beyond the scope of this introductory text. There are variations from person to person in the degree of response to the three ranges of sensitivity. As a result, colors seen by different observers are not identical. Usually these differences are small but in some cases they are large. In extreme cases, some people are *color blind*. While there are different types of color blindness, the most common is *red–green* color blindness. Such people cannot distinguish between red lights and green lights.

In order to specify color and predict responses to mixtures of colorants, a mathematical model of a "standard human observer" was established by an international committee of experts known by the acronym of its French name as CIE. Figure 16.12 shows a graph of three functions that model color vision. At any given wavelength, the standard observer will have the same response to the ratio of the values of $\bar{x}$, $\bar{y}$, and $\bar{z}$ at that wavelength as he/she will to monochromatic light of that wavelength. Tables of these *CIE Color Matching Functions*, $\bar{x}$, $\bar{y}$, and $\bar{z}$, are available as a function of wavelength with bandwidths of 1, 10, and 20 nm [3].

Within the range of higher illumination, the eye adapts to changes in levels of illumination. For example, if an area of white is surrounded by black, the white will look whiter than if the black were not present. The eye is adapting to the level of light reflected by the combined black and white and, therefore, responds more

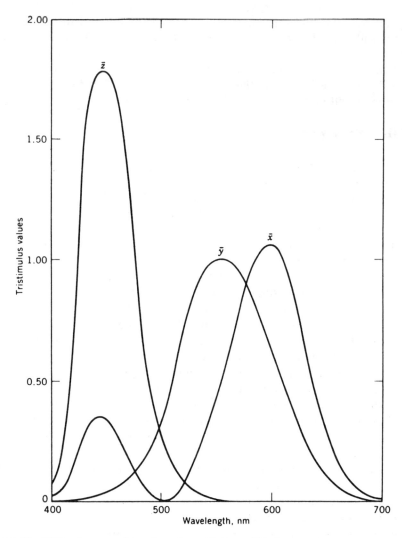

Figure 16.12. CIE color matching functions, $\bar{x}$, $\bar{y}$, and $\bar{z}$, for equal energy spectra. (From Ref. [1], with permission.)

to the white in the presence of the black. Similar effects will occur when two strongly colored fields of view are adjacent to each other. Yellow surrounded by blue-green will look more orange than it would if the blue-green were absent. In general, if one looks at small color chips in selecting a coating, the color will be different than if that coating is applied to a large surface because of the effect of the surrounding area on the color. Many other effects result from such interactions between the eyes and combinations of color surfaces (see Ref. [5] for further discussion).

16.6. INTERACTIONS OF LIGHT SOURCE, OBJECT, AND OBSERVER

As stated earlier, color depends on the interaction of three factors—light source, object, and observer. If any one of these change, the color will change. If we have observed an object under a light source with the energy distribution of standard source A and shift to a different light source, for example, standard source C (similar to D_{65}), the color will change. Light source A has relatively lower emission in the blue end of the spectrum and relatively higher emission in the red end. The light reaching the eye when the object is illuminated with source A will have more red light and less blue light than the light reflected from the same object illuminated with source C. The color is different. This situation is illustrated in spectral sets a and b of Figure 16.13, which show the different responses as product spectra of

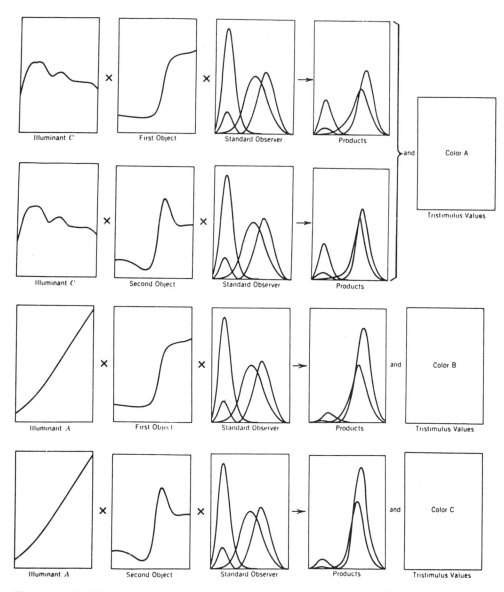

Figure 16.13. Illustration of light source effects and metamerism. (From Ref. [1], with permission.)

light source times object times observer for the same object with light sources A and C, respectively.

If the chemical compositions and the physical states of the colorants in two coatings are the same, the reflectance spectra will be identical, and the coatings will match under any light source. It is possible for two materials with different colorant compositions, and hence different reflectance spectra, to be the same color under a certain light source. However, such a pair will not match under light sources with significantly different energy distributions. This phenomenon is called "metamerism." Note that in the first case of the spectral match, the two panels change color in the new light source but it is the same change in both cases. In the case of a metameric pair, the color is the same with one light source; the colors of both panels also change when the light source is changed but the extent of change is different between the two panels. This situation is also illustrated in Figure 16.13, which shows responses of a second object with illuminants C (spectral set c) and A (spectral set d). The color of the first and second objects (a metameric pair) under illuminant C is identical, but the colors are different under illuminant A.

The only way to be sure that a pair of coatings will match under all illuminations is to utilize colorants with identical chemical composition and physical states. Unfortunately, this requirement cannot always be met (see Section 16.9).

16.7. COLOR SYSTEMS

The human eye can discriminate thousands of colors. However, it is difficult for a person to tell another person what colors he/she sees. A color description such as "light grayish blue-green" illustrates this difficulty. Many systems have been developed to permit definitive identification of colors. There are two types of systems, ones that use visual color samples in specific arrangements and others that identify colors mathematically. All color systems share the common denominator that they require at least three dimensions to include all the possible colors.

The most widely used visual color system is the Munsell Color System. Thousands of carefully prepared and selected color chips are classified in a three-dimensional system. The dimensions of the Munsell System are called "hue," "value," and "chroma." Hue refers to the dimension of color described as blue, blue-green, green, greenish-yellow, yellow, yellowish-red, red, reddish-purple, purple, purplish-blue, back to blue. The value dimension characterizes the lightness of a color as compared to a series of gray samples—a "gray scale." The value 0 is assigned to pure black and the value 10 to pure white. A light blue has a high value while a dark blue of the same hue has a low value. The chroma dimension refers to the degree of difference between a color and a gray of the same value. A bright red would have a high chroma, whereas a grayish-red of the same hue and value would have a low chroma.

The color chips in the Munsell System are prepared so that there are equal visual differences between all pairs of adjacent chips. The chips are all systematically labeled; for example, a chip labeled G5/6 would be a green with a value of 5 and a chroma of 6. One could look at a set of Munsell chips and know the color some one meant when they made such a designation. There are two critical limitations to this statement. First, the light source must be specified. Chip G5/6 will give a different color under source A than under source D_{65}. Second, as mentioned earlier, surface roughness affects color, so comparisons have to be made at equal gloss levels. Two sets of Munsell chips are available, one with high gloss and the other with low gloss; but significant errors can result in comparing either of these with other materials that are semigloss.

The other very widely used color system is the CIE Color System. It is based on mathematical descriptions of light sources, objects, and a standard observer. As mentioned earlier, light sources are specified by their relative energy distributions, objects are specified by their reflectance spectra (or transmission spectra), and the observer by the CIE standard human observer tables. For color analysis, the light reflected (or transmitted) from (or through) an object is measured with a spectrophotometer. Since in most cases, the reflection is diffuse, it is essential to use a spectrophotometer with an integrating sphere so that all of the light being reflected will be sampled, not just that at some narrow angle. The data can be reported in analog form as a curve or in digital form as tables of the fraction of light reflected (or transmitted) as a function of wavelength.

To identify the color resulting from the interaction of a light source, an object, and a standard observer, one uses the data for these three dimensions to calculate the *tristimulus values*, X, Y, and Z:

$$X = \sum_{380}^{770} \bar{x}_\lambda E_\lambda R_\lambda$$

$$Y = \sum_{380}^{770} \bar{y}_\lambda E_\lambda R_\lambda$$

$$Z = \sum_{380}^{770} \bar{z}_\lambda E_\lambda R_\lambda$$

Note that the tristimulus values will be different for the same object and the same observer when the energy distribution E of a different light source is used. This is as it should be since we know that colors change as the light source changes. The tristimulus values uniquely and unequivocally define colors. For example, $X = 14.13$, $Y = 14.20$, and $Z = 51.11$ is a definitive description of a color, but what color? Unfortunately, even experts often cannot say by looking at the numbers. This set of tristimulus values is for a blue but few, if any, people could look at them and tell you it is a blue, much less whether it is a grayish-blue that approaches being a purplish-blue, which happens to be the case.

The X and Y tristimulus values can be converted to *chromaticity values*, x and y, by normalization.

$$x = \frac{X}{X + Y + Z} \qquad y = \frac{Y}{X + Y + Z}$$

The chromaticity values of each wavelength of the spectrum can be calculated and plotted against each other to form the *CIE spectrum locus* (see Fig. 16.14). The ends of the locus are connected by a straight line called the "purple line." There are no purples in the spectrum, in CIE color space all the hues of purple

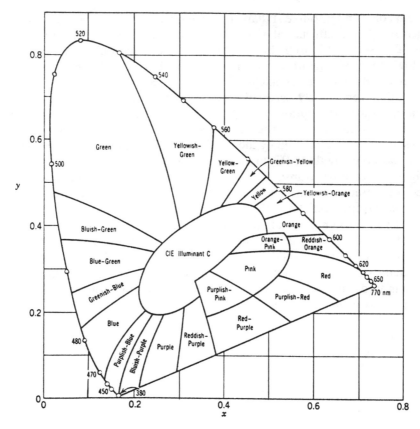

Figure 16.14. CIE Chromaticity Diagram showing the location of various hues. (From Ref. [1], with permission.)

lie along this line. As shown, the plot can be divided into color areas, so that one can look at the x and y values and have a reasonable idea of the shade of color designated. The third dimension rises vertically from the plot. It is the Y tristimulus axis, luminance; $Y = 100$ (or 1, in some conventions) only at the x,y values of the light source. The Y at the spectrum locus line approaches 0. At x,y points between the spectrum locus and the point of the source, Y will always be less than 100. As the value of Y gets larger, the gamut of possible colors narrows, as seen in Figure 16.15.

If one draws a line from the point of the source through the point of a sample extended out to the spectrum locus, the wavelength at the intercept is called the *dominant wavelength* of the color. This dimension corresponds to the hue dimension in the Munsell System, but the scale is different. If this extrapolation intercepts the purple line, then the line is extrapolated as a straight line in the opposite direction and the intercept of that line with the spectrum locus is called the *complementary dominant wavelength*. If one divides the distance from the source point to the sample point by the total distance from the source point to the spectrum locus (some workers express it as a percentage), one obtains the "purity." Purity is the same dimension as chroma in the Munsell System, but the scale is different.

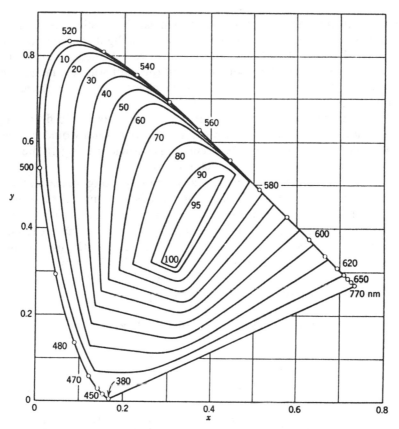

Figure 16.15. Topographical diagram of three dimensional CIE color space with light source *C*. (From Ref. [1], with permission.)

The vertical *Y* dimension, *luminosity*, is a gray scale as in the Munsell value dimension, but again the scale is different.

Figure 16.15 shows a topographical diagram of three dimensional CIE color space with light source *C*. All real colors that can be seen with light source *C* fall inside of this color space. In the Munsell System, the boundaries are limited not by reality of color but by the color purity of the pigments available to make reference chips.

As shown by looking at Figures 16.14 and 16.15, CIE color space is not visually uniform. For example, a small difference in the values of *x* and *y* in the blue part of the color space represents a substantial difference in color, whereas in the green part of color space the same difference in *x* and *y* values would represent little difference in color. In this sense, Munsell color space is preferable since the differences are visually uniform. However, color calculations of the types needed for instrumental color matching are feasible with CIE color space but not with Munsell space. If the comparisons are to be visual, one commonly uses the Munsell System. If the comparisons are to be mathematical and to include all possible color space, one uses the CIE system.

If one has a metameric pair of coated panels, the tristimulus values and the chromaticity coordinates would be the same with the light source under which the panels match. They would not be the same, however, if calculated with the energy distributions from another light source. When the light source changes, the X, Y, and Z, and the x and y values of both panels change, but the changes would be to different degrees.

For the most accurate data, reflectance measurements would be taken at each wavelength and the values over the range of 380 to 770 nm would be used in the summation. For most purposes, the accuracy is sufficient using 16 measurements at 20-nm intervals from 400 through 700 nm.

There have been many attempts to transform CIE color space mathematically to a visually equal color space. While progress has been made, the objective has not been reached. The reason for making such a transformation is not just to *look nice* but to have the mathematical differences between colors have the same visual difference throughout color space. Then the differences could be used as meaningful measures for specification purposes. Some progress has been made and color differences (ΔE) can be calculated using the CIE 1976 equations, where L^* corresponds to Y, u^* corresponds to x, and v^* corresponds to y; n is the source.

$$L^* = 116 \, (Y/Y_n)^{1/3} - 16; \qquad (Y/Y_n > 0.01)$$

$$u^* = 13L^* \left[\left(\frac{4X}{X + 15Y + 3Z} \right) - \left(\frac{4X}{X + 15Y + 3Z} \right)_n \right]$$

$$V^* = 13L^* \left[\left(\frac{9Y}{X + 15Y + 3Z} \right) - \left(\frac{9Y}{X + 15Y + 3Z} \right)_n \right]$$

$$\Delta E_{u^*,v^*} = [(\Delta L^*)^2 + (\Delta u^*)^2 + (\Delta v^*)^2]^{1/2}$$

These equations still do not represent fully uniform color space. If specifications are written for a line of colors specifying a fixed $\pm$ range for ΔE for the whole series, the requirements will be more stringent for some colors than for others. Even if there were color difference equations available that were visually uniform, there would still be difficulty using them for specifications. By using such a specification, the color would be permitted to vary equally in any direction from the central standard. However, it is common for people to be more concerned about deviations in one direction in color space as compared to other directions. For example, there is commonly a greater tolerance for whites to be off in the blue direction than for them to be off in the yellow direction.

16.8. COLOR MIXING

There are two broad types of color mixing: *additive* and *subtractive*. In additive mixing, the primary colors are red, green, and blue. Additive mixing is involved in theatrical stage use of overlapping colored spot lights, as well as in color television, where three colors of dots (red, green, and blue) are projected near each other on the screen. The lights from nearby dots are *added* when we look at them to give colors that depend on the ratio of the three colors in nearby dots. In additive color mixing, equal amounts of blue and green light give blue-green (cyan); similarly, blue and red light give purple (magenta); green and red light give yellow; and equal addition of all colors gives white light. With appropriate light sources, all color lights can be made. Note that we see as yellow the combination of all wavelengths in the green, yellow, orange, and red segments of the spectrum.

In almost all cases, however, color mixing encountered in the coatings field is not additive but subtractive mixing. We use colorants that absorb (subtract) some of the light of some wavelengths from the white light. If we add a second colored pigment to a coating containing one colored pigment, we subtract more of additional wavelengths; we cannot add back any missing wavelengths by using a further colorant. The primary colors in subtractive mixing are cyan, magenta, and yellow. If we mix equal quantities of an ideal cyan and an ideal magenta, the result will be blue. That is, the cyan absorbs red and the magenta absorbs green, so that blue remains. Similarly, cyan and yellow yield green; magenta and yellow yield red; and a mixture of equal amounts of all three ideal absorbing colorants would absorb all light and yield black. Idealized transmission (or reflectance) spectra of the primary colors, which illustrate their complementary nature, are provided in Figure 16.16.

16.9. COLOR MATCHING

A very large fraction of all pigmented coatings is color matched. The customer originally chose some color he/she liked for the refrigerator, automobile, or other product, and the coating formulator was given a sample of a material having this

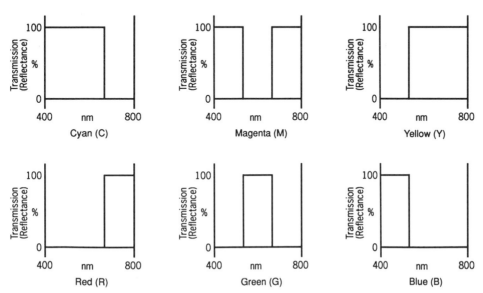

Figure 16.16. Transmission (or reflectance) spectra of cyan, magenta, and yellow colorants, together with their complementary colors.

color and was asked to match the color using a coating formula appropriate for the refrigerator, automobile, or other product. After the formulator establishes the appropriate ratio of the appropriate pigments and after the customer approves the color match, then the factory must color match batch after batch of the coating to meet this standard.

16.9.1. Information Requirements

Before starting the initial laboratory color match, the color matcher needs a substantial amount of information:

1. *Metamerism.* Is a spectral (nonmetameric) match possible? That is, can exactly the same colored pigments (including white and black) be used in establishing the match as were used in making the customer's sample? If not, the customer must realize that any match will be metameric; that is, the colors will match under some light source but not under others. For example, if the sample is a dyed fabric, the color of the textile cannot be exactly matched under all lights by any coating, since the colorants cannot be identical. If, as another example, the customer has been using a coating made with one or more pigments containing lead compounds and wishes to be supplied with a lead-free coating only a metameric match will be possible.

2. *Light Sources.* If the match is to be metameric, the customer and the supplier must agree on the light source(s) under which the color is to be evaluated. If there is to be more than one light source, a decision should be made whether it is more desirable to have a very close match under one light source without regard to how far off the match might be under other light sources, or to have a fair color match under several light sources but not a close one under any individual light source.

3. *Gloss and Texture.* The color of a coating depends on its gloss and texture. Some of the light reaching the eye of an observer has been reflected from the surface of the film and some from within the film. The color seen by the observer will vary depending on the ratio of these two types of reflected light. At most angles of viewing, more light is reflected from the surface of a low gloss coating than from the surface of a high gloss coating. It is impossible to match the colors of a low and high gloss paint at all angles of viewing. Therefore, there must be agreement as to the gloss of the coating and, if the gloss of the standard is different from the gloss desired for the new coating, the angles of illumination and viewing must be agreed upon. It is impossible to make even a metameric match of the color of some fabric sample with a paint at all angles of viewing since both the colorants and the surface textures have to be different. When paint manufacturers advertise on TV that their paint stores can do this, they are engaging in misleading advertising.

4. *Color Properties.* Obviously, colorants have to be chosen that will permit formulation of a coating that can meet the performance requirements. Need the coating have exterior durability, resistance to solvents, resistance to chemicals like acids and bases, resistance to heat, meet some regulation for possible toxicity, and so on?

5. *Film Thickness and Substrate*. Since in most cases, the coating will not completely hide the substrate, the color of the substrate will affect the color of the coating applied to it and the extent of that effect will vary with the film thickness. This variable is particularly important in applications like can coatings and coil coatings where relatively thin coatings are applied (< 25 μm). A thin coating that was color matched over a gray primer will not match over a red primer. A coating set up for one coat application on aluminum will not match the color standard if it is applied instead over steel.

6. *Baking Schedule*. Since the color of many resins and some pigments used in coatings is affected by heating, particularly at high temperatures, the color of the coating will be affected by the time and temperature of baking. It is also necessary to specify what the color requirements are for overbaking.

7. *Cost*. The color matcher should know the cost limitations in advance. There is no sense making an excellent color match with pigments that are not economically feasible for the particular application. An important element affecting cost, which is often not considered seriously, is how close are the tolerance limits permitted in production? This is such an important factor that it is described separately.

8. *Tolerance*. How close a color match is needed? In some cases, such as coatings for exterior siding or automobile top coats, very close color matches are absolutely necessary. For many others, close matching is not needed, but some customers, in their ignorance, set tight tolerance limits anyway. Overly tight tolerances raise cost without performance benefits.

For coatings that are going to be produced over time with many repeat batches, the most appropriate way to set color tolerances is to have an agreed on set of limit panels. For example, for a deep yellow coating, these tolerances would be the greenness and the redness limit, the limit of brightness and darkness, and so on. Since panels can change with age, spectrophotometric measurements should be made of the standard and the limit panels and CIE tristimulus values calculated. The customer would agree that any batch of coating that gave a color falling within this volume of color space would be acceptable.

As mentioned earlier, while ΔE color differences, calculated from the CIE 1976 color difference equations, have been used for setting color tolerances, they have severe limitations. It is undesirable to use them as a general statement of tolerance for a series of colors. Furthermore, as noted earlier, they permit variation equally in any direction from the standard color, which is generally undesirable.

16.9.2. Color Matching Procedures

There are two broad approaches to matching colors: visual and instrumental. In visual color matching, experienced color matchers look at the sample to be matched and from their experience select a combination of pigment dispersions that they think will permit matching of the color. A formula is mixed and the coating applied. The color matcher looks at the resulting color in comparison to the standard and decides what further addition of either one or more of the same pigment dispersions or perhaps dispersions of different pigments are needed to adjust the color match. This process continues until a satisfactory combination of pigment dispersions has been selected. In matching a new color, even highly experienced color matchers usually need at least 3 "hits," and commonly it will take 8, 10, or more *hits*, sometimes requiring changes in one or more of the pigments originally chosen.

It should be noted that precise color matching must be based on dry coating films. It can be useful to compare samples of wet paints to estimate progress towards a match, but since color changes considerably during application, film formation, and drying, decisions have to be based on dry films prepared under conditions approximating the way the coating will actually be used.

The laboratory color matcher not only has the responsibility of selecting the colorants and their ratios to make a color match, but he/she must also attempt to make the color match so that it will be as easy as possible to produce efficiently batch after batch in the factory. It is not possible for pigment manufacturers to produce successive batches of color pigments that are exactly the same color and color strength. Furthermore, there will be batch-to-batch variation in dispersions made from these pigments. Yet, the formula should be set up so that, in the factory, it will be possible to adjust the formula somewhat to permit matching the color standard in spite of these variations. For this reason, it is desirable to use at least four pigments (counting black and white, but not inerts, as pigments) in making the original match. This will provide the four degrees of freedom necessary to move in any direction in three-dimensional color space. Sometimes use of single pigments cannot be avoided, but it is undesirable, especially in applications where close color matches are needed. The factory should be discouraged from changing from one type of pigment to another because the result will be a metameric match.

Visual color matching is a highly skilled craft requiring years of apprenticeship to master. It is still practiced, but the trend is toward computerized instrumental color matching systems.

Spectrophotometric curves of the original standard sample can be used in an analytical mode to help identify component colorants in the standard, hence simplifying pigment selection. Organic pigments can often be identified by examining the absorption spectra of solutions of colorants from a sample of coating [9].

More importantly, instrumental color data bases along with computer programs can be used to help select the colorants and their ratio both in the original color matches in the laboratory and to provide information as to the amount of the different pigment dispersions to be added in the factory to match production batches. Establishing an instrumental color matching program requires a major effort. The data base must be set up carefully. The pigments to be used must be made into dispersions and then multiple single color coatings with different concentrations of color pigment dispersion in a series of appropriate white coatings formulas must be made and applied. The reflectance values must be measured at 16 wavelengths, and for critical situations values at 35 wavelengths may be needed.

Complete discussion of computer color matching is beyond the scope of this introductory text. See Ref. [1] for an introduction to computer color matching and the references given in it for more detailed discussions. The process can be illustrated with a description of the simplest case: A case where all the colors to be matched have a high level of hiding, are pastel colors, and have the same surface reflectance as the coatings on which the computer data base has been accumulated. In such cases the equations used for calculation can be substantially simplified. If surface reflection is ignored and full hiding is assumed, the Kubelka-Munk equations given in Figure 16.10 simplify to

$$\frac{K}{S} = \frac{(1 - R_\infty)^2}{2R_\infty}$$

In this simplified form, the ratio of absorption coefficient K to scattering coefficient S is used. The reflectance of the standard sample is measured and the ratio of K/S is calculated for each wavelength. The term R_∞ signifies that the reflectance does not change as film thickness is increased further. For a mixed colorant system, the K/S of the system is related to the K/S of the components by the following equation:

$$\left(\frac{K}{S}\right)_{mix} = \frac{c_1 K_1 + c_2 K_2 + c_3 K_3 + \cdots}{c_1 S_1 + c_2 S_2 + c_3 S_3 + \cdots}$$

This equation requires the availability of separate K and S values, but for the simple case of pastel colors, it is reasonable to assume that the scattering is dominated by the white base. If one assumes that the scattering results only from the white, the K/S of the system can be calculated from the simpler equation where only K/S values are needed.

$$\frac{K}{S_{mix}} = c_1 \left(\frac{K}{S}\right)_1 + c_2 \left(\frac{K}{S}\right)_2 + c_3 \left(\frac{K}{S}\right)_3 + \cdots + \left(\frac{K}{S}\right)_w$$

The $(K/S)_w$ is the K/S of the white base paint to be used. The K/S of the colorants are those of known concentrations of colorant dispersions in the same white base. Since the K/S ratios are constant only over a relatively narrow range of concentration, they must be determined in the concentration range in which they will be used.

For each wavelength, the computer can select colorants and colorant concentrations such that the K/S of the mixture will equal that of the standard to be matched. The calculations are done iteratively so that the closest possible K/S ratios are obtained at all 16 (or 35) wavelengths simultaneously. Furthermore, the reflectance values of the standard and various combinations of possible colorants can be used to calculate tristimulus values using two or more standard light source energy distributions. In this way, colorants can be selected that will give nonmetameric matches and/or that will exhibit relatively low degrees of metamerism.

For darker colors, the equations and computer programs used are more complex but the principle is the same. Matching of metallic colors has been the most difficult to computerize since the colors have to match at multiple angles. The reflectances of the samples and all of the data base colorants must be determined at multiple angles.

Other information on the colorants such as cost, properties, and regulatory restrictions can be built into the data base. The computer can then calculate a series of alternative colorant combinations selected for lowest cost, least degree of metamerism, excluding pigments with inadequate exterior durability, and so forth.

The formulator can then choose the most appropriate combination to start a particular color matching assignment.

The formulator then makes up an experimental batch of paint, applies it to the appropriate substrate, bakes or dries it, and measures the reflectance values. The result is seldom a satisfactory color match on this first attempt. Reflectance values of the first hit are instrumentally compared with those of the standard, and the computer calculates the amounts of the various colorants that need to be added to the batch so that the color will match. The number of hits required to make color matches has been found to be significantly less using computer color matching than the number required by even experienced visual color matchers.

The same program is used in production of factory batches of colored coatings. Of course, in this case, the colorants to be used have already been selected and must not be changed. The laboratory formula is used for the initial mix, but a fraction of each of the colorants is held back. (If the factory batch of some pigment dispersion were stronger than standard, too much of that colorant would be put in.) The batch is mixed, a panel coated, and its reflectance measured. The data is then used to calculate what additions have to be made to the batch to obtain a color match. If necessary, the process is repeated.

The savings in time, and hence in cost, using computer color matching can be very large, but the cost of establishing and maintaining the data base is substantial. Since the colors of pigment dispersions vary from batch to batch of pigment and of pigment dispersion, measurements of each batch are compared with the standard data base values. Computer programs can be written to correct for small differences in the colors from batch to batch. The growing data base is valuable for productivity improvement and quality assurance programs.

The applicability of any computer color matching program will be limited. In our simple example, we said that the white base paint had to be constant. Actually the scattering efficiency of different whites and their surface reflectances will be different. Obviously, if two whites are similar, the data base information obtained using one white can be used with the other white. How similar do they have to be? The answer to this is pragmatic; if color matching is accomplished in fewer hits than it would be without the program, the whites are similar enough. Similarly, our example was restricted to pastel colors. How dark a pastel color can be handled by the system? Again, the answer is pragmatic; if the color matching efficiency is increased you use it.

GENERAL REFERENCES

Anonymous, *ASTM Standards on Color and Appearance Measurement*, 3rd ed., American Society of Testing and Materials, Philadelphia, 1991.

F. W. Billmeyer, Jr., and M. Saltzman, *Principles of Color Technology*, 2nd ed., Wiley, New York, 1981.

G. Wysecki and W. S. Stiles, *Color Science*, 2nd ed., Wiley-Interscience, New York, 1982.

D. B. Judd and G. Wysecki, *Color in Business, Science, and Industry*, 3rd ed., Wiley, New York, 1975.

REFERENCES

1. F. W. Billmeyer, Jr., and M. Saltzman, *Principles of Color Technology*, 2nd ed., Wiley, New York, 1981.

2. F. W. Billmeyer, Jr., "Color," in *Encyclopedia of Chemical Technology*, 3rd ed., Vol. 6, Wiley, New York, 1979, pp. 523–548.
3. G. Wyszecki and W. S. Stiles, *Color Science*, 2nd ed., Wiley-Interscience, New York, 1982.
4. L. M. Greenstein, "Pearlescence," in T. C. Patton, Ed., *Pigment Handbook*, Vol. 3, Wiley, New York, 1973, pp. 357–390.
5. D. B. Judd and G. Wysecki, *Color in Business, Science, and Industry*, 3rd ed., Wiley, New York, 1975.
6. P. B. Mitton, "Opacity, Hiding Power, and Tinting Strength," in T. C. Patton, Ed., *Pigment Handbook*, Vol. 3, Wiley, New York, 1973, pp. 289–339.
7. Z. W. Wicks, Jr., and W. Kuhert, *J. Paint Technol.*, **47** (610), 49 (1975).
8. E. Cremer, *Prog. Org. Coat.*, **9**, 241 (1981).
9. R. Kumar, F. W. Billmeyer, Jr., and M. Saltzman, *J. Coat. Technol.*, **57** (720), 49 (1985).

_____

CHAPTER XVII

Gloss

Gloss is another important property of coatings that is affected by the interaction between light and a coating. In some cases, high gloss is desirable; in others, low gloss is wanted. Whatever the choice, controlling the reproducibility of the gloss is essential.

17.1. TYPES OF GLOSS

Gloss is a complex phenomenon. Individuals frequently disagree on gloss differences. Partly because of the difficulty of visual assessment, little progress has been made in developing useful mathematical treatments or measurements of gloss. Unfortunately, many people working on coatings assume that there are appropriate mathematical treatments and that the measurements that are made every day by thousands of technicians are much more meaningful than they really are.

One problem has been that people do not have a clear definition of gloss in mind. They assume they know what it means without thinking through their ideas. There are several types of gloss. *Specular gloss* is the type most often considered in coatings. A high gloss surface reflects a large fraction of the light that is reflected from the surface at the specular angle, that is, the angle of reflected light equal to the angle of the incident light beam (both angles being measured from a perpendicular to the reflecting surface). Note that gloss is *not*, as seems to be commonly assumed, directly related to the fraction of light reflected from the surface. The fraction of light reflected at a coating surface increases as the angle of illumination increases (see Fig. 16.5). At most angles of illumination, surface reflection from a low gloss surface is higher than from a high gloss surface. In considering gloss, people visually compare the amount of light reflected at the specular angle with the amounts reflected at other angles. If the contrast in reflection is high, the gloss is said to be high.

If the surface is rough, the angle of incidence of a beam of light will not be the same as the geometric angle of the surface with the light beam. Light will be reflected at the specular angles between the light beam and the individual rough facets of the surface. If the surface has many such facets oriented at all possible angles, the beam of light will be reflected in all directions. That is, such a surface would be a *diffuse reflector* and would be said to have a very low gloss. It is called a "matte" surface; it is also called a "flat." Flat is unfortunate terminology, since a perfectly smooth (flat) surface, gives high gloss; whereas, a microscopically rough surface gives what we call a flat. At intermediate surface roughness, the gloss is intermediate and is classified as semigloss.

Surface roughness can result from a variety of causes. If one sands the surface of a high gloss coating, the gloss will be reduced due to the small surface scratches giving a rough surface. If the pigment volume concentration in the dry coating is high enough, the surface will be rough due to the protrusion of pigment particles making the surface of the film uneven. If the particles of pigment are agglomerated into clusters, that is, flocculated, the gloss is likely to be lower than the gloss of the same coating in which the pigment dispersion remained stable during application and film formation. If the surface of the film cross-links before the lower layers of the film, wrinkling is likely to occur. That is, when the lower layers do cross-link, shrinkage occurs leading the top layer to pucker up in folds—to *wrinkle*. A wrinkled pattern on the surface leads to low gloss since the surface is no longer smooth.

A closely related phenomenon to specular gloss is what has been called distinctness-of-image (DOI) gloss. A perfect specular reflector is a perfect mirror with an image that exactly mimics the original. If a surface gives perfectly diffuse reflection, no mirror image can be seen. At intermediate stages between perfect specular and perfect diffuse reflection, the image will be more and more blurred as the ratio of specular/diffuse reflection decreases. There is a further effect of larger surface irregularities, which lead to distortion of the image. Commonly, one sees some degree both of blurring and distortion.

"Sheen," as the term is used in the coatings industry, refers to the reflection of light when a surface of a low gloss coating is viewed from an angle near the grazing angle. A high gloss coating will (see Chapter 16) reflect a very high fraction of light whose angle of incidence approaches grazing at that specular angle. The reflection from a low gloss, low sheen coating viewed at the grazing angle is low. A low gloss paint is said to have a high sheen if there is substantial light reflection at the grazing angle. The effect is easy to see but difficult to describe; there is no glare (as when a gloss paint is illuminated and viewed at a grazing angle) but a "soft" relatively high reflection.

Luster is another type of gloss effect. (To illustrate the problem of defining gloss, the dictionary definition of luster is gloss, sheen.) Luster is "directional gloss." For example, some woven fabrics are much *glossier* when viewed looking at the weave parallel to the warp than when viewed parallel to the woof. It is common for such fabrics to appear to have very different gloss effects when looking at the folds in a draped fabric. If the contrasts are great, the fabric is said to have a high luster. Similar effects are sometimes observed in textured coating surfaces.

Still another factor that can be said to be a form of gloss is *haze*. When light enters a hazy film, it is scattered to some degree, that is, diffusely reflected. This diffuse light can be reflected back out of the film and reach the eye of the observer. The result is, thus, similar to having some of the light reflected at nonspecular angles at the surface. The contrast between the fraction of light reflected at the specular angle to that reflected at nonspecular angles is reduced. In pigmented coatings, it may be difficult to distinguish visually between reduced gloss resulting from haze and reduced gloss resulting from poor pigment dispersion. It is always desirable for the formulator to make up a batch of coating with no pigment to verify the clarity of the dried, cross-linked, pigment-free film.

Another phenomenon with a similar effect is called *bloom*. If some liquid component of the coating film is not truly soluble in the resin binder, it can separate from the body of the film in small droplets. These can come to the surface making the surface uneven, hence diffusing light beams striking the surface and reducing gloss. The superficial appearance is much like haze, but bloom can be wiped off the surface with a cloth that is damp with a solvent for the blooming material; commonly, the bloom will eventually reappear.

17.2. VARIABLES IN SPECULAR GLOSS

A way of understanding some of the variables in specular gloss is to use schematic diagrams based on hypothetical measurements by an idealized goniophotometer. A goniophotometer is an instrument in which a beam of light can be directed at a surface with any angle of incidence. When an angle of incidence has been fixed, the instrument can measure the amount of light reflected at any angle of reflection in the same plane as the incident beam of light. The instrument is shown schematically in Figure 17.1. Figure 17.1a shows a cross section through the instrument through the plane of the beam of light.

Figure 17.1b shows a view looking down on the instrument from above. This drawing shows that the only light detected by the photometer is that which is reflected in the plane of the beam of light. Actually, no instrument can measure reflected light at all angles (e.g., one cannot measure all of the light reflected in all planes at one angle). There is no way to build an instrument that would measure reflected light at an angle of 90°. In our idealized instrument, the beam of light is narrow enough so that only a point on the surface is illuminated. Furthermore, the ideal photodetector has been made so that it can detect a beam of reflected light with a beam width approaching zero.

If we measured the light reflected from a perfect mirror in our goniophotometer, all of the light would be reflected at the specular angle and only in the same plane as that of the beam of light. If, on the other hand, the sample is a perfect diffuse reflector, light will be reflected at all angles and in all planes. Line S_o in Figure 17.2 shows photocurrent as a function of the angle of viewing for a perfect diffuse reflector illuminated at 45° as "measured" by an ideal goniophotometer. At first glance, the results do not seem rational but it must be recalled that the photometer can measure all the light reflected at 0° and only a fraction approaching zero of the light reflected in all planes at 90°.

Also shown in Figure 17.2 is the ideal response curve u for a semigloss white coating. Comparison of these two curves illustrates three important factors in considering specular gloss interpretation. First is the *relative brightness A* of the sample compared to the perfect diffuse reflector. Here A is the ratio of light intensities in the perpendicular direction, u_o/S_o. The second factor, *height–of–gloss h*, which is calculated from the *peak height p* as follows:

$$h = \frac{p - u_{45}}{S_{45}} = \frac{p}{S_{45}} - \frac{A}{100}$$

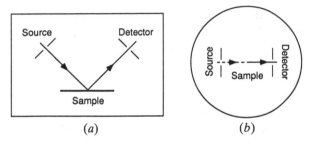

Figure 17.1. Simplified schematic representation of a goniophotometer.

The third factor, the *distinctness of image* δ aspect of gloss is defined as follows, where P is the total area under the sample curve; U is the area of diffuse background reflection from the sample; and S is the area under the standard diffuse reflector curve.

$$\delta = \frac{h}{F} \qquad F = \frac{P - U}{S}$$

If A is large, an observer will consider the gloss of the sample to be lower than that of a sample with a lower A value. If h is small, an observer will consider the gloss of the sample to be lower than that of a sample with a higher h value. It is revealing to compare a white gloss coating with a black gloss coating. A will be much larger and h will be smaller for the white coating than for the black coating, since the white coating scatters the light that enters the film so that this light is diffusely reflected. On the other hand, the black coating absorbs almost all the light entering the film, so that there is little diffuse reflectance from within the film. If the surfaces of the two coatings are equally smooth, so that surface reflectance is equal, the black coating will be viewed as being much glossier than the

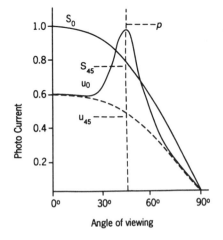

Figure 17.2. Schematic representation of the photocurrent reflected as a function of the angle of viewing. (From Ref. [1], with permission.)

white coating, even though it reflects much less of the total incident light. It is impossible to make a white coating with as high a gloss as that of a high gloss black coating. Obviously, the possible gloss of colored coatings is intermediate between these extremes, the darker the color, the higher the potential gloss. We say in Chapter 16 that gloss affected color, and now we see that color affects gloss. If the value of δ is high, the surface acts as a mirror giving a clear image. As the value of δ decreases there will be more and more blurring of the specular image.

Gloss is also affected by the distance between the observer and the object. If the observer is close enough so that he/she can visually resolve irregularities on the surface, he/she will say that a surface is a rough, high gloss surface. On the other hand, if the observer is far enough away from the same surface so that he/she cannot visually resolve the irregularities, he/she will say that it is a low gloss smooth surface. A coating with a fine wrinkle pattern will have a low gloss when looked at without magnification, but under a microscope it can be seen to be a high gloss wrinkled surface.

17.3. FACTORS AFFECTING GLOSS

The major factor controlling gloss of coatings is their pigmentation. The roughness of the surface varies with the volume of pigment in the dry coating film. These volume considerations are discussed in some detail in Chapter 21. For now it is enough to visualize that, as the pigment content increases, more pigment particles will be near the surface. As the film shrinks with solvent loss, pigment particles will cause irregularities in the surface.

In solvent-borne, high gloss coatings, the pigment content of the top micrometer or so of the dry film contains little, if any, pigment. This means that there is little effect of the first few percent of pigment volume on the gloss of the coating. This unpigmented layer results from the motions within the film as the solvent evaporates. Initially, as the solvent evaporates, convection currents are set up within the film and both the resin solution and the dispersed pigment particles move freely. As solvent evaporates, the viscosity of the film increases and the movement of the pigment particles becomes sluggish. The movement of resin solution continues longer so that the top surface contains relatively little pigment. Since large particles stop moving before small particles and flocculated pigment dispersions have larger particle size than stable dispersions, it is sometimes possible that a coating with a somewhat flocculated pigment dispersion will exhibit higher gloss than a well-stabilized system since there can be a thicker clear layer at the top of the film. Reference [2] discusses effects of pigment particle size and clear layer thickness on specular gloss.

In some solvent-borne coatings (e.g., furniture lacquers) it is desirable to have a low gloss but still a high degree of transparency. This is accomplished by using a small quantity of very fine particle size silicon dioxide (SiO_2) as a pigment. The combination of small particle size and low refractive index difference results in minimal light scattering as long as the concentration is kept low. When the solvent evaporates from such a lacquer, the SiO_2 particles keep moving until the viscosity of the surface of the film becomes high. This results in the presence of a higher than average degree of pigmentation in the top of the film, reducing gloss with relatively low pigment content.

Latex paints generally exhibit lower gloss than solvent-borne coatings at equal levels of pigmentation in the dry film. There probably are several reasons:

1. Latex paints have both the resin particles and pigment particles as dispersed phases. During the drying of the film, there is not the same opportunity for separation as was discussed in a solvent coating. Consequently, there is not a corresponding thin layer of pigment-free resin at the top of the dry film as there commonly is with solvent solution coatings. Latexes with smaller particle size do tend to give somewhat higher gloss in the final films than are obtained with larger particle size latex. Some difference in segregation may occur since particles with larger particle size and higher density will stop moving first as the film dries.

2. Another restriction to the gloss of latex paints occurs because the pigment-free dry films are not completely transparent, as is usually the case with the films from the solvent solution paints. They are hazy and reduce gloss of a pigmented film. This haze may be due to the presence of dispersants and water-soluble polymers that are not completely soluble in the latex polymer films.

3. It is also fairly common to find that surfactants can bloom onto the surface of the latex coating. As we have seen, gloss could be reduced.

4. It is generally more difficult to achieve good leveling with latex paints than with solvent-borne paints. Surface roughness from poor leveling can certainly affect gloss. Brush marks that do not completely level will reduce gloss. Analogous to the effect of wrinkling already mentioned, the effect of brush marks will depend on how close the observer is to the surface. If one is close enough to resolve the brush marks, the paint can look like a high gloss paint with a wavy surface. From a greater distance, the gloss will appear to be low. Factors affected leveling of latex paints, as well as other factors affecting the gloss of latex paints, are discussed in Section 35.3

Since the fraction of light reflected at a coating surface increases as the refractive index increases, the gloss of high gloss coatings tends to increase as the refractive index of a coating increases, since the contrast between reflection at the specular angle and other angles is greater. Refractive index differences between different binders are small, so that while detectable, this effect is generally very small as compared to the effects of surface roughness [2].

Another important aspect of the gloss of coatings is that gloss changes during the life of an applied coating film. The most common initial sign of failure, when coatings are exposed outdoors, is loss of gloss. In some cases, the surface of the film embrittles and then cracks as the film expands and contracts. Generally, this mechanical failure is progressive, and after initial loss of gloss there is film erosion. In other cases, erosion happens first, and loss of gloss becomes evident only after erosion is deep enough to cause protrusion of pigment particles; this type of gloss loss is often observed in lightly pigmented, glossy coatings. Erosion of the binder can proceed to a stage where pigment particles are also freed from binder on the surface and can rub off readily; this phenomenon is called "chalking." When the surface of a glossy coating chalks, the gloss of the surface, of course, drops drastically.

In some cases, the gloss of a low gloss paint surface will increase with use. For example, it is fairly common for the surface of flat wall paints to increase if they are rubbed, for example, near a light switch. This type of increase in gloss is called burnishing; means of minimizing burnishing are discussed in Section 35.2.

17.4. GLOSS MEASUREMENT

No fully satisfactory method for measuring gloss is available, and no satisfactory rating scale for visual observation has been developed. While all people will agree as to which film is the glossiest, if the gloss difference is large, they will frequently disagree as to ranking, if the difference is not large. Even the same observer experiences difficulty in rank ordering a series of panels consistently, if the differences in gloss are small.

Obviously, it is desirable to have some objective means of measuring gloss. Instruments have been developed but one must exercise considerable caution in their use. For research purposes, goniophotometers are probably the most useful. For quality control purposes for high gloss coatings, DOI meters are often the most useful.

For lower gloss coatings some form of specular glossmeter is all that is available. With sufficient care, they can be used for quality control purposes and for following loss of gloss with aging. For specification purposes specular glossmeters are not appropriate, and at this stage one must rely on standard panels. The customer selects three panels of each color, one having the desired gloss and the others representing the upper and lower acceptable limits. These panels become the standards for visual comparisons of production batches.

In Section 17.2 we discussed an idealized goniophotometer. In practice, one has to accept less than this ideal. The beam diameter cannot actually approach zero because there must be sufficient light intensity so that there will be sufficient reflected light at all angles to give a measurable response with available photodetectors. In actual instruments, there will be a light source shining upon a slit aperture at some fixed distance from the sample surface. The reflected light also goes through a slit aperture at a fixed distance to shine on a photodetector. In research instruments, both the angle of illumination and the angle of viewing can be varied independently. These instruments are expensive and relatively difficult to maintain. They are used for research [2] and, in the case of the National Institute of Science and Technology (formerly the National Bureau of Standards), for calibrating standards for less sophisticated glossmeters.

The most widely used instruments are specular glossmeters, also called reflectometers. They are simplified goniophotometers where one measures a response only at the specular angle. Those most commonly used in the coatings industry can make measurements when the angles of incidence (and viewing) are 20°, 60°, and 85°. A schematic diagram is shown in Figure 17.3.

The first step in using the instrument is to calibrate it with two standards: one with high gloss and the other with a lower gloss. If the second standard does not give the *standard* result after the instrument is set with the first standard, something is wrong; most commonly, one or both of the standards is dirty or scratched. Other possible problems include: panel misalignment, deterioration of the light source,

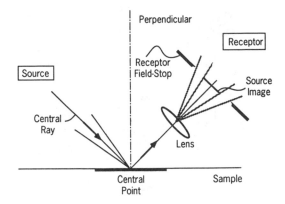

Figure 17.3. Schematic drawing of a glossmeter. (From Ref. [1], with permission.)

or malfunction of the photometer. The standards must have been calibrated for use at the angle that has been selected. Black and white standards are available. As is evident from the earlier discussion, reflection at the specular angle will not be the same from a white and a black standard with equal surface roughness; in the case of the white, there will be diffuse reflectance at the angle of measurement as well as surface reflectance near the angle of reflectance used. Since it is not feasible to have standards of all different colors, white standards are used for light colors and black standards for dark colors. (One should always report which set of standards was used.)

In normal practice, one first measures at 60°. If the reading obtained is over 70, readings should be made at 20° rather than at 60° since the precision will be higher nearer the midpoint of the meter reading. It is common to read low gloss panels at both 60° and 85°. Readings at 85° may have some relationship to sheen. It is essential to report the angle at which the reading is made.

Multiple readings are taken on each panel and analyzed statistically. This will reduce the probability that local surface irregularities or dirt particles are affecting the meter reading. It has been said that slight brush marks do not affect readings in instruments with slits, if the direction of the brush marks is parallel to the plane of incidence. If results are to be compared between two laboratories, it is essential to check the compatibility of their instruments. This is best done by measuring at least three black and three white standards on each of the instruments to be used in comparisons. It has been said that readings are reproducible, on carefully calibrated instruments, to ±3%. This may be true in the high gloss end of the range but it is patently untrue in the low gloss range, if for no other reason than that readings can only be made to one gloss unit. It has been suggested that the error is better expressed in terms of units, such as ±2 gloss units rather than as a percentage [3]. Obviously, this is a high range of error in the lower gloss range, but this is an inherent property of this type of instrument.

There is considerable confusion as to what the numbers mean. Commonly, though undesirably, they are expressed as percentages. It is better to call them gloss units or just meter readings. They are *not*, as many people seem to believe, the percent of light reflected at the surface. They are closer to being the percent

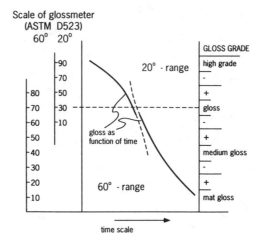

Figure 17.4. Schematic representation for transition of gloss readings according to Freier [4]. (From Ref. [1], with permission.)

of light reflected by the sample at that angle in that instrument compared to the reading that would be obtained if a perfect mirror were measured. As noted earlier, the total reflectance from a black matte surface is much higher at most angles of illumination and viewing than from a high gloss black surface. The point can be emphasized by comparing the meter readings of the same panels at 60° and 20°. As shown in Figure 17.4, the meter readings are lower for the same panel when the setting is 20° compared to 60°.

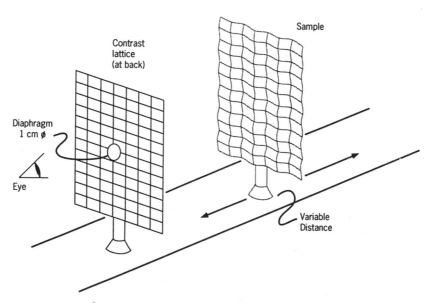

Figure 17.5. Schematic of a DOI meter. (From Ref. [1], with permission.)

Figure 17.4 illustrates a procedure used when it is desired to follow the loss of gloss when a high gloss panel is exposed to exterior weathering. Initial readings are taken at 20°, then at both 20° and 60°, and for the lower gloss range only at 60°. Different scale units in the 20° and 60° systems of ASTM D 523 (*left*) provide for continuous slope at the transition. Glossmeter readings are correlated with designations of gloss grade as observed visually (*right*).

Distinctness-of-image meters (Fig. 17.5) rely on using the sample as a mirror. The reflection of a grid on the surface of the panel is compared visually to a set of photographic standards ranging from a nearly perfect mirror image to a blurred image where the grid cannot be detected. One reports the comparison of the degree of blurring and also qualitatively a statement about distortion.

GENERAL REFERENCES

H. K. Hammond, "Gloss," in *Paint Testing Manual*, 13th ed., American Society for Testing Materials (ASTM), Philadelphia, 1972, pp. 15–21.

U. Zorll, *Prog. Org. Coat.*, **1**, 113 (1972).

REFERENCES

1. U. Zorll, *Prog. Org. Coat.* **1,** 113 (1972).
2. L. A. Simpson, *Prog. Org. Coat.,* **6,** 1 (1978).
3. S. Huey, *Off. Digest*, **36,** 344 (1964).
4. H.-J. Freier, *Farbe Lack*, **73,** 316 (1967).

Pigments

Pigments are insoluble, fine particle size materials used in coatings for one or more of five reasons: to provide color, to hide substrates, to modify the application properties of the coating, to modify the performance properties of films, and/or to reduce costs. In general terms, pigments are divided into four broad classes: white, color, inert, and functional pigments.

Dyes are soluble colored substances, while pigments are insoluble materials used as colloidal dispersions. Dyes are used only in specialized coatings applications, such as in stains for wood furniture (see Section 34.1). Usually it is undesirable to have a soluble colorant in a coating. Some pigments are called "lakes." The original meaning of lake was a dye that had been converted into a pigment by irreversible adsorption onto some insoluble powder. The term lake is now sometimes used when a colored pigment is blended with an inert pigment; when the pigment is essentially 100% pigment, it is sometimes called a "toner."

Particle size is an important factor that affects the color strength, transparency or opacity, exterior durability, solvent resistance, and other properties of pigments. For any given pigment, the manufacturer selects the most appropriate compromise for the particle size and designs his/her process to produce that average particle size consistently. The final step in the manufacture of most pigments is precipitation from water. The conditions of precipitation determine the particle size of the pigment. The precipitated pigment is filtered and the filter cake is dried. In the process of drying, it is common for the pigment particles to become more or less cemented together in aggregates. The coating manufacturer generally receives pigment from the pigment manufacturer as a dry powder of aggregates and must disperse these aggregates in a way that breaks up the aggregates to their original particle size and makes a stable dispersion of the pigment. Pigment dispersion is the subject of Chapter 20.

18.1. WHITE PIGMENTS

A very large fraction of all coatings contain a white pigment. White pigments are used not only in white coatings but also in a substantial fraction of other pigmented coatings to give lighter colors than would be obtained using color pigments alone.

Furthermore, many color pigments give transparent films, and the white pigment provides a major part of the hiding power of the coating. As discussed in Section 16.2.3, the ideal white pigment would absorb no visible light and would have a high scattering coefficient. Since a major factor controlling scattering efficiency is the difference in refractive index between the pigment and the binder, refractive index is one of the critical properties of a white pigment.

18.1.1. Titanium Dioxide

By far the most important white pigment used in coatings is TiO_2. Two different crystal types are used: *rutile* and *anatase*. Rutile is used in much larger volume, primarily because it gives about one-third greater hiding power than anatase; the refractive index of rutile is 2.76 compared to 2.55 for anatase. On the other hand (see Figure 18.1), rutile TiO_2 absorbs some violet light, whereas anatase TiO_2 absorbs almost no light. While a coating pigmented with only rutile TiO_2 looks white, when compared next to a coating pigmented with anatase TiO_2, the rutile pigmented coating has a yellowish hue.

The color of rutile TiO_2 coatings can be adjusted by tinting. If one adds a small amount of a dispersion of a violet pigment, such as carbazole violet, the white coating obtained will be less yellow. The violet pigment strongly absorbs wavelengths of light other than violet and hence reduces the difference in reflection of the violet wavelengths compared to all the others. This makes the coating a very light gray but, unless there is a higher reflectance white present for comparison, it is seen as white, not gray. At the same time, there is the added advantage that the hiding power is further increased due to the light absorption. Phthalocyanine blue is also used as a tinting pigment. Phthalocyanine blue does not give as *white a white* as carbazole violet but is less expensive and gives a slightly greater im-

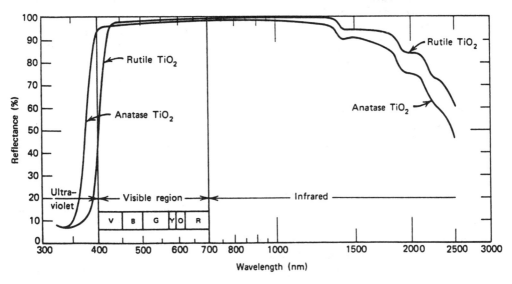

Figure 18.1. Reflectance spectra of rutile and anatase TiO_2 coatings. (*Source*: W. A. Kampfer in Ref. [1], 1st ed., Vol. I, p. 2, with permission.)

provement in hiding. The most common tinting pigment used in whites is a carbon black (see Section 18.2.4). Carbon black is less expensive than either the violet or blue pigments and gives a greater increment to hiding. Since it absorbs all wavelengths of light approximately equally, it does not reduce the yellowness of the white but does provide a low cost method of increasing hiding.

Anatase TiO_2 also absorbs less near-UV radiation than rutile TiO_2. In white pigmented UV cure coatings, this lesser absorption is a distinct advantage since it gives less interference with absorption of the UV radiation by many photoinitiators. Absorption of radiation by the photoinitiator is the essential first step for UV curing (see Chapter 32).

Another important difference between rutile and anatase TiO_2 is their photoreactivity. Anatase TiO_2 pigmented coating films fail quicker on outdoor exposure by *chalking* than rutile TiO_2 pigmented films. Chalking is the formation of loose pigment particles on the exposed coating surface resulting from the erosion of the binder as a result of photodegradation (see Section 25.7). Chalking is usually undesirable, although "self-cleaning" white house paints are deliberately formulated with anatase TiO_2 so they will chalk readily. When the surface of the film erodes, dirt accumulations are also eroded so that the paint surface stays whiter.

In most exterior applications, chalking must be minimized, and rutile TiO_2 is used. However, while rutile is less photoreactive than anatase, it is still sufficiently reactive to reduce exterior durability. This problem has been minimized by treating the surface of the TiO_2 particles during pigment manufacture. The most common surface treatments are silica (SiO_2) and/or alumina (Al_2O_3); zirconium dioxide and antimony trioxide, among others, are also used together with silica and/or alumina. A typical surface treatment for exterior grades of rutile is about 6 wt% silica and 4% alumina. An ideal surface treatment would be a continuous layer on all of the TiO_2 particles preventing any contact between the TiO_2 and the binder. Perfection has not been achieved but substantial progress has been made on this process engineering challenge, and the best chalk-resistant grades of rutile TiO_2 show virtually no acceleration of chalking. The best treated anatase pigments accelerate chalking more than treated rutiles but much less than the untreated anatase pigment.

Since (see Fig. 18.1) TiO_2, especially rutile, strongly absorbs UV radiation, it competes with the binder in a coating for UV absorption. If chalk-resistant grades are used, the reduced UV absorption by the binder can give enhanced exterior durability.

The surface treatment can affect other properties of the pigments besides chalking. In coatings containing Class I MF resins, sulfonic acids or amine salts of sulfonic acids are used as catalysts. As discussed in Section 6.3.1.1, alumina surface treated TiO_2 can neutralize part, or perhaps even all, of the catalyst over time leading to reduction in cure rates.

Specially treated TiO_2 dry pigments have been developed for use in water-borne coatings. These are surface treated so as to permit production of more stable dispersions in water.

Titanium dioxide dispersions in water, called slurries, are increasingly being used in the manufacture of latex paints. The slurries are manufactured by the TiO_2 producer and shipped in tank cars to replace dry pigment shipped in multiwall paper sacks. The price on a pigment basis is about the same, but there are substantial savings in material handling.

A wide variety of TiO_2 pigments is offered by each TiO_2 manufacturer. Some are designed for maximum exterior durability, some are designed to provide maximum gloss, some are more easily dispersed in water systems, whereas others are more easily dispersed in solvent systems. The TiO_2 content of the various types of pigments varies widely, from as high as 98% to as low as 75%. Generally, the hiding varies with the TiO_2 content, so the amount of TiO_2 needed for some level of hiding will vary with pigment type. It is important for formulators to be aware of the differences among the types of pigments. Numerous paint performance problems and much economic waste have resulted from the use of the wrong type of TiO_2 and from indiscriminate substitution of one type for another.

18.1.2. Other White Pigments

Until the late 1930s, white lead, basic lead carbonate, $2PbCO_3 \cdot Pb(OH)_2$, was widely used as a white pigment. Its relatively low refractive index, 1.94, means that its hiding power is low, only about one-eighth that of rutile TiO_2. When TiO_2 became available commercially (anatase in the 1920s and rutile in the 1930s), the use of white lead dropped rapidly. White lead is somewhat soluble in water and hence presents a serious toxic hazard. Many children suffer from lead poisoning attributed to eating paint containing white lead that peels from the walls and woodwork in older residential buildings. As a result, the lead content of any paint sold to retail consumers through interstate commerce in the United States is limited to 0.06% of the dry weight of the paint whether or not the lead is there as a water-soluble compound. Actually, because of its low refractive index and corresponding high cost for hiding, the use of white lead virtually stopped decades ago, and by far the largest source of lead in the U.S. environment has been from the emissions of vehicles burning leaded gasoline. The regulation has primarily served to eliminate the use of lead driers and other lead pigments such as lead chromate, a questionable accomplishment.

Zinc oxide (ZnO) was formerly used as a white pigment, but because of its low refractive index, 2.02, it cannot compete for hiding power with TiO_2. As discussed in Section 18.4, ZnO is used in exterior house paints as a fungicide and in some can linings as a sulfide scavenger. Zinc oxide should not be used in primers since it is somewhat water soluble and can, therefore, induce blistering by osmotic pressure when water vapor penetrates through the top coats into the primer coat. Other white pigments, important at one time but now largely obsolete, are zinc sulfide (ZnS) with a refractive index = 2.37, and lithopone ($ZnS/BaSO_4$).

Small air bubbles scatter light and are another important white pigment. In some cases, cost reductions are possible by taking advantage of the scattering by air bubbles in coating films. The refractive index of air, 1, provides a difference of refractive index of about 0.5 in a typical binder (refractive index, ~ 1.5). While this difference is small compared with the difference of 1.25 with rutile TiO_2, it still provides significant light scattering and hence hiding. The most common way of incorporating air bubbles in a coating film is by use of such high levels of pigmentation that, after the solvent evaporates, there is not enough resin to adsorb on all the pigment surfaces and to fill all the interstices between the pigment particles. Air voids result, which increase the hiding. Such coatings are said to be formulated at pigment volume levels above the *Critical Pigment Volume Concen-*

tration (CPVC). This concentration is defined as the volume fraction of pigment in the dry coating film for which there is just enough binder to provide a complete adsorbed layer on the surfaces of the pigment particles and to fill all the interstices between the randomly close-packed pigment particles in the film (see Chapter 21). Coatings with PVC > CPVC have air voids and, hence, increased hiding. If, as is commonly the case, interfaces between TiO_2 and the air are developed, the very large refractive index difference of 1.75 is especially effective in increasing hiding. There are, of course, disadvantages to this approach since the voids make the film porous and markedly reduce the protective properties of the paint films. In some cases, such as ceiling paints, advantage is taken of the added hiding since requirements for other properties are not severe for ceiling paints.

Particles containing air bubbles can take advantage of the hiding from the air–resin interface without imparting porosity to the films. For example, high T_g latexes containing water trapped within the particles are available and are finding increasing usage in latex paints as a partial replacement for TiO_2 [2]. When the paint is applied and the water evaporates, the conventional latex particles of the binder coalesce to form the film, but the high T_g latex particles do not coalesce. Water diffuses out of the particles leaving air bubbles within the particles that add to the hiding. The *hiding* latex particles are prepared by sequential emulsion polymerization [3]. Initially, a low T_g latex copolymer of acrylic acid and acrylic esters is prepared; these particles are highly swollen with water, especially at alkaline pHs. These core particles are encapsulated in shells of high T_g cross-linked polymer such as a styrene–divinylbenzene copolymer. The core is swollen with water; when the water diffuses out of the particles, the rigid outer shell prevents collapse so that air voids are left in the core.

Another air containing pigment has been developed in which there are TiO_2 particles inside air bubbles in a resin particle matrix [4]. This gives a very high refractive index difference and, therefore, efficient hiding. The particle size is large, so use of such pigments is limited to fairly thick films of low gloss paint. The use of such pigments is discussed in Chapter 35.

18.2. COLOR PIGMENTS

A very wide variety of color pigments is used in coatings. Reference [1] provides a detailed discussion of the chemistry, properties, economics, and uses for most of those in commercial use; Ref. [5] gives a more condensed coverage of organic pigments. In this book, brief descriptions of some of the most important color pigments are provided. First, it is appropriate to summarize the considerations involved in selecting color pigments for a particular coating application.

Color. Obviously, the first criterion is the color of the pigment. Pigment suppliers provide technical bulletins that contain color chips showing the color that can be obtained with each pigment. Generally, there will be two or three color chips for each pigment. The *mass tone* is displayed by a color chip in which the pigment is used as the sole pigment in the coating. Then, there will be one or two additional color chips showing the color obtained when the pigment is used together with TiO_2 in different ratios. These are called the *tint colors*. In some cases, color chips showing the effect of an aluminum pigment are also provided.

Color Strength. Some colorants are *strong* and others are *weak* corresponding to the absorption coefficients of the pigments. It may well be more economical to use an expensive (i.e., high cost per unit volume), strong pigment than a weak, low cost per unit volume pigment.

Opacity or Transparency. Depending on the end use, it may be desirable to use a pigment that increases hiding by both scattering and absorbing radiation or it may be important to select pigments that scatter little, if any, light in the coating film so that a transparent color can be obtained.

Ease of Dispersion. Some pigments are more easily dispersed than others. If, as is seldom the case, everything else were equal, one would select the most easily dispersed pigment. Many pigments are given proprietary surface treatments by their manufacturers to enhance their ease of dispersion. Pigment dispersion is a complex process involving several factors; it is discussed in detail in Chapter 20.

Exterior Durability. Some pigments are much more sensitive to photodegradation than other pigments. Pigment manufacturers provide some preliminary data that can be useful in selecting pigments that may be appropriate for use outdoors. However, there can be significant variation in exterior durability depending on the combination of pigment and resin. Therefore, for critical applications, the durability of specific formulations must be determined (see Chapter 25 for further discussion of exterior durability and testing for exterior durability).

Heat Resistance. In some coatings, resistance to change of color on heating is not an important consideration, but with coatings that are baked or exposed to high service temperatures it can be critical. A few pigments (e.g., yellow iron oxides) undergo chemical changes with heat; a more common problem is that some pigments will slowly sublime at elevated temperatures.

Chemical Resistance. When in use many coatings are exposed to chemicals, most often to acids and bases, and the pigments must resist color change under those conditions. For example, automobiles are exposed to acid rain, and home laundry machines are exposed to detergents, which are alkaline.

Water Solubility. In most applications, pigments with any significant water solubility are best avoided because they leach out of the coating film, resulting in loss of properties. The presence of water-soluble substances in a primer can lead to blistering of the films when water permeates through the film. Furthermore, especially in latex paints, care must be taken to assure that the amount, if any, of water-soluble impurities is not such as to affect stability of the dispersion of the latex particles. The stability of colloidal aqueous dispersions can be particularly affected by the presence of very low concentrations of polyvalent salts. On the other hand, pigments that help protect against corrosion by passivation (see Section 18.4) must be slightly soluble in water to be effective. This important subject is discussed in Chapter 27.

Solvent Solubility. Partial solubility of pigments in some solvents can be a problem. For example, if one is making a red coating for a bicycle that will have a white stripe coated on the red coating, it is critical to use a red pigment that will not dissolve in solvents used in the white striping coating. If the red pigment is somewhat soluble, the stripes may turn out pink rather than white. Such pigments are said to "bleed."

Moisture Content. Most "dry" pigments adsorb small-to-moderate amounts of water on the particle surfaces. The adsorbed water can result in serious problems with water-reactive binders such as polyisocyanates.

Toxic and Ecological Hazards. Some pigments are toxic or carcinogenic; such pigments must be handled with appropriate care. Filter face masks should always be worn when handling dry powder pigments; it is dangerous to inhale fine-particle dusts even if they are chemically inert. In most cases, there is little or no hazard once the pigment is incorporated into a coating. However, there are some cases in which toxicity even in a paint film can be important. For example, as mentioned earlier, white lead is no longer permitted in architectural coatings in the United States. Lead pigments are also prohibited in coatings for children's toys and furniture. Some chromate pigments are known to be human carcinogens. In some European countries their use in coatings is prohibited; currently in the United States there is no such ban. Regulations can be expected to become increasingly restrictive. Another factor expected to affect pigment choice is the increasing restriction on disposal of materials containing *heavy metals* in land fills.

It is often suggested that every chemical that is known to be toxic or is even suspected to be carcinogenic should be banned, but life is not that simple. Risk cannot be eliminated, and society is often faced with choosing among different kinds of risks. For example, does the safety benefit of yellow highway striping outweigh the risk from using lead chromate pigments?

Cost. While cost is clearly a critical consideration, as noted earlier, one cannot tell which is the least expensive pigment by just looking at the price per unit weight. The critical question is how much will the final coating cost, and this will be affected by color strength, as well as unit cost. Since in most cases, coatings are sold on a price per unit volume basis, density can have a major effect on costs.

18.2.1. Yellow and Orange Pigments

Inorganic Yellows and Oranges. Iron oxide yellows, FeO(OH), are low chroma brownish yellow pigments. They give opaque films with good hiding and high exterior durability; chemical and solvent resistance are excellent. The pigments are generally easily dispersed and are comparatively inexpensive. When heated above 150°C, they gradually change color to a low chroma red since they dehydrate to form iron oxide red (Fe_2O_3). Most iron oxide yellows now in use are made synthetically but some natural ore *ochre* pigments are still used. In some cases, particularly with the natural pigments, the presence of soluble iron and other metal salts can affect the stability of coatings that cure by a free radical mechanism. Extremely fine particle size iron oxide pigments are also available; they are used when transparency is required. This effect of particle size on opacity is an excellent example of the relationship between light scattering and particle size (see Section 16.2.3).

Chrome yellow pigments are bright, high chroma yellows. Medium chrome yellow pigments are predominantly lead chromate ($PbCrO_4$). Greenish yellow colorants, called primrose yellow and lemon yellow, are cocrystals of lead chromate with lead sulfate. Redder yellows, that is, chrome oranges, are cocrystals of lead chromate with PbO. Still redder oranges, molybdate oranges, are cocrystals of lead chromate with lead molybdate ($PbMoO_4$) and lead sulfate. On the basis of their relatively high color strength, chrome yellows are relatively low cost, although higher than iron oxide, at least partly because their density is higher. Chrome yellows discolor on exterior exposure to form lower chroma yellows but the exterior durability is adequate for many outdoor uses. They are bleed and heat resistant. Due to the lead content, their use is not permitted in consumer paints in the United States. Their use in industrial applications is declining because of concern about the use of lead compounds. The major current use is in traffic striping paint for the familiar yellow lines on highways. Use of chromates for this purpose has been banned in many European countries but continues in the United States. While fully satisfactory alternatives to yellow chromates remains a research challenge, use of lead and chromium free yellow pigments is increasing.

Titanium yellows are derived by introducing other metal ions into the lattice of anatase TiO_2 crystals followed by calcining to convert to the rutile crystal structure. Greenish yellow shades are based on introducing antimony and nickel; reddish yellow shade grades contain antimony and chromium. Opaque films result from their use in coatings. The resistance of the color to exterior exposure, chemicals, heat, and solvents is excellent. However, only relatively weak yellow colors can be produced so cost is high, and the range of colors that can be made is restricted.

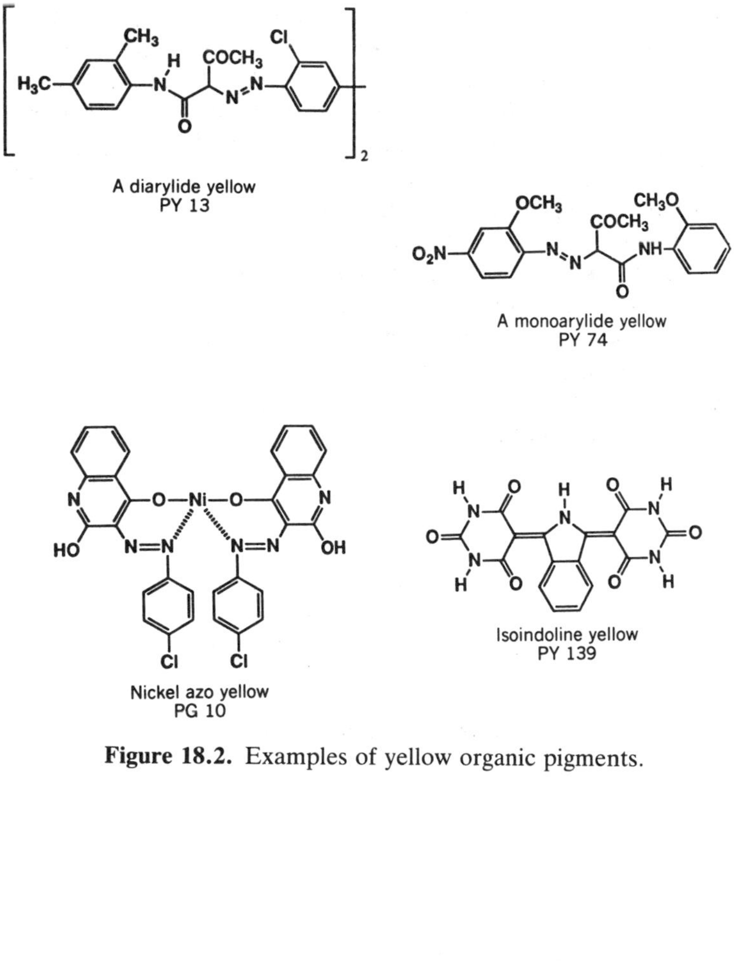

A diarylide yellow
PY 13

A monoarylide yellow
PY 74

Nickel azo yellow
PG 10

Isoindoline yellow
PY 139

Figure 18.2. Examples of yellow organic pigments.

Organic Yellows and Oranges. Chemical structures of representative pigments are shown in Figure 18.2. The code designations in this and subsequent figures are from the standard classification system developed by the Society of Dyers and Colourists in the United Kingdom and the Association of Textile Chemists and Colorists in the United States. The P in the code stands for pigment, the next letter designates the hue, and the number is chronologically assigned.

Diarylide yellows are bisazo pigments derived from 3,3'-dichlorbenzidene, for example, PY 13 shown in Figure 18.2. They have very high color strengths and high chroma. The hue and photostability are controlled by the number, positions, and structure of substituent groups on the aromatic ring of the anilide portion of the molecule. Even the most photochemically stable diarylide yellow pigments fade on exterior exposure, especially when used in tints. On the other hand, their solvent, heat, and chemical resistances are excellent. Due to the high strength and comparatively low density, their cost is relatively low. They are widely used in interior coatings when the bright yellow color is needed in tints and in such applications as coatings for pencils. Grades with excellent transparency can be made. Diarylide yellows are the major yellows used in printing inks, especially in the yellow inks used in four color process printing.

Monoarylide (monoazo) yellow pigments, such as PY 74, also have high chromas. Color strengths are high but lower than those of diarylide yellows. They exhibit relatively poor bleed resistance and they sublime when exposed to high temperatures. However, their lightfastness is significantly better than the diarylide yellows, although still inferior to inorganic yellow pigments. Some grades have sufficient light resistance for use in outdoor paints. Some are beginning to replace chrome yellows in traffic paints, although they cost more and do not perform quite as well.

Nickel azo yellow is a relatively weak, very greenish-yellow—in fact, as seen from the designation PG 10 in Figure 18.2, it can be classified as a green. Nickel azo yellow exhibits excellent exterior durability and heat resistance. It gives transparent films and is used predominantly in automotive metallic coatings. There have been some situations where bleeding into stripes has been reported so that specific applications must be checked for bleeding.

Vat yellow pigments, as exemplified by isoindoline yellow PY 139, give transparency with excellent exterior durability and heat, chemical, and solvent resistance. They are expensive and used only when their outstanding properties are required (e.g., in automotive metallic coatings).

Benzimidazolone orange pigments offer excellent light fastness and resistance to heat and solvent. They have become widely used as replacements for molybdate orange.

Benzimidazolone orange
PO 36

18.2.2. Red Pigments

Inorganic Reds. Iron oxide red (Fe_2O_3) gives the familiar barn red color. It is a low chroma red with excellent properties and low cost. In contrast to iron oxide yellows, iron oxide reds are thermally stable. When particle size is optimal for scattering, they provide a high degree of hiding. There are also very fine particle grades available that provide transparent films. Excellent exterior durability makes the transparent grades suitable for use with aluminum in metallic auto top coats.

Organic Reds. Toluidine red pigment, PR 3, is a moderate cost, bright red azo pigment with high color strength, good exterior durability in deep colors, good chemical resistance, and adequate heat resistance to permit use in baking enamels. As shown in Figure 18.3, it is an azo derivative of β-naphthol. Toluidine red is soluble in some solvents and gives coatings that are very likely to bleed when another coating is put on top.

Bleed resistance of azo pigments can be achieved by the presence of carboxylic acid salts. For example, 2-hydroxy-3-naphthoic acid (BON) can be coupled with diazo compounds. Permanent Red 2B is an example of such bleed resistant high chroma red azo pigments. It is available as the calcium, barium, or manganese salt. The somewhat higher cost manganese salt shows better exterior durability than the calcium or barium salts. A variety of related azo pigments, which exhibit somewhat different shades, are also available. This general class is the largest volume of organic red pigments used in coatings and inks. However, many of them are sensitive to bases, hence they may not be suitable for some latex paints.

Naphthol reds are a large family of azo pigments with various substituents (Cl, OCH_3, NO_2, etc.) on the rings of the generic structure shown in Figure 18.3. They

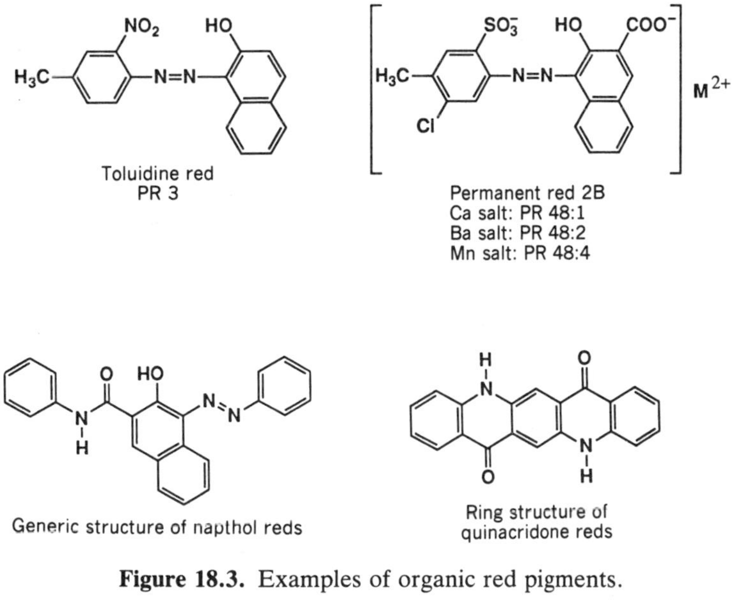

Figure 18.3. Examples of organic red pigments.

are more resistant to bases, soap, and acid than the permanent reds, and also have fairly good exterior durability and solvent resistance.

Quinacridone pigments are nonbleeding, heat and chemical resistant, and give outstanding exterior durability even in light shades. However, their cost is high. Depending on substitution and crystal form, a variety of orange, maroon, scarlet, magenta, and violet colors are available. They find considerable use in metallic automotive top coats.

18.2.3. Blue and Green Pigments

Inorganic Blues and Greens. Iron blue pigment, ferric ammonium ferrocyanide, $FeNH_4Fe(CN)_6$, is an intense reddish shade of blue with fairly good properties. It was the first synthetic pigment, and its introduction in the eighteenth century led to a fashion for bright blue and green colors. Examples have been reproduced in the wall paints in George Washington's home at Mount Vernon. Fairly substantial volumes of iron blue are still used, although since the 1930s, it has been increasingly supplanted by phthalocyanine blues, which have greater color strength.

Cocrystals of various ratios of iron blue and chrome yellow are called chrome greens. The popularity of these green pigments has decreased because of the high color strength of phthalocyanine green. Also, they can no longer be used in architectural coatings in the United States because of the high lead content.

Organic Blues and Greens. The principal blue and green pigments in current use are copper phthalocyanine (CPC) pigments, commonly called phthaloblue and phthalogreen. They exhibit outstanding exterior durability, bleed and chemical resistance, are heat stable, and have high tinting strengths. Although their cost per pound is fairly high, the high tinting strength and quite low density mean that their use cost is moderate. While the structures are complex (Fig. 18.4), CPCs can be synthesized from inexpensive raw materials: phthalic anhydride, urea, and a copper salt.

Phthaloblue is available in three crystal forms: alpha, beta, and the seldom used gamma. The beta form has a relatively green shade of blue and is stable. The alpha

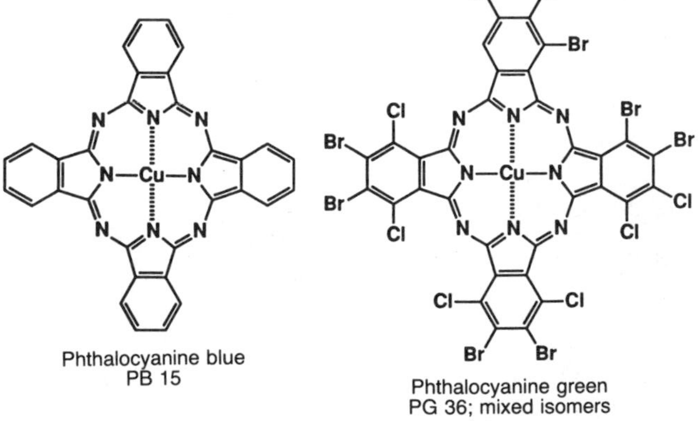

Phthalocyanine blue
PB 15

Phthalocyanine green
PG 36; mixed isomers

Figure 18.4. Representative phthalocyanine pigments.

form is redder but not as stable; in some cases, when some grades of the alpha form are used, there can be serious problems of change of color and strength during storage of coatings or during baking. More stable alpha form pigments are now available; these incorporate, as an additive, long-chain amide derivatives of sulfonated phthalocyanine blue. The additive not only stabilizes the crystal form but also minimizes problems with flocculation of dispersions. Some grades of phthalocyanine blue are slightly chlorinated; these have greener blue shades.

Phthalogreens are made by halogenating copper phthalocyanine to produce mixtures of isomers in which many of the 16 hydrogen atoms of CPC have been replaced with chlorine or with mixtures of chlorine and bromine (Fig. 18.4). Phthalogreens with 13–15 chlorine atoms, but no bromine, have blue-green shades with the lower chlorine content pigments being the bluest greens. Partial replacement of chlorine with bromine shifts the color toward a yellow-green shade. The yellowest shades have a high ratio of bromine/chlorine, an example being PG 36, which has, on average, about 9 bromines and 3 chlorines.

18.2.4. Black Pigments

By far the majority of black pigments used in coatings are carbon blacks. They absorb UV radiation as well as light, and with most binders black colors are the most stable on exterior exposure. Carbon blacks are made by a variety of processes of partial combustion and/or cracking of petroleum products or natural gas. Depending on the process, the particle size and, therefore, the degree of jetness (intensity of blackness) varies. High color channel blacks have the smallest particle sizes, with diameters of 5 to 15 nm, and have the greatest jetness. They are used as the sole pigment when intense, gloss black coatings are desired. Furnace blacks are lower in cost, give less jet blacks, and have larger particle sizes. Various grades are available with average diameters of 50 to 200 nm. Lampblacks have still larger particle size, on the order of 0.5 μm, and have much lower color strengths than other carbon blacks. They are primarily used in making gray coatings. They are much preferable to high color blacks for this purpose; if a small excess of high color black is added by mistake, it may be necessary to increase the batch size 50% or more to make up for the mistake. The shade of gray obtained changes much more slowly with the addition of lampblack making color matching easier.

While all carbon blacks are predominantly elemental carbon arranged in polynuclear six-membered rings, the chemical structure of the surfaces varies depending on the raw material and process used in their manufacture. The surfaces are generally quite polar and in some cases are fairly highly acidic. There are also variations in the porosity of the particles depending on the process.

Black coatings can present difficulties in formulating coatings, especially when channel blacks are used. Due to the small particle size, smaller than any other pigment, the ratio of surface area/volume of the pigment particles is very high. The large surface area leads to adsorption of very high ratios of resin on the pigment particles—they commonly adsorb many times their volume of polymer, substantially increasing their volume and hence giving high viscosity at relatively low pigment loading levels. Due to the polarity of the surface and the large surface area, they can selectively adsorb polar additives such as catalysts from a coating formulation. For example, in drying oil and oxidizing alkyd formulations, the metal salt driers slowly adsorb on carbon black during storage so that the coating dries more and more slowly.

18.2.5. Metallic and Other Flake Pigments

The most important metallic pigments are aluminum flake pigments. They are produced by milling finely divided aluminum metal suspended in mineral spirits in steel ball mills to make thin flakes. A wide variety of particle size pigments is available. There are two major classes of aluminum pigments used in coatings: "leafing" and "nonleafing" pigments.

Leafing aluminum pigments have been surface treated (e.g., with stearic acid) so that they have a very low surface tension. When a coating containing leafing aluminum pigment is applied, as a result of their low surface tension, flakes orient at the surface. This gives a bright metallic appearance and, furthermore, acts as a barrier to permeation of oxygen and water vapor through the film. As a result of this barrier development, leafing aluminum pigments are used in corrosion protection top coats for steel structures.

Nonleafing aluminum pigments have higher surface tensions and do not come to the surface of the coating. As discussed in Section 16.4, the formulations using nonleafing aluminum are designed to maximize the fraction of flakes oriented within the film parallel to the surface of the film. Films made with transparent color pigments and nonleafing aluminum change shade with the angle of viewing. They are most widely used in metallic top coats for automobiles. Nonleafing aluminum pigments have proprietary surface treatments designed to minimize sensitivity to acid in the environment. The aluminum pigments used in water-reducible coatings must be treated to minimize the chemical reaction of water with the aluminum.

Bronze, nickel, and stainless steel pigments with platelets similar to those of aluminum pigments are also available, but are used in lower volume. Bronze alloy flakes have *gold* colors; depending on the alloy composition shades from greenish-yellow to reddish-golds are available. Generally, they are surface treated so that they will leaf. Due to the presence of copper, bronze alloy pigments will change color on outdoor exposure to a blotchy, muddy green appearance. Gold and bronze colored automotive metallic coatings are pigmented with nonleafing aluminum along with transparent yellow and red pigments.

While pearlescent pigments are not metal pigments, it seems appropriate to describe them here. Pearlescent pigments are platelets of mica with thin surface treatment layers, which serve to give interference reflection of light striking the pigment surfaces. As a result, at some spots on the surface of the platelets, some wavelengths of light are strongly reflected and others are transmitted. At other spots, where the film thickness of the treatment layer is different, different wavelengths will be reflected and transmitted [6]. The result is a mother-of-pearl effect. These pigments are fairly widely used in automotive coatings to give an effect related to that obtained with aluminum flake but with additional color effects.

18.3. INERT PIGMENTS

Inert pigments absorb little, if any, light and have refractive indexes close enough to those of binders that they give very inefficient light scattering when used as pigments. Several synonymous terms are used: inert pigments, inerts, fillers, and extenders. Their function is primarily to adjust the rheological properties of fluid coatings and the gloss and mechanical properties of coating films. Commonly, but not always, they are inexpensive and their use helps reduce coatings cost. The principal function of inert pigments is often to occupy volume in the coating film. As discussed at length in Chapter 21, many film properties are importantly controlled by the volume of pigment in the films. References [1] and [7] give detailed discussions of the multitude of inerts available.

Calcium carbonate ($CaCO_3$) pigments are widely used. The lowest cost grades are simply ground limestone or the mixed calcium magnesium carbonate ore, dolomite. Synthetic calcium carbonate pigments are more costly, but they are whiter. Pigments are available with a variety of average particle sizes. In some applications, the reactivity of calcium carbonate with acids makes carbonate pigments undesirable, especially in exterior paints where degradation of film properties may be accelerated with acid rain. Calcium carbonate pigments should not generally be used in exterior latex paints. Water and carbon dioxide can easily permeate through the film of a latex paint, some calcium carbonate dissolves forming calcium bicarbonate, which is water soluble and can permeate back out of the film. On the film surface, the water evaporates and the reaction reverses leaving a "frosting" of insoluble calcium carbonate deposited on the film. Frosting is especially noticeable on dark color paints.

A very wide range of clays (aluminum silicates) are used as inert pigments. They are available in various particle size ranges. Cost is frequently related to whiteness. As seen in Section 18.4, bentonite and attapulgite clays are used to modify the viscosities of coatings. Mica (aluminum potassium silicate) has a platelet structure and can be useful in reducing the permeability of films to oxygen and water vapor when it orients parallel to the surface of the film.

Magnesium silicate minerals are also used as inert pigments. Talcs of various crystal structures affect the film strength of coatings differently. Some talcs are platey and will reduce vapor permeability, others are fibrous and may be particularly effective in film reinforcing. Asbestos is a very fibrous magnesium silicate, which is no longer used because it causes lung cancer when fibers are inhaled.

Silicon dioxide pigments are another important class of inerts. Ground natural silicon dioxide, sand, is used in a variety of particle sizes. A fascinating example of SiO_2 pigments is diatomaceous earth, also called fuller's earth. This material is composed of fossil skeletons of diatoms as shown in Figure 18.5. The large ratio of surface area/volume affects the properties (see Chapter 21). Very fine particle size synthetic silicon dioxide pigments are used to reduce the gloss of clear coatings and to impart shear thinning flow properties to coatings as briefly discussed in the following section.

Barytes, barium sulfate, has been widely used as an inert pigment especially in automotive primer formulations. It is said to provide a *harder* primer. The density of barytes is high (4.5) roughly twice that of most other inerts. In at least some cases, it is probable that the use of barytes was based on weight comparisons with other inerts rather than volume comparisons as should be done. While barytes is not expensive on a weight basis, it is more expensive than most other fillers on a volume basis.

While most inert pigments are inorganic minerals, organic materials can also be

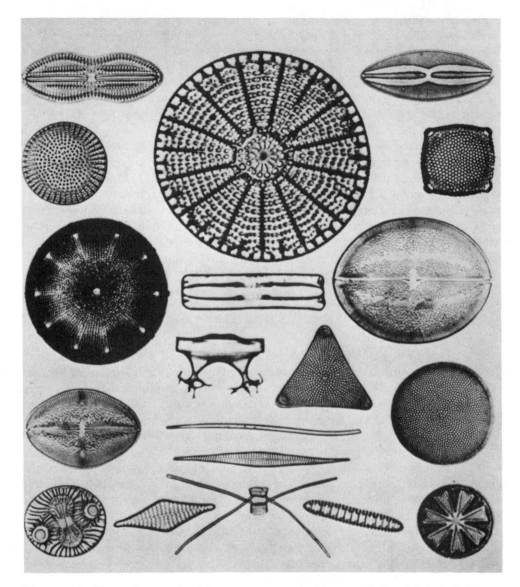

Figure 18.5. Photomicrograph of diatomaceous earth. (*Source*: H. Kranich in Ref. [1], 1st ed., Vol. I. p. 147, with permission.)

used as inert pigments. For example, powdered polypropylene is insoluble and acts as an inert pigment. High T_g latexes such a polystyrene latex can be used as an inert pigment in latex paints. Synthetic fibers, such as aramid fibers, have been shown to be effective in increasing the mechanical strength of coating films.

18.4. FUNCTIONAL PIGMENTS

Functional pigments are used to modify the application characteristics, appearance, or film properties of coatings. An important example is corrosion-inhibiting pigments. Complex zinc chromate pigments, red lead, zinc phosphate, and many others are used in primers to inhibit corrosion of steel by passivation of anodic areas. In contrast to other pigments, they must be somewhat soluble in water in order to function. Zinc metal pigment provides corrosion protection by galvanic action and is used in a class of primers known as zinc-rich primers. Detailed discussion of these pigments is deferred until Chapters 27 and 36. References [1] and [8] provide considerable information.

"Flatting pigments," are another important type of functional pigments used to reduce gloss. While low gloss is often attained by formulating coatings with a high volume ratio of pigment in the dry film, sometimes this is not a desirable approach. For example, in lacquers for wood furniture it is essential to have the transparency of the dry film as complete as possible so that the beauty of the wood grain is not concealed. Fine particle size silicon dioxide is widely used in such coatings as a flatting pigment. As discussed in somewhat more detail in Section 17.3, during solvent evaporation from the coating film, convection currents carry the fine particles to the surface so that the pigmentation at the surface of the dry film is high enough to reduce gloss. The combination of low total pigmentation, small particle size, and small refractive index difference means that there is little reduction in transparency. Powdered polypropylene has been used in a somewhat analogous fashion.

Some pigments are used as biocides. For example, zinc oxide is widely used as a fungicide. Barnacles, algae, and other organisms can grow on the outer hulls of ships, in order to minimize this growth "antifouling paints" are used. Cuprous oxide and organotin pigments have been extensively used in antifouling paints, although their use is now constrained by environmental concerns (see Section 36.2 for a discussion of antifouling paints and the trend towards nontoxicant approaches to solving the problem of barnacle growths).

Zinc oxide (ZnO) is used in linings for cans used for packing vegetables, such as corn, that evolve some hydrogen sulfide during cooking. The ZnO reacts with the H_2S to form white zinc sulfide, preventing black stains resulting from the formation of tin sulfide by reaction between the H_2S and the tin oxide layer on the tin surface of the tinplated cans.

Antimony oxide (Sb_2O_3) is a white pigment whose refractive index of 2.18 is too low and whose cost is too high to warrant its use simply for hiding. Its major use is in fire retardant coatings. When a combination of Sb_2O_3 and a chlorinated or brominated polymer is heated to high temperature, combustion byproducts are generated that suppress flame propagation.

Another important class of functional pigments is viscosity modifiers. In general terms these are pigments that will increase the low shear viscosity of the coatings either to inhibit pigment settling during storage of the coating and/or to reduce sagging after application of the coating. Important examples used in solvent-borne coatings are quaternary ammonium salt treated bentonite clays [9] and fine particle size silicon dioxide. Attapulgite clay is used in water-borne systems. The effect of such pigments on flow is discussed in Chapter 19.

GENERAL REFERENCES

P. A. Lewis, *Organic Pigments*, Federation of Societies of Coatings Technology, Philadelphia, 1988.
P. A. Lewis, Ed., *Pigment Handbook*, 2nd ed., Vol. I, Wiley-Interscience, New York, 1989.

REFERENCES

1. P. A. Lewis, Ed., *Pigment Handbook*, 2nd ed., Vol I, Wiley-Interscience, New York, 1989.
2. D. M. Fasano, *J. Coat. Technol.*, **59** (752), 109 (1987).
3. J. W. Vanderhoff, J. M. Park, and M. S. El-Aasser, *Polym. Mat. Sci. Eng.*, **64** 345 (1991).
4. R. W. Hislop and P. L. McGinley, *J. Coat. Technol.*, **51** (658) 63, 75 (1979).
5. P. A. Lewis, *Organic Pigments*, Federation of Societies for Coatings Technology, Philadelphia, 1988.
6. L. M. Greenstein, "Pearlessence," in *Pigment Handbook*, Vol. 3, T. C. Patton, Ed., Wiley-Interscience, New York, 1973, pp. 289–339.
7. D. H. Solomon and D. G. Hawthorne, *Chemistry of Pigments and Fillers*, Wiley-Interscience, New York, 1983.
8. A. Smith, *Inorganic Primer Pigments*, Federation of Societies for Coatings Technology, Philadelphia, 1989.
9. S. J. Kemnetz, A. L. Still, C. A. Cody, and R. Schwindt, *J. Coat. Technol.*, **61** (776), 47 (1989).

APPENDIX _____

Sources

BOOKS ON COATINGS

Anonymous, *Annual Book of ASTM Standards, Paints, Related Coatings, and Aromatics*, Vols. 6.01, 6.02, and 6.03. American Society for Testing and Materials, Philadelphia. Revised annually. Provides standard procedures for carrying out tests. It is critical to remember that these are, in general, quality control tests whose precision has been checked not performance predicting tests. Many of the tests have not been validated against actual performance.

Anonymous, *Paint/Coatings Dictionary*, FSCT, Blue Bell, PA, 1979. Contains 464 pages of definitions and other information. Some are not too scientifically rigorous but it can be a useful guide to the jargon of the industry.

D. A. Bate, *The Science of Powder Coatings*, 2 Vols., Selective Industrial Training Associates, London, 1990.

J. J. Brezinski, Ed., *Manual on Determination of Volatile Organic Compounds in Paints, Inks, and Related Coating Products*, ASTM Manual Series: MNL. 4, ASTM, Philadelphia, 1989. Useful. Periodic new editions have been promised.

L. J. Calbo, Ed., *Handbook of Coatings Additives*, Marcel Chicago Society for Paint Technology, *An Infrared Spectroscopy Atlas for the Coatings Industry*, 2nd ed., Federation of Societies for Coatings Technology, Blue Bell, PA, 1983. Contains 1400 spectra of coatings materials.

E. M. Corcoran, A. G. Roberts, and G. G. Schurr, Eds., *H. A. Gardner and G. G. Sward Paint Testing Manual*, 13th ed., ASTM, Philadelphia, 1972. Very useful review of paint testing. A new edition is in preparation.

Federation of Societies for Coatings Technology Monograph Series on Coatings. New Series began in 1986; monographs continue to appear at a rate of about four per year. Many are excellent. Federation of Societies for Coatings Technology, Blue Bell, PA.

FATIPEC (Federation d'Associations de Techniciens des Industries des Peintures, Vernis, Emaux et Encres d'Imprimerie de l'Europe Continentale) Congress Books. Contain copies of the hundreds of papers presented at FATIPEC biannual congresses. Many interesting and valuable papers, the majority are in English, all with at least an English abstract.

R. Holman, *UV & EB Curing Formulation for Printing Inks, Coatings, and Paints*, Scholium International, Port Washington, NY.

P. Oldring and G. Hayward, *Resins for Surface Coatings*, 3 vols. Scholium International, Port Washington, NY.

Organic Coatings, Science and Technology, American Chemical Society, Washington, annually. These are the proceedings of The International Conferences in Organic Coatings Science and Technology held near Athens, Greece or in Switzerland each summer. Have included excellent review papers.

S. P. Pappas, Ed., *UV Curing: Science & Technology*, 2 vols., Technology Marketing Corp., Norwalk, CT, Vol I, 1978, Vol. II, 1985.

S. P. Pappas, Ed., *Radiation Curing: Science and Technology*, Plenum, New York, 1992.

T. C. Patton, *Paint Flow and Pigment Dispersion*, 2nd ed., Wiley-Interscience, New York, 1979. Engineering flavor; now somewhat outdated but still very useful.

Proceedings of Symposia on Water-Borne and Higher Solids Coatings. Sponsored by Department of Polymer Science, University of Southern Mississippi and Southern Society for Coatings Technology annually in New Orleans. Contain some useful papers. Can be ordered by mail from USM.

Steel Structures Painting Council, Pittsburgh, publishes bulletins and books on maintenance coating. For example: *Steel Structures Painting Manual, Vol. I, Good Painting Practice and Vol. II, Systems and Specifications.*

G. E. Weismantel, *Paint Handbook*, McGraw-Hill, New York, 1981. Written mainly from the end-users point of view for application of architectural paints. Much practical information on application.

BOOKS ON POLYMER SCIENCE—GENERAL (Not specific to Coatings)

Springer-Verlag, New York. Collections of critical reviews. Many are useful. For example, K. Dusek, Ed., *Epoxy Resins and Composites I, II, III, and IV*, 1985 and 1986.

K. E. J. Barrett, Ed., *Dispersion Polymerization in Organic Media*, Wiley, London, 1975. The source book on NAD polymer systems.

F. W. Billmeyer, Jr., *Textbook of Polymer Science*, 3rd ed., Wiley-Interscience, New York, 1984. Comprehensive popular textbook.

H.-G. Elias, *Macromolecules*, 2nd ed., Plenum, New York, 2 Vols., 1984, 1985. More advanced readable textbook.

D. C. Blackley, *Science and Technology of Polymer Colloids*, G. W. Poehlein, R. H. Otteweill, and J. W. Goodwin, Eds. NATO ASI Series, M. Nijhoff Publishers, Brussels, 1983.

J. Brandrup and E. B. Immergut, Eds., *Polymer Handbook*, 3rd ed., Wiley-Interscience, New York, 1989. Reference work for polymer and polymerization data.

D. H. Everett, *Basic Principles of Colloid Science*, Royal Society of Chemistry, Letchworth, UK, 1988.

K. C. Frisch, *Advances in Urethane Science and Technology*, Periodic Series, Technomic Publishers, Westport, CT.

H. Lee and K. Neville, *Handbook of Epoxy Resins*, McGraw-Hill, New York, 1966. Severely out-of-date but useful.

L. W. Nielsen, *Polymer Rheology*, Marcel Dekker, New York, 1977. Most readable book on subject.

F. Rodriguez, *Principles of Polymer Systems*, 3rd ed., Hemisphere Publishing Co., New York, 1988.

G. Odian, *Principles of Polymerization*, 3rd ed., Wiley-Interscience, New York, 1991. Excellent treatment of polymer synthesis and kinetics of polymerization.

L. H. Sperling, *Introduction to Physical Polymer Science*, Wiley-Interscience, New York, 1986. Readable introductory text on polymer chemistry from a physical chemistry point of view. Complements Odian.

K. J. Saunders, *Organic Polymer Chemistry*, Chapman & Hall, New York, 1988.

V. I. Yeliseeva, *Emulsion Polymerization*, I. Piirma, Ed., Academic, New York, 1982.

Symposium Series and *Advances in Chemistry Series*, American Chemical Society. Many cover polymer topics and some are directly relevant to coatings. For example: R. W. Tess and Poehlein, Eds., *Applied Polymer Science*, ACS Symposium Series No. 285, 1985. Contains 54 review articles a majority relevant to coatings. R. A. Dickie and F. L. Floyd, Eds., *Polymeric Materials for Corrosion Control*, ACS Symposium Series No. 322, 1986; R. A. Dickie, S. S. Labana, and R. S. Bauer, Eds., *Cross-Linked Polymers: Chemistry, Properties and Applications*, ACS Symposium Series No. 367, 1988. ACS Books, Washington, DC.

BOOKS—COLOR and APPEARANCE, and PIGMENTS

F. W. Billmeyer, Jr. and M. Saltzman, *Principles of Color Technology*, 2nd ed., Wiley-Interscience, New York, 1981, Useful general introduction. Contains good bibliography.

R. S. Hunter, *The Measurement of Appearance*, Wiley-Interscience, New York, 1975. Only book whose coverage includes discussion of gloss. Useful but not authoritative.

D. B. Judd and G. Wysecki, *Color in Business, Science and Industry*, 3rd ed., Wiley-Interscience, New York, 1975. Excellent

T. C. Patton, Ed., *Pigment Handbook*, 3 vols., Wiley-Interscience, New York, 1973. P. Lewis, Ed., 2nd ed. Vol. I, Wiley-Interscience, New York, 1988.

J. D. Sanders, *Pigments for Inkmakers*, Scholium International Inc., Port Washington, NY, 1990.

G. Wysecki, *Color Science, Concepts and Methods, Quantitative Data and Formulae*, 2nd ed., Wiley-Interscience, New York, 1982. The authoritative reference work. Extensive tables of standard values.

ENCYCLOPEDIAS

"Encyclopedia of Polymer Science and Engineering," 2nd ed., Wiley-Interscience, New York, 1985–1990.

"Kirk–Othmer Encyclopedia of Chemical Technology," 4th ed., Wiley-Interscience, New York, being issued starting in 1991.

ABSTRACTS

World Surface Coatings Abstracts, Paint Research Association, Pergamon Press, Helmsford, NY. Invaluable. Covers most literature and many patents relating to coatings. About 10,000 abstracts per year. Well organized.

Chemical Abstracts, American Chemical Society, Columbus, OH. Extensive coverage and indexing. Section 42 covers coatings but abstracts of interest in the coatings field are also in several other sections. Less convenient than WSCA but more comprehensive. One can subscribe to individual sections and to on-line computer search services.

JOURNALS—SCIENTIFIC

A word of caution: In general, refereeing of papers published in coatings journals tends to be spotty, resulting in publication of some atrocious papers. Just because it is in print does not make something true.

Color—Research and Application, Wiley, New York.

Farbe und Lack, Curt R. Vincens Verlag, Hannover, Germany. In German with summaries in English. Often has excellent papers.

Journal of Coatings Technology, Federation of Societies for Coatings Technology, Blue Bell, PA. Formerly named *Journal of Paint Technology* and before that the *Official Digest*. Leading U.S. journal in the field. Since paginations start anew with each issue reference must include issue number as well as page and year.

Journal of the Oil and Colour Chemists Association, Wembley, UK.

Polymeric Materials: Science and Engineering Proceedings, ACS, semiannual. These are the papers published at the *spring* and *fall* meetings of the ACS, Division of Polymeric Materials Science and Engineering. Formerly *Organic Coatings and Plastics Chemistry*. While the papers are not refereed, some very useful papers. Frequently the first publication of new developments.

Progress in Organic Coatings. Elsevier Science Publishers, Lausanne, Switzerland. The outstanding review journal in the field. Most articles in English. Recently began accepting research papers, potentially diluting its value.

Surface & Coatings Technology, Elsevier Science Publishers, Lausanne, Switzerland. Inorganic coatings and surface studies.

Most general scientific journals covering polymers include papers relevant to coatings science.

JOURNALS—TRADE

American Paint Journal. St. Louis. Weekly. Large circulation among paint industry people, newsy, good coverage of industry happenings.

Coatings, Kay Publishing Co., Oakville, Ontario, Canada. Canadian counterpart of Modern Paint and Coatings.

Industrial Finishing, Hitchcock Publishing, Wheaton, IL.

Journal of Protective Coatings and Linings, Steel Structures Painting Council, Pittsburgh. Good source of information about industrial maintenance coating.

Modern Paint and Coatings. Atlanta—formerly *Paint and Varnish Production*, Atlanta, GA.

Paint & Coatings Industry, Canoga Park, CA. Industrial coatings spin off of Western Paint and Decorating.

Paint & Resin, Rickmansworth, UK.

Polymer Paint and Colour Journal, Redhill, UK.

Resin Reviews, Rohm & Haas Co., Philadelphia. Published quarterly. Articles, generally based on Rohm & Haas products, presented in technically useful form.

Western Paint & Decorating, Canoga Park, CA. Trade sales focus. Covers the West Coast regulatory developments.

MARKET and PRODUCT INFORMATION

Chemical Economics Handbook, SRI International, Palo Alto, CA. Contains useful information on coatings. Expensive.

Kline Guide to the Paint Industry, Charles H. Kline & Co., Fairfield, NJ. A new edition of this market research report appears every few years.

Paint Redbook, Communication Channels, Inc., Atlanta, GA. Annual. Lists suppliers of raw materials and equipment manufacturers.

Powder Coating Institute, 1800 Diagonal Road, Suite 370, Alexandria, VA, 22314. Market and technical bulletins on powder coatings.

U.S. Department of Agriculture, Forest Products Laboratory, Madison, WI, publishes bulletins on painting wood and wood products. Some very useful.

U.S. Department of Commerce, Bureau of Census. *Current Industrial Reports—Paint, Varnish, and Lacquers*, M-88-F, issued monthly. Provides statistics on production of various classes of coatings.

Technical Bulletins. Resin, solvent, pigment, and additive suppliers issue technical bulletins on their products. Many of them provide excellent background information and comparisons as well as suggested uses and starting formulations. One must expect some bias in such publications—there is a wide range of the extent of bias. One must learn from experience (and by talking to others), which company's claims are most reliable and which have to be discounted substantially. In the case of resin suppliers, remember that in most cases, their preliminary evaluations are of clear coatings or at most white gloss coatings.

SAFETY and TOXIC HAZARDS

Material Safety Data Sheets. In the United States all producers of chemicals and chemical formulations are required to provide "MSDS Sheets (sic)" to their customers. MSDSs contain reliable safety information on each substance and formulation.

N. I. Sax and R. J. Lewis, Sr., *Dangerous Properties of Industrial Materials*, 7th ed., 3 Vols., Van Nostrand-Reinhold, New York, 1989. A comprehensive encyclopedia of toxic and other hazards of industrial chemicals.

CONFERENCES, CONVENTIONS, and SYMPOSIA

American Chemical Society meetings are held each spring and fall. Papers of direct interest in coatings are generally in the meetings of the Division of Polymeric Materials, Science and Engineering but also in several of the other division's papers.

FATIPEC Congresses are held on even numbered years in successive European cities. Many papers presented.

Federation of Societies for Coatings Technology (North America) Annual Meeting in late October or early November each year. Technical papers presented, and many are subsequently printed in the Journal of Coatings Technology. There is always a trade exhibit called the "Paint Show" with extensive displays by raw material and equipment manufacturers. Technical bulletins are available. Each of the 26 individual societies hold monthly meetings at which one or two papers are presented, most commonly by raw material suppliers. Several Societies also have short symposia annually.

Gordon Conference on The Chemistry and Physics of Coatings and Films. Held on odd-numbered years in late summer. By invitation only (you ask for an invitation). Roughly a dozen and a half papers on research in progress. No proceedings or summaries are published. Ample time for discussion of papers and for informal discussions with the hundred odd other participants.

International Conference on Organic Coatings Science and Technology. Organized by Prof. A.V. Patsis, SUNY—New Paltz, NY. Held each summer near Athens, Greece—1991 conference was held in Switzerland, but plans are to return to Athens.

Oil and Colour Chemists Association Convention. Held on odd numbered years in the United Kingdom. Smaller version of the FATIPEC Congress.

Society of Manufacturing Engineers, PO Box 930, Dearborn, MI. Organizes symposia oriented toward industrial end-users. Symposium topics include powder coatings, radiation curing and, more generally, "finishing." Proceedings of symposia are often published.

Water-Borne and Higher Solids Coatings Symposium organized by Southern Society for Coatings Technology and Polymer Science Department at University of Southern Mississippi. Held just before Mardi Gras in New Orleans each spring.

REGULATIONS

A. B. Waldo and R. deC. Hinds, Esq., *Chemical Hazard Communication Guidebook*, Executive Enterprises, New York, 1988. Outlines U.S. Occupational Safety and Health Administration, Environmental Protection Agency and Department of Transportation requirements.

Index